THE EXPERIMENTAL STUDY OF FOOD

SECOND EDITION

THE EXPERIMENTAL STUDY OF FOOD

ADA MARIE CAMPBELL
The University of Tennessee, Knoxville

MARJORIE PORTER PENFIELD
The University of Tennessee, Knoxville

RUTH M. GRISWOLD
Late of Indiana University

HOUGHTON MIFFLIN COMPANY BOSTON
Dallas Geneva, Illinois Hopewell, New Jersey Palo Alto London

Printed in the U.S.A.

Library of Congress Catalog Card Number: 78-69535

ISBN: 0-395-26666-1

TO OUR STUDENTS

CONTENTS

PREFACE

This book has two main purposes: to present the scientific basis for an understanding of the nature of food, and to promote the principles of experimental methodology as applied to food. The book is intended for use in the first course in experimental foods.

The experimental approach is an emphasis of most courses for which this book will be used. As in the first edition, some detailed planned experiments are included at the ends of chapters. Although objective measurements should be made on products whenever possible, the lack of all or some of the equipment for making such measurements need not eliminate the experiments from the class schedule. Some experiments require only the simplest equipment, and sensory evaluation is nearly always feasible. Some suggested exercises are included without detail to provide ideas for experiments and yet to permit planning experiments to relate them to the interests and resources of specific groups and individuals. Some planning of this type is an important part of a course in experimental food science. Procedures described in the detailed experiments can be applied in many of these exercises.

Incorporation of recent research findings and recognition of specific technological advances related to food have been major goals during the revision process. Recent changes in available market forms of foods have been noted throughout. Specific chemical additives have been mentioned in the discussions of various foods. Developments in these and other aspects of food science are continuous and sometimes far-reaching. An example is the currently doubtful status of certain food additives. The food science student needs to maintain contact with both scientific and lay literature and to be alert to developments that affect the consumer.

In addition to updating, this edition represents both expansion and deletion. Increased emphasis has been placed throughout on relationships between chemical and physical properties. Chapter 2 has been expanded to include sections on water, proteins, and hydrocolloids. The chemistry of sugars has been expanded in Chapter 13. Chapter 15 has been reorganized and expanded to reflect current trends in sensory evaluation. The Appendix contains sample evaluation scales from which an appropriate scale can be selected whenever sensory evaluation is done in connection with an experiment. Current concern of educators regarding the writing skill of their students is recognized in the inclusion of a sample portion of a report in Chapter 17. On the other hand, market orders for instructors have not been retained because their usefulness is limited by the inevitable variation in portions of experiments selected and in numbers of replications carried out. Chapter summaries have given way to expanded text because repetition seemed less important than inclusion of new material. Some topics that are covered fully in elementary textbooks have been somewhat de-emphasized herein.

Chapters 1–13 contribute to both goals of the book—presenting a scientific basis and promoting principles of methodology for food study—and Chapters 14–17 to the second. Many chapters are self-contained and can be read in any sequence. The two introductory chapters provide a foundation for effective use of the remainder of the book; thus they should be read first and consulted frequently thereafter. Chapters 11 and 12 are somewhat dependent on several preceding chapters that deal with ingredients of flour mixtures. Although Chapters 15 and 16 are located as they are because of their relation to the carrying out of an individual study, they contain information useful also during the conduct of class experiments.

Some background in organic chemistry is assumed, as is some fundamental knowledge about food. The subject matter presentation draws heavily not only on the principles of related sciences, but also on published food research, both early and recent. The reference lists though far from complete are meant to be representative. It is hoped that the student will gain an appreciation for the tremendous contribution of early workers in the field, as well as for the advances resulting from more recent work, and will recognize the dynamic nature of the field. The perceptive student will observe that new information results in new questions to be answered and that new approaches to food study become possible as technological advances occur.

The student is encouraged to read both textbook material and research literature critically, to examine evidence presented, and to evaluate statements made and conclusions drawn. In this way the student will realize that reading should be an active, evaluating process, not a passive, accepting one.

Two major changes in form have been incorporated into this revision. The bibliographic entries and the literature citations now follow the form of the Institute of Food Technologists, and the metric system is used almost exclusively. Both the Celsius (metric) and Fahrenheit scales are used for oven temperatures.

Suggestions made by users of the first edition have been helpful during the revision process and are gratefully acknowledged. Students in many classes have provided inspiration through their questions, their comments, their needs, and their interests. Their contributions are immeasurable.

The revisers and the publisher are grateful for the assistance of those who made professional reviews of the manuscript. Among these reviewers, to whom we give thanks, were Eva Medved, Kent State University, and Rose Mirenda, Hunter College.

A.M.C.
M.P.P.

1. Introduction to Food Experimentation

A scientific approach to the experimental study of food is presented in this book. The experimental study of food is concerned with why foods are handled and prepared as they are, how and why variations in treatment influence the quality of food, and how this knowledge can be used to improve the quality of food products. The scientific approach includes three basic steps: defining the problem or arriving at a hypothesis; testing the hypothesis in a carefully planned experiment; and reporting and interpreting the results. The scientific study of food is an exciting field of investigation. Adequate answers to some food-related problems have been found, often by applying sciences, such as chemistry, physics, and biology, that are basic to the study of foods. Answers to other problems are only partial or not yet known. Incomplete answers and the unknown offer challenges to the food scientist. Food science students can find challenging questions to study in an experimental food science class.

This book is divided into two parts. Part I includes information about various food groups and also includes experiments designed to illustrate the points made in the text and to answer questions that students may have regarding the foods and the effects of various treatments on food. This chapter is an introduction to the laboratory work of Part I. Part II explores methods used in food science and should serve as a guide to the student who is starting an independent problem in experimental foods or who wants to expand the experiments outlined in Part I.

The experiments in Part I contain several of the same sections as a published research report. The review of the literature is supplied in the text. As knowledge in food science changes rapidly, however, the most current literature should also be reviewed. The purpose of the experiment is stated, and then the experimental methods, including sensory and sometimes objective methods of evaluating the product, are described. The other essential parts of the research report—the data, the discussion of the results, the conclusions, and the reference list—should be incorporated into the student's notebook. The reference list should give credit to the sources of information used in preparation of the report. For reference lists in the text, the Institute of Food Technologists' (IFT) style has been used. Details of this style are found in the style guide published by IFT (1977). For class reports, either this style or one of the styles used in other professional journals may be selected. It is important, however, that the same style be used throughout a report.

Experiments at the ends of the chapters include basic formulas and procedures followed by suggested experimental variations. The product from the basic formula serves as a control, in other words, a basis for comparison of experimental products. Variations not included in the experiment description may better serve the needs and interests of a class. For example, food-related problems of current concern and interest may serve as a basis for class or individual projects—interesting projects often are suggested by class members.

Carefully controlled experiments are necessary if the results are to be meaningful and, in the case of more extensive research studies, worthy of publication. The experiments in the text are designed to be controlled experiments; in other words, one factor is varied while all other conditions that might affect the results are controlled as much as possible. If more than one factor is altered, it is difficult to determine the cause of any changes that might be observed in the quality of the products. It is beyond the scope of this book to discuss the principles of experimental design and statistical analysis that must be applied in studies in which more than one factor is varied. Even in the most carefully planned experiment, unexpected variables may arise. Temperature and humidity of the laboratory are difficult to control but may influence the results of an experiment. Faulty measurements, misinterpretation of instructions, and variation in individual techniques are other possible unexpected variables that should be considered. A record of any unplanned variation should be made in the laboratory notebook so that it may be considered when the results are evaluated and discussed. An unplanned variable, if its presence is known, may be most helpful in explaining the results of an experiment. Research workers have sometimes changed the direction of work because of the accidental occurrence of a condition that gave a clue to the solution of a problem. The procedures used to control experimental conditions will be discussed in this chapter. These procedures should be studied carefully and followed as experiments outlined in the following chapters are conducted.

CONTROLLING INGREDIENTS

The uniformity, temperature, and quantity of each ingredient must be carefully controlled. If possible, sufficient ingredients for the entire experiment should be purchased at one time and should be well mixed prior to sampling. Such ingredients as flour, sugar, and shortening are easily mixed and stored in quantity, but perishable ingredients such as milk and eggs are usually obtained daily. Greater uniformity can be obtained by the use of reconstituted nonfat dry milk or reconstituted dried egg whites in place of their fresh counterparts. A few foods, such as meat, present special sampling problems that must be considered carefully in planning the experiment. Suggestions for sampling meat are discussed in Chapter 5.

Ingredients should be at the same temperature for each replication of an experiment. When used for baking, most ingredients should be at room temperature. The initial temperature of meat to be heated should be uniform in a controlled study because it affects the heating rate (Heldman, 1975).

Measuring

At the time of this revision, the United States was in the midst of a voluntary conversion from the customary system of measurement, previously referred to as the *English system,* to the *metric system.* Therefore, it was anticipated that this text would be used during periods when both systems would be in use and that discussions including both would be appropriate in some portions of the book. Therefore, both systems are referred to occasionally as an aid to transition. The customary system is retained in instances where it would of necessity be used for some time. For example, oven temperatures are given in customary as well as metric units in the Experiments and Suggested Exercises. Some discussion of customary household measuring utensils is included because it is conceivable that they will also be used for some time.

For the customary system, household measuring devices were designed for quick and easy use with enough accuracy and precision[1] for home use. The American Standards Association, in accordance with the recommendations of the American Home Economics Association (AHEA), established standards for the capacity of the household measuring devices (ASA, 1949). These utensils were volume-measuring devices. Even though some food scientists (Miller and Trimbo, 1972) have recommended the use of a weighing system for household measurements, the AHEA has recommended that a volume system for both dry and liquid ingredients in metric units be adopted (Anon., 1976). Standard sizes and tolerances for metric measuring devices for household use have been recommended by an American National Standards Institute Committee sponsored by the AHEA (1977). Recommended standard volumes include 1000 ml and 500 ml subdivided every 50 ml; 250 ml subdivided every 25 ml; and 125, 50, 25, 15, 5, 2, and 1 ml (AHEA, 1977).

Canadian groups have established standard sizes for household measuring devices for use in that country (Metric Committee, 1975). For liquid measurements, a 250-ml measure with 25-ml graduations has been accepted. For dry ingredients, 250-, 125-, and 25-ml measures designed for leveling have been approved. The third set of Canadian

[1]*Accuracy* refers to whether the results agree with the true result. Thus, an accurate 100-ml pipet delivers 100 ml. *Precision* refers to the agreement among repeated determinations and is not necessarily accompanied by accuracy. For example, repeated careful measurements with a 100-ml pipet that delivers only 98 ml might be precise but not accurate. Precision under these circumstances means wasted effort. Both terms may refer to either equipment or technique.

measures was designed for measurement of small amounts of dry and liquid ingredients and includes 1-, 2-, 5-, 15-, and 25-ml measures.

New standard measuring devices in the United States will have standardized capacities with tolerances like those applied to customary cups and spoons (AHEA, 1977). A tolerance of 5% has been accepted for these measuring utensils. Thus, the 1-cup measure with a standard capacity of 236.6 ml could actually have a capacity of 224.8 to 248.4 ml. The 250-ml measure to be adopted for use in the United States will have to have a capacity within the range of 237.5 to 262.5 ml.

Household measuring utensils such as a customary 1-cup liquid measure lack precision because their diameters at the point of measurement are larger than the diameters of more precise measuring devices, such as 100-ml pipets, 100-ml burets, or 100-ml graduated cylinders. A difference of a few drops in the volume of the liquid measure shows very clearly in a pipet or buret, somewhat less clearly in a graduated cylinder, and far less clearly in a household measuring device. Therefore, a measuring utensil no larger than required is selected when possible. For instance, a 10-ml graduated cylinder will measure 9 ml more precisely than will a 100-ml cylinder. Similarly, a 250-ml measure is used instead of a 500- or 1000-ml measure for amounts of more than 100 ml but less than 250 ml.

Errors in measuring food by volume may be caused by the manner in which measuring devices are used. Liquid measurements should be read at eye level and the position of the bottom of the meniscus noted.

The way in which a food is handled for measurement will influence the precision of measurement. For example, if the measuring cup is dipped into the flour, the flour will weigh from 130 to 158 g; if the flour is spooned into the cup without sifting, the weight will range from 121 to 144 g; and if the flour is sifted and then spooned, the weight will range from 92 to 120 g (Arlin et al., 1964). The factors that affect the weight of a cup of flour were studied by Grewe (1932a, b, c). Shortening is packed into the cup to avoid air bubbles. The weight of a measured cup of food may reflect variations in the food itself as well as in the way in which it is measured.

For experimental work, the amount of each ingredient in a formula is usually controlled by weighing, except that liquids sometimes are measured in a graduated cylinder. The volume and weight measurements shown in Table 1.1 and in more extensive tables published by the

Table 1.1 Weight and volume measurements of selected food materials

Food material	Weight of 1 cup, g
Corn sirup, dark or light	328
Cornstarch	125
Cornmeal	
White, degerminated	140
White, self-rising, wheat flour added	141
Yellow	
Degerminated	151
Stone ground	132
Cream	
Half-and-half,	242
Whipping	232
Fats and oils	
Butter	227[a]
Lard	205
Margarine, regular	225
Margarine, soft	208
Oil, cooking	209
Shortening, hydrogenated	187
Flours	
Barley, unsifted, spooned	102
Oat, unsifted, spooned	
Coarse grind	120
Fine grind	96
Potato, unsifted, spooned	179
Rice, unsifted, spooned	
Brown	158
White	149
Rye	
Dark, stirred, spooned	127
Light	
Unsifted, spooned	101
Sifted, spooned	88
Soy, full-fat, unsifted, unspooned	96
Triticale, unsifted, spooned	119[b]
Wheat	
All-purpose, hard wheat	
Sifted, spooned	115[b]
Unsifted, spooned	125[b]
All-purpose, soft wheat	
Sifted, spooned	98[b]
Unsifted, spooned	116[b]
Bread	
Sifted, spooned	117
Unsifted, spooned	123
Unsifted, dipped	136

Table 1.1 *continued.*

Food material	Weight of 1 cup, g
Cake	
Sifted, spooned	99
Unsifted, spooned	111
Gluten	
Sifted, spooned	136
Unsifted, spooned	135
Self-rising	
Sifted, spooned	106
Unsifted, spooned	127
Whole wheat, stirred, spooned	120
Honey, strained	325
Milk	
Whole, fresh, fluid	241
Skim or buttermilk	244[a]
Nonfat dry	
Regular	134
Instant	74
Molasses	309
Sugar	
Brown, packed	211
Brownulated	152
Confectioner's	
Sifted	95
Unsifted	113
Granulated	196

	Weight of 1 tsp, g
Baking powder	
Phosphate	3.8[a]
SAS-Phosphate	3.0[a]
Tartrate	2.8[a]
Baking soda	4.1[c]
Cream of tartar	3.1[c]
Salt, free-running	5.5[a]

[a]Adams, 1975.
[b]Class data, Department of Food Science, Nutrition, and Food Systems Administration, University of Tennessee, Knoxville.
[c]AHEA, 1975.
SOURCE: Except as noted, the data are drawn from Fulton, L., Matthews, E. and Davis, C. 1977. Average weight of a measured cup of various foods. Home Econ. Res. Rpt. 41, USDA, Washington, DC.

AHEA (1975) and the United States Department of Agriculture (Adams, 1975; Fulton et al., 1977) make it possible to convert one type of measurement to the other.

Weighing

Trip balances Of the types of balances available for weighing foods, trip balances probably are most often used in experimental foods classes because they are relatively inexpensive. A balance with 10- and 200-g riders can be used to weigh objects up to 210 g without added weights. A set of weights will increase the capacity of a trip balance to 2 kg.

When food is to be weighed into a container, the container can first be weighed, but more often it is counterbalanced. *Counterbalancing* containers is a means of avoiding the necessity of weighing an empty container and making a calculation to determine the weight of the food. Counterbalancing speeds the weighing process. Before counterbalancing is begun, the balance must be adjusted to zero by adjusting the screw weights on the right until the pointer rests on the center line of the pointer scale when the riders are on zero. The container that will be used for the food is placed on the left platform of the balance. Another, lighter, container is placed on the right platform. Sufficient shot is poured into the right container to balance the two sides. To weigh the sample, the desired weight is placed on the right by moving the riders and adding additional weights if needed. Food is added to the container on the left until the two sides balance.

With any type of balance, time can be saved by weighing all the foods to be used for a series of experiments at one time. Rigid freezer containers are convenient for weighing because they are light and can be covered if the food is to be stored before use. To weigh small amounts of food, papers of equal weight can be placed on the two pans of the balance. Because no sifting or careful packing is necessary, it is often possible for experienced workers to weigh foods more rapidly than they could measure them by household methods. To avoid damage, a balance always is picked up by the base, with a pan supported by a hand.

The suitability of a balance for weighing small quantities depends on its sensitivity. *Sensitivity* can be defined as the weight necessary to shift the equilibrium position of the pointer by one scale division. It is likely to decrease as more weight is placed on the balance. For trip balances with a load of 100 g, the sensitivity is about 0.1 g, which means that a cup of flour (96–115 g) can be weighed with an error of one part per thousand. This very small error can be contrasted with the error of about 10% that might occur if flour were measured with household methods. The sensitivity of the trip balance is not satisfactory, however, for very small amounts such as 1/4 teaspoon baking powder or soda, which weighs about 1 g. When such an amount is weighed to the nearest 0.1 g, the accuracy is only about 10%. A torsion balance, with a sensitivity of about 0.02 g, or an automatic, top-loading, direct reading balance with a sensitivity of at least 0.01 g is more satisfactory than the trip balance for weighing small quantities. If neither of these types of balances is available, careful measurements with graduated cylinders for liquids or household measuring devices for solids may afford greater precision than the trip balance.

Torsion and other balances Torsion balances are satisfactory as to sensitivity and ease of use, but usually are more costly than trip balances. Various sizes, differing in capacity and in sensitivity, are available. It is convenient to have one of the larger sizes in a foods laboratory for weighing objects that are too large for the trip balance. Spring-type scales have the advantage of speed but their sensitivity, which depends on their capacity and construction, usually is not sufficient for weighing ingredients for small recipes. Dietetic scales also are spring- type scales, but their sensitivity is only about 0.5 g so they are not commonly used for experimental foods work.

CONTROLLING TECHNIQUES

Variations in the techniques used to prepare samples for experimental work are often more difficult to control than variations in the quality and amount of food used, especially for batters and doughs. For food research, mixing is usually controlled by use of electric mixers timed with stop watches. The same mixer, or identical ones, should be used throughout the experiment. If mixers are not available for class experiments, some other method of control may be substituted. Timing hand mixing or counting the number of strokes is an alternative. Counting may be more reliable than timing since mixing speed varies with the worker. When strokes are counted, however, they should be as nearly the same type or strength as possible. Standardization of techniques may take practice.

The effects of individual differences in technique can be reduced by class work if a team of students does an experiment that includes several variations and the basic formula. The work is divided as it would be for an assembly line, so that one student does the same step on each product. In making biscuits, for example, one student might weigh the ingredients and bake the biscuits; a second student might mix; and a third might knead, roll, and cut. If some steps are time consuming or tiring, they can be divided. For example, if a product is to be kneaded two hundred strokes, one student can knead the first hundred strokes, and another student the second hundred.

Identical equipment should be used for each variation in an experiment. If instrumental methods are used for evaluation of the food, the same instrument should be used each time in order to avoid variability due to differences between instruments. Conditions under which instruments are used also should be constant. Obviously, each product should be labeled at all times during preparation and evaluation.

EVALUATION OF CLASS EXPERIMENTS

The outlines for experiments at the ends of chapters include suggestions for both objective and subjective evaluations of the quality of products. These two types of evaluation techniques are described in more detail in Chapters 15 and 16.

In some experiments in Part I, food is evaluated by objective methods that employ some measure of quality other than the human senses. Objective tests are less subject to human variability or error than are sensory tests and are of value if related to the palatability of the product. Experiments in Part I may be adapted to make use of other objective tests if the specialized equipment for appropriate testing is available. Results of such tests are recorded on the chalkboard or duplicated so that all class members may use the results in writing laboratory reports.

Each class member should have the opportunity to evaluate the samples for each experiment. The principles of sensory evaluation of food as discussed in Chapter 15 should be used in arranging samples for class evaluation. Students can be responsible for getting their own white plates, flatware, and water before taking samples of the products displayed in white containers. All samples should be randomly coded or at least coded with the identifying letters and numbers used in the experiment outline. The students record their observations in their notebooks using a scoring system selected from the Appendix or one devised by the class. Before judging begins, it is helpful to discuss the qualities of a desirable product. After judging, discussion of the results is a useful learning experience. It may be helpful to summarize the responses of the total class to get a broader idea of the quality of the products.

During the actual judging, it is important to avoid talking because judges are easily distracted and prejudiced. Care should be exercised so that the number of samples is not too large for careful judging. If freezer space is available, it is sometimes convenient to freeze samples like bread for testing at a later date when more time is available.

REPORTING THE RESULTS

Recording Data

Experimental work is not valuable unless a written record of the results is made. This record is best made with ink in a bound notebook as the experiment is being done. Copying from pieces of paper should be avoided because of possible errors, wasted time, and the danger of losing the papers. If data that have been written in a notebook are to be omitted for any reason, a single line is drawn through them with a note about the reason for the omission. Results must not be rejected only because they are not as expected.

The necessary parts of an experiment report are: objective or purpose, methods, results, discussion of results, conclusions, and reference list. If the purpose and the method have been published, as with the experiments in this text, a page reference to them is sufficient unless modifications have been made. Modifications of a published experiment must be described fully so that the reader of the report can repeat the experiment if desired. Data are easier to locate and interpret if they are in tabular form rather than text form. Tables should be prepared in advance of the laboratory session as suggested in the experiments, allowing space for descriptive terms and remarks, as well as for scores. A carefully worded title for each table is desirable so that the reader of the report can understand how the data were collected and what they mean.

Interpreting Data

Before drawing conclusions, it is necessary to interpret the data. A study of the purpose and plan of the experiment and of the tabulated results will show the comparisons that can be made. As this comparison is being done, one of the first questions is whether any differences that may be observed in the data are real differences resulting from the treatments, or whether they are due to chance variation. In research this question frequently is answered by the use of statistics. For classroom experiments, one practical standard is that a real difference is greater than the difference between replications of the same experiment. Common sense also indicates that a result is more likely to be significant if results of all tests made on the product point in the same direction and if a gradual change in properties accompanies a gradual change in an ingredient or a procedure.

Variation cannot be eliminated entirely from an experiment. Uncontrollable variation is referred to as experimental error and contributes to variation in the results of a study. Experimental error must be minimized if real differences are to be identified. Replication or repetition of an experiment is important in minimizing experimental error. Repeated determinations or measurements on a single sample also are important. These are called *replicates,* or specifically, *duplicates* or *triplicates* if the determination is done two or three times.

After evaluating any differences observed in the main portion of the experiment, it is wise to look for additional relationships and interpretations of the data in order to make the most of the experimental work that has been done. It is equally important, however, to avoid drawing conclusions and assuming relationships that are not warranted by the data and the scope of the experiment.

Drawing Conclusions

After the data are interpreted, conclusions can be drawn. These must be based on the observations made in the laboratory, not on the reading or any feeling of what the results ought to be. If the results of an experiment do not agree with published work, differences in experimental conditions often can be found. Exploration of inconsistencies is likely to lead to an explanation. The conclusions must be limited by the conditions of the experiment. Thus, a method that proved best for cooking frozen broccoli may not be best for other frozen vegetables or even for fresh broccoli. It is good practice to use the phrase "under the conditions of this experiment" in interpreting results in order to avoid any temptation to generalize more than the experiment justifies.

SUGGESTED EXERCISES

1. Evaluate the accuracy of household measuring devices. Accuracy may be checked by weighing or by measuring the volume of the water held by the measuring device. Use the process of counterbalancing for weight measurements. For volume measurements, use graduated cylinders with capacities of 100 and 250 ml. Repeat each determination at least three times. Compare the average values to the standard capacities for the utensils.

   For most class work, it can be assumed that the weight of water in grams equals its volume in milliliters. This is strictly true only at 3.98°C, the temperature of maximum density for pure water. At higher temperatures, 1 ml of water weighs less than 1 g. Thus 250 ml of water weigh 249.6 g at 20°C, and 249.3 g at 25°C. For such precise work as calibrating burets and pipets, corrections should be made for variations in the density of water. Consult the *Handbook of Chemistry and Physics* (CRC, 1977) for information on the density of water at various temperatures.

   Do any of the utensils exceed the acceptable tolerances? How do the variations of your individual determinations from your averages compare with the difference between your average and the standard capacity?
2. Evaluate the accuracy of recommended household methods of measuring various foods. Before beginning, check the accuracy of the measuring utensil that you plan to use. Follow the procedure outlined in suggestion 1. Measure the selected food as described in the listing below. Weigh the measured quantity. Don't forget to use the

technique of counterbalancing in order to speed the process. Repeat the measurement at least five times. If possible, make each observation on a portion of food not measured previously. Average your observations. Determine the deviation of each observation from the average.

a. Sift all-purpose flour once and spoon lightly into the measuring utensil. Avoid shaking. Level with the straight edge of a spatula.
b. Stir whole-grain flours and meals lightly but do not sift, then measure like all-purpose flour.
c. Sift white sugar only if it is lumpy. Without shaking, fill the measuring device until it overflows, then level with the straight edge of a spatula.
d. Pack brown sugar firmly into the measuring utensil, then level with a spatula.
e. Press solid fat firmly into the container until it is full, then level with a spatula.

How does your average compare with the published weight? Are the deviations from the average important in household measurements? Were deviations greater for some foods than for others? Why should ingredients be weighed rather than measured by volume for experimental foods work?

3. Study the effects of the following variables on measurement of the weight of a common unit (cup or 250-ml measure depending on the system in use) of flour. Do each experiment five times (ten if time permits), average the results, and determine the deviation from the average. Compare both the averages and the variability for the experiments.
   a. Sift three times before measuring.
   b. Sift directly into a cup instead of using the standard method of sifting and then spooning into the cup.
   c. Shake the cup slightly to level the flour instead of using the standard method.
   d. Spoon unsifted flour into the cup instead of using the standard method.
   e. Dip the measure into unsifted flour instead of using the standard method.
   f. Compare different types of flour.
   g. Compare five (ten if time permits) measurements taken by the

same person with single measurements taken by five to ten people.

Why is it necessary to sift flour before using?

4. Study one of the experiments in Part I for variations in amount and type of ingredients, equipment, procedures, techniques, and experimental conditions. Notice what factor is being studied in each section of the experiment. Decide which of the variables must be controlled and which will be difficult or impossible to control.
5. Calibrate the ovens in the laboratory with either thermocouples and temperature recorder or an oven thermometer. Check at 149, 177, 204, and 232°C (300, 350, 400, and 450°F).
6. Read an assigned food research paper in a current journal and be able to discuss the following questions:
   a. What was the purpose of the experiment? (What were the treatments?)
   b. What variables were controlled?
   c. What uncontrolled variables were taken into account in the interpretation of the results?
   d. What objective and sensory methods of evaluation were used?
   e. Did the investigators achieve their objectives?
   f. What new problems became apparent as a result of this experiment?
   g. What suggestions do you have about the procedures used in the experiments?

REFERENCES

Adams, C. F. 1975. Nutritive value of American foods. Agric. Handbook 456, Agricultural Research Service, USDA, Washington, DC.

AHEA. 1975. "Handbook of Food Preparation." American Home Economics Assoc., Washington, DC.

AHEA. 1977. "Handbook for Metric Usage." American Home Economics Assoc., Washington, DC.

ASA. 1949. American standard dimensions, tolerances, and terminology for home cooking and baking utensils. Z61.1, American Standards Assoc., New York.

Anon. 1976. Metric, rehabilitation committees take action at annual meeting. AHEA Action 3(2): 3.

Arlin, M. L., Nielson, M. M. and Hall, F. T. 1964. The effect of different methods of flour measurement on the quality of plain two-egg cakes. J. Home Econ. 56: 399.

CRC. 1977. "Handbook of Chemistry and Physics," 58th ed., Ed. Weast, R. C. CRC Press, Inc., Cleveland.

Fulton, L., Matthews, E. and Davis, C. 1977. Average weight of a measured cup of various foods. Home Econ. Res. Rpt. 41, USDA, Washington, DC.

Grewe, E. 1932a. Variation in the weight of a given volume of different flours. I. Normal variations. Cereal Chem. 9: 311.

Grewe, E. 1932b. Variation in the weight of a given volume of different flours. II. The result of the use of different wheats. Cereal Chem. 9: 531.

Grewe, E. 1932c. Variation in the weight of a given volume of different flours. III. Causes for variation, milling, blending, handling, and time of storage. Cereal Chem. 9: 628.

Heldman, D. R. 1975. Heat transfer in meat. Proc. 28th Ann. Recip. Meat Conf., p. 314, Natl. Livestock and Meat Board, Chicago.

IFT, 1977. Style guide for research papers. J. Food Sci. 42: 1417.

Metric Committee. 1975. Style guide for metric recipes. KIP 5He. Canadian Home Economics Assoc., Ottawa, Ontario.

Miller, B. S. and Trimbo, H. B. 1972. Use of metric measurements in food preparation. J. Home Econ. 64(2): 20.

2. Introduction to Food Science

Throughout most of this text, the chemical and physical nature of food will be studied, along with the chemical and physical changes that occur when foods are exposed to various conditions. The relationship between chemical and physical properties will be emphasized. A brief orientation to some underlying scientific principles is presented in this chapter. It will provide some students with needed background and others with a review.

WATER

Water, the universal chemical substance so important to the properties of food, will be considered first. Not only is water a component of all foods, but it contributes significantly to the physical differences among foods and to the changes that foods undergo. Food structure depends on water; try to imagine making a food emulsion, a crystalline candy, or a frozen dessert without water. Water is important chemically as a reactant in hydrolysis, which has tremendous effects on food properties. Chemical reactions not involving water directly also are important to food properties, both positively and negatively, and water plays a role in occurrence of reactions. It dissolves other chemicals and provides the mobility required to bring reactants together. On the other hand, water may interfere with chemical reactions by diluting reactants so that interaction becomes difficult. Some compounds do not react without first being ionized in water. Substances that are not truly soluble in water are dispersed as larger particles in water in many food dispersions.

Water can exist in three physical states: gas (vapor), liquid, or solid. Heat transfer, either from or into water, is necessary for its conversion from one state to another. Conversion from water vapor to liquid or from liquid to solid (ice) is *exothermic:* Heat must be removed. Conversion from solid to liquid or from liquid to vapor is *endothermic:* Heat must be supplied. The amount of heat that must be transferred into a given amount of water to produce a change in physical state or temperature equals the amount of heat that must be removed in order to reverse the change. The amounts of energy that must be supplied or removed to change only the physical state of water at the solid–liquid level and at the liquid–gas level are relatively large: 80 calories/g for the heat of fusion and 540 calories/g for the heat of vaporization. The amount of energy that must be supplied or removed to change the temperature without affecting the physical state also is large: The heat capacity (specific heat) is about 1 calorie/g/°C, and varies slightly with the temperature of the water.

The properties of water are closely related to its chemical nature. Although the positive charges equal the negative charges in the water molecule, the asymmetric arrangement of the atoms represents an imbalance of the charges. The oxygen atom with its two negative charges is on one side of the molecule and the two hydrogens, each with its one positive charge, are on the other side. This polar nature of the water molecule is responsible for its hydrogen-bonding ability. Each hydrogen atom is attracted to the oxygen of another water molecule and each oxygen atom can attract two hydrogen atoms. Thus each water molecule may be connected, through hydrogen bonding, to four other water molecules, and these in turn may be hydrogen bonded to other water molecules. The actual extent of their clustering depends on the temperature. There is so much hydrogen bonding in ice that an extremely ordered, rigid structure exists. When the ice is melted and the water heated, hydrogen bonds are distorted and broken until, when steam is formed, there are none. Individual hydrogen bonds are weak compared with covalent bonds, but their net effect can be great. Although this discussion is concerned with hydrogen bonding between water molecules, other types of compounds, such as those having hydroxyl, carboxyl, or amino groups, also may participate.

The rather large energy requirements mentioned above are significant with regard to the function of water as a medium for cooling and heating. Water in the form of ice is particularly effective as a cooling medium because, in order to melt, it must absorb a large amount of energy. If it obtains that energy as heat from food, the food is cooled. For example, an ice cream freezer has an inner container made of metal and an outer container made of wood or another poor conductor; thus, the melting of ice in the ice-salt mixture between these containers removes heat mostly from the mixture to be frozen, rather than from the surrounding air. Water is effective as a heating medium because to become hot it must absorb a large amount of energy from the source of heat. The energy then is available for transfer to food.

Some of the water in food is bound. *Bound water* is immobilized on the surfaces of other components, mostly proteins and carbohydrates, and is so closely associated with those substances that it behaves differently from free water. In contrast to water that is free, bound water is not available as a solvent; it is resistant to crystallization and evaporation; it will not freeze or boil; and it can be pressed from food only with tremendous pressure. The proportion of bound water present affects the response of a food to various treatments, and is itself affected by treatments.

The physical constants of water include the following: freezing point, 0°C; boiling point, 100°C; and density, 1 g/cm^3 at 4°C. The density of water decreases below and above 4°C; it is 0.958 at 100°C (CRC, 1977). Much more information about water is provided by Eisenberg and Kauzmann (1969).

FOOD DISPERSIONS

Water is the usual dispersion medium in foods, though certain food constituents, notably fat-soluble vitamins and some flavor compounds, are dissolved in fat. Fat is also the dispersion medium in water-in-oil emulsions. Solutions will be discussed first, then colloidal dispersions.

Solutions

A *solution* is a homogeneous mixture of the molecules and/or ions of different substances, one of which usually is water. The particles of the dissolved substance are so small and have so much kinetic energy that they can make room for themselves between water molecules. In a sugar solution, the particles that are distributed uniformly among water molecules are sugar molecules. In a solution of sodium chloride, which ionizes into Na^+ and Cl^-, the particles that are uniformly mixed with water molecules are ions. The dissolved substance, whether a nonionizing compound or an ionizing compound, is the *solute,* and the medium in which it is dissolved is the *solvent.*

Solubility is the amount of solute that can be dissolved in a given amount of solvent at a given temperature. Solubilities of inorganic and organic compounds are given in tables of physical constants (CRC, 1977) as g/100 g solvent at specified temperatures. The effect of temperature on solubility varies with solutes. Increasing temperature greatly increases the solubility of some solutes and has less effect on the solubility of others. The extent of the solubility of a water-soluble compound at a given temperature depends on the attraction of its molecules or ions to water as compared with the attraction of its molecules or ions to one another.

The *concentration* of a solution is the amount of the solute dissolved in a specified amount of the solvent or solution and is expressed in several ways, including the following: percentage of solute by weight; percentage of solute by volume if the solute is a liquid; molarity, or the number of moles (gram molecular weights) of solute dissolved in a liter of solution; and normality, or the number of gram equivalent weights of solute dissolved in a liter of solution. The details of these methods of expressing concentration can be reviewed in an elementary chemistry textbook. The important point here is to be specific in identifying the concentration of a solution used in work with food. Note that an excess of solute lying undissolved in a solution does not contribute to the concentration. Unsaturated, saturated, and supersaturated solutions will be discussed in Chapter 13.

Several properties of solutions are particularly important in food preparation. Among these are the colligative properties of solutions. *Colligative properties* are those that depend on the number of solute particles in a given amount of solvent and do not depend on the nature of the solute (aside from the effect of its molecular size on the maximum number of particles that can be in a given amount of solvent). The colligative properties to be discussed are vapor pressure, boiling point, freezing point, and osmotic pressure.

Vapor pressure It is obvious to anyone who has boiled water that a liquid vaporizes at its boiling point. What is less obvious, but still apparent if one leaves a container of water standing uncovered for a long time at room temperature, is that vaporization also occurs at lower temperatures, though more slowly. If the container is closed, vaporization occurs, but condensation of the vapor back into the water also takes place. *Vapor pressure* is the pressure exerted by the gas above the liquid when equilibrium exists—that is, when evaporation and condensation occur at the same rate. Vapor pressure increases with increasing temperature. It decreases with the addition of a nonvolatile solute because the solvent constitutes a smaller proportion of the liquid, and the nonvolatile solute occupies space without contributing to vapor pressure. The greater the concentration of solute, the greater is the reduction in vapor pressure.

Boiling point The *boiling point* of a liquid is the temperature at which the vapor pressure just exceeds the external pressure, which is atmospheric pressure under ordinary conditions. The greater the reduction in vapor pressure resulting from the presence of a nonvolatile solute, the higher is the temperature required to raise the vapor pressure sufficiently to permit boiling. The extent of the effect is directly proportional to the number of solute particles in a given quantity of solvent. One mole of a nonvolatile nonionizing solute raises the boiling point of a liter of water 0.52°C. The effect of nonvolatile nonionizing solutes on the boiling point of water is applied in the use of a thermometer for achieving the desired extent of concentration of a jelly or candy mixture.

One mole of a nonvolatile ionizing solute raises the boiling point of a liter of water 0.52°C per ion formed. Sodium chloride forms two ions and calcium chloride forms three ions. A mole of sodium chloride, therefore, raises the boiling point of a liter of water 1.04°C, and a mole of calcium chloride raises the boiling point of a liter of water 1.56°C.

The three preceding sentences assume complete ionization; this is not necessarily the case, but the maximum effect is approached.

Volatile solutes actually lower the boiling point because of their contribution to, rather than hindrance of, vaporization. With the increased vapor pressure, a lower temperature than 100°C permits water to boil.

References to 100°C as the boiling point of water are based on standard atmospheric pressure, 760 mm. At reduced atmospheric pressures, such as at high altitudes, pure water boils at a lower temperature than at standard atmospheric pressure and a given solution boils at a correspondingly lower temperature than does the same solution at standard atmosphere. Conversely, in a pressure cooker, a greater external pressure must be overcome and the boiling point is elevated.

Freezing point Solutes depress the freezing point of water. The extent of their effect on a liter of water is 1.86°C per mole of a nonionizing solute. Ice crystals will begin to form at −1.86°C in a sucrose solution containing 342 g sucrose/liter. The corresponding effect is 1.86°C per mole of ions formed in the case of an ionizing solute. Therefore, a mole of sodium chloride lowers the freezing point of a liter of water 3.72°C, and a mole of calcium chloride lowers the freezing point of a liter of water 5.58°C. The effects do not differ for volatile and nonvolatile solutes.

Osmotic pressure *Osmosis* is the passage of water molecules through a semipermeable membrane that is separating either solutions that differ as to concentration of dissolved particles, or pure water and a solution. If pure water and an aqueous solution are on opposite sides of the membrane, water molecules pass through the membrane in both directions, but much more rapidly from pure water to the solution than vice versa. The more rapid passage is in the direction in which equalization of concentrations can be approached. The rate at which osmosis occurs depends on the difference in solute concentration on the two sides of the membrane. Osmosis is particularly rapid when the difference in concentration is large; it is retarded as equilibrium is approached.

The occurrence of osmosis causes a change in the relative volumes of the two liquids. The volume of the solution that becomes more dilute increases. *Osmotic pressure* is the pressure required to prevent that increase in volume, and is measurable. Considerable osmosis occurs in living tissues. Membranes are changed after death, but selective permeability is still evident in some fruits that have an intact, firm skin, such as plums. Plums cooked in sirup having a sugar concentration greater than that of the plum are likely to shrivel because of the passage of water through the skin into the sirup.

Any general chemistry textbook will provide further information on the subject of solutions.

Colloidal Dispersions

The dispersed particles in a *colloidal dispersion* are sufficiently large that their numbers in a given volume of dispersion medium cannot be great enough to permit effects on colligative properties comparable to those observed in solutions. On the other hand, the particles are not large enough to be observed with an optical microscope. The particles in colloidal dispersions are aggregates of molecules and their size falls into a rather wide range, 1–100 nm. Many substances can exist, depending on the conditions, in true solution, in colloidal dispersion, or in suspension. (In suspensions, the particles are large enough to be observed with an optical microscope.) Sugar is an example. When the sirup for a crystalline candy has been concentrated, the sucrose still is in solution in the form of molecules. In the crystallization process, crystal growth involves increasing particle size first to colloidal dimensions and then to suspended crystals. Some of the sucrose remains in solution, as will be discussed in Chapter 13.

Foams and emulsions, because of their properties, frequently are considered as a special class of colloidal dispersion, even though their dispersed particles are large enough to be observed with an optical microscope. Gels also are frequently discussed with colloidal systems.

Terminology The *dispersion medium* is the continuous phase, sometimes called the *external phase.* The *dispersed substance* (the colloid) is the dispersed, discontinuous, or internal phase. It consists of colloidal particles. A *sol* is a solid-in-liquid colloidal dispersion in the liquid state. When gelatin has been hydrated with cold water and then dispersed with hot water, it is a sol. Sols in which water is the continuous phase are *hydrosols.* Most food sols are hydrosols. Sols often are changed to gels. A *gel* is a liquid-in-solid dispersion. A *foam* such as whipped cream is a gas-in-liquid dispersion; a foam such as a marshmallow is a gas-in-solid dispersion. An *emulsion* is a liquid-in-liquid dispersion. All combinations of physical states are possible except gas-in-gas. Those combinations mentioned above are most characteristic of food dispersions.

Several terms apply to the colloid itself rather than to the colloid system. A *lyophilic colloid* has an affinity for the dispersion medium. When the dispersion medium is water, the lyophilic colloid more specifically is *hydrophilic.* A *lyophobic* (or hydrophobic) *colloid* does not have an affinity for the dispersion medium. Therefore, it can be dispersed only through special treatment, as with a protective colloid. A *protective colloid* is a material that is lyophilic and can be adsorbed on the surface of the lyophobic colloidal particle, thus conferring lyophilic dispersal properties. Colloidal particles frequently are called *micelles,* particularly those formed by condensation (see next paragraph).

Preparation Colloids can be prepared by reducing the size of large particles (dispersal) or by increasing the size of small particles until they are of colloidal size (condensation). Certain substances, such as gelatin and dried albumen, form colloidal dispersions without any special treatment to reduce their particle size. Reduction of particle size of other substances may be brought about chemically, as by hydrolysis, or mechanically, as by grinding, mixing at high speeds, or forcing a liquid dispersion through a very small opening. The growth of crystals from molecules is a condensation process.

Stabilization Frequently it is desirable to maintain a colloidal dispersion as unchanged as possible. Stability necessitates preventing aggregation of particles. There are several mechanisms of stabilization.

Solvation This mechanism, called *hydration* when water is the dispersion medium, is characteristic of hydrophilic colloids. Hydrophilic colloids have groups that bind water. The effect is a layer of water that surrounds each colloidal particle and moves with the particle. Such particles do not aggregate unless they are destabilized. Most food colloids are hydrophilic, and thus their dispersions are stabilized by solvation. *Caseins* are hydrophobic individually, but they occur in milk combined into micelles that are structured in such a way that hydrophilic groups are concentrated at the particle surface.

Adsorption of surface-active substances A protective colloid is an example of a surface-active substance. Though a substance contributing this type of stabilizing effect does not have to be a colloid, it does have to be able to orient itself on the particle surface in such a way that a protective film is formed to prevent aggregation. Such behavior is characteristic of foam stabilizers and emulsifying agents. Surfactant properties are discussed further in Chapter 8.

Presence of electric charge on particle surfaces Sulfate, phosphate, carboxyl, and amino groups may be present and may be charged. When all the particles carry the same net charge, they repel one another. The effect is stabilization of the dispersion. A substance that does not have charged groups might adsorb either positive or negative ions preferentially. Whether the adsorbed ions are positive or negative, their like charges cause the particles to repel one another.

Lyophilic colloids may be stabilized simultaneously by hydration and electric charge. Pectin is an example of such a colloid.

Destabilization *Destabilization* is aggregation into particles of larger-than-colloidal dimensions. Colloidal dispersions stabilized by hydration are destabilized by competitive removal of the water layer. In the making of jelly, the sugar contributes to gel structure in that it dehydrates the pectin particles. The effectiveness of many surface-active substances can be reduced by heating. The stabilizing role of an electric charge can be eliminated by neutralization of the charge. For example, the acid in jelly neutralizes the charge on the pectin particles. With stabilizing forces overcome, interactions between particles become possible and gels may form, flocculation may occur, or aggregates may precipitate, depending on the specific substances and the prevailing conditions. In the case of gel formation, bonding occurs between dispersed particles to such an extent that a solid rigid structure is formed, with water trapped in the meshes of the solid network. Note that there is a reversal of phases in gelation. A solid-in-liquid dispersion is changed to a liquid-in-solid dispersion.

Many sources of information on the general principles of colloid chemistry are available, including the summarization by Whitney (1977).

Some specific colloidal substances Proteins represent a major group of colloids in food and are discussed below. Hydrocolloids, which have come into prominence in food applications in recent years, will also be discussed briefly.

Proteins Specific proteins will be discussed throughout the book, but some general information will be provided here. *Amino acids* are the structural units of proteins. The characteristic feature of the amino acids that differentiates them from fatty acids is the amino group on the α-carbon:

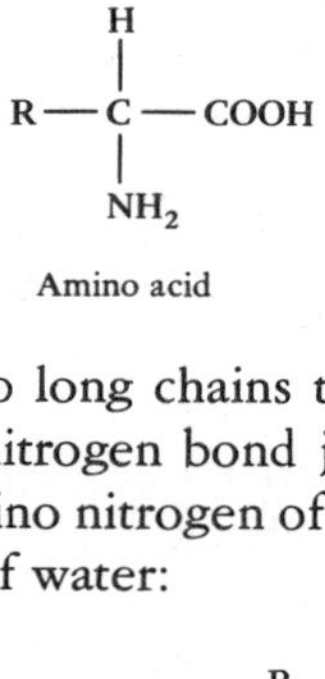

$$\begin{array}{c} \text{H} \\ | \\ \text{R}-\text{C}-\text{COOH} \\ | \\ \text{NH}_2 \end{array}$$

Amino acid

Amino acids are joined into long chains through peptide linkages. A *peptide linkage* is a carbon-nitrogen bond joining a carboxyl carbon of one amino acid and an α-amino nitrogen of another amino acid with the splitting off of a molecule of water:

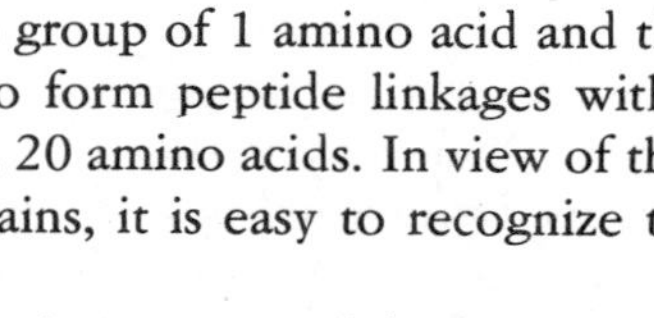

$$\begin{array}{ccc} \text{H} & \text{O} & \text{R} \\ | & // & | \\ \text{R}-\text{C}-\text{C} & \quad \text{OH} \quad & \text{HC}-\text{COOH} \\ | & & | \\ \text{NH}_2 & & \text{HN} \\ & & \text{H} \end{array}$$

Peptide linkage

Note that the amino group of 1 amino acid and the carboxyl group of the other are free to form peptide linkages with other amino acids. There are more than 20 amino acids. In view of that fact plus the great length of protein chains, it is easy to recognize that the potential for variation is large.

A protein molecule is represented diagrammatically as a backbone in which —C—C—N— is a repeating unit and the distinctive portions of the individual amino acid molecules, represented by R, extend as side chains:

```
     H     O     R     H     O     R     H
     |     ‖     |     |     ‖     |     |
     N  H  C     C     N  H  C     C     N
   /  \ | /  \  /|\  /  \ | /  \  /|\  /  \
        C     N  H  C     C     N  H  C
        |     |     ‖     |     |     ‖
        R     H     O     R     H     O
```

Backbone structure

The repeating —C—C—N— segment of a polypeptide backbone consists of an α-carbon, a carboxyl carbon, and an α-amino nitrogen. The specific amino acids and their sequence in the polypeptide chain constitute the primary structure of a protein. Only quite recently have protein chemists been able to establish the primary structure of many proteins. The magnitude of their task is apparent if one considers that κ-casein, not one of the larger food protein molecules, has 169 amino acid residues.

If there were no structure in proteins beyond the primary structure, all proteins would be monomeric extended chains. However, proteins differ in their physical and biological properties to a greater extent than can be explained by differences in primary structure. Individual proteins in their native state have characteristic conformations based on secondary, tertiary, and quaternary structure. Secondary structure results from interactions between amino acid residues that are rather close to one another in the sequence, resulting in quite regular helical or folded structures. The classical example is the α-helix with hydrogen bonding between carboxyl oxygen atoms and amino nitrogen atoms:

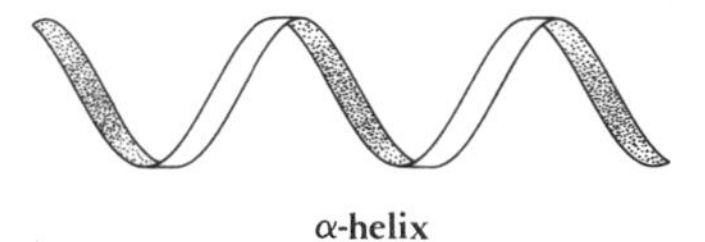

α-helix

The helical conformation does not necessarily exist over the entire chain:

Chain with helices

The chain above is quite extended in spite of the helices and is not characteristic of native proteins. Further structure is provided at the tertiary level. Tertiary structure results from interaction between amino acid residues that are farther apart in the polypeptide chain than those involved in secondary structure:

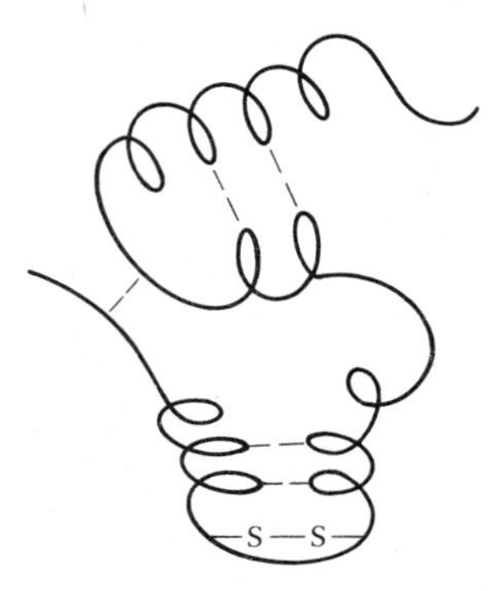

Tertiary structure

Tertiary interactions are less regular than secondary. Both secondary and tertiary structures result from intrachain bonding. Quaternary structure results from interaction between polypeptide chains, or interchain interaction:

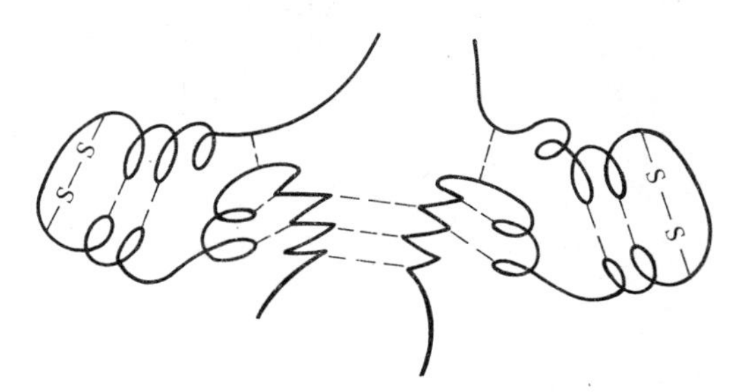

Quaternary structure

Conformation is the net result of the above levels of structure. For each protein there is a three-dimensional shape, or conformation, that is most stable under normal conditions of temperature, ionic strength, and pH; this is the "native conformation." The particle of native protein represents a low level of free energy.

The nature and sequence of the amino acid residues in a protein molecule determine the kinds of intrachain and interchain bonding that are possible contributors to secondary, tertiary, and quaternary structure. Some of the amino acids have unsubstituted aliphatic hydrocarbon chains. Others have chains with substituents such as hydroxyl (—OH) groups, additional carboxyl (—COOH) groups, additional amino ($—NH_2$) groups, sulfhydryl (—SH) groups, or ring structures. There is much potential for hydrogen bonding because of the —OH, $—NH_2$, —COOH, and some other groups. Ionic bonding is possible because the —COOH groups can be negatively charged and the $—NH_2$ groups can be positively charged. Covalent bonding can occur when cysteine, with its —SH group, is present; when two —SH groups get close enough together under oxidizing conditions, a disulfide (—S—S—) bond forms between them. Disulfide bonding can occur between portions of a single chain and also between chains. Hydrophobic bonding, or more accurately *hydrophobic interaction,* involves the hydrophobic (nonpolar) unsubstituted hydrocarbon chains or rings of certain amino acids. Such groups do not have an inherent attraction for one another, but have a common dislike for water. They come into intimate contact with one another in their attempt to avoid water. Not surprisingly, these groups tend to be oriented toward the interior of a protein molecule since proteins occur in aqueous environments. Hydrophobic interaction probably has a large influence on the folding pattern. The native conformation of a protein molecule is stabilized, as well as formed, by the various types of interactions between amino acid residues. With many proteins, native structure does not extend beyond the tertiary level. With others, association between monomers is characteristic.

Protein denaturation is a common occurrence in food. Sometimes it is desired and sometimes it is not. *Denaturation* consists of nonproteolytic changes in the three-dimensional native protein structure; in other words, denaturation does *not* include breaking of peptide bonds. Some protein chemists also exclude the breaking of disulfide bonds from the definition of denaturation. In any case, the process results in the unfolding of molecules.

Denaturation is not an all-or-nothing phenomenon. It may involve conformational change that is so slight as to be practically undetectable by most techniques. When denaturation is sufficient to cause detectable effects, those effects can be quite variable, but they always include decreased solubility. Viscosity, another property that is important to food properties, usually is increased. Heat is the primary factor contributing to protein denaturation in food. Salts and pH also can have important effects.

Renaturation is reversal of denaturation. It is more likely to be partial than complete. The heat resistance of the enzymes catalase and peroxidase is a reflection of protein renaturation.

Coagulation follows denaturation. Once extensive denaturation has occurred, new cross-bonding between chains can occur extensively because the chains are more extended and interaction is easier. More groups are exposed and molecules can approach each other closely over larger portions of their chains. With increased cross-bonding, flocculation occurs, and finally a visible coagulum appears. If the coagulating mixture is being stirred, the coagulum is manifest as increased viscosity caused by small curds. If the mixture is undisturbed during coagulation and conditions are otherwise favorable, the coagulum is continuous, resulting in a gel. A gel such as an egg gel, formed through protein coagulation, is not a reversible structure.

Gelation of a gelatin sol occurs through hydrogen bonding after the gelatin chains have been dispersed and the sol has been cooled without agitation. This solidification is not an example of coagulation, even though gelatin is a protein. The hydrogen bonds in a gelatin gel can be broken by heat and the sol re-established. Cooling then can result in gelation. This case of gelation, which does not involve protein coagulation, is reversible.

Further information concerning proteins is available from many sources, including Aurand and Woods (1973). Whitaker (1977) has discussed protein denaturation extensively.

Hydrocolloids *Hydrocolloids,* or gums, are natural or synthetic in origin. In both cases, they consist of complex polysaccharide derivatives. Many natural hydrocolloids are obtained from sea plants. Alginates, carrageenan, and agar are examples. Other natural hydrocolloids are obtained from land plants. Examples are tree exudates such as gum arabic, gum karaya, and gum tragacanth, as well as seed gums such as guar gum and locust bean gum. Synthetic hydrocolloids—for example, carboxymethyl cellulose (CMC)—are derivatives of cellulose. Xanthan gum is a biosynthetic hydrocolloid. It is synthesized by the bacterial species *Xanthomonas campestris* in a medium containing glucose and certain salts.

Hydrocolloids differ as to whether they have charged groups, and, for those that do, the extent of charge. Charged groups in various hydrocolloids include sulfate and carboxyl groups. The presence of such groups affects the response to ions, such as Ca^{++}, in food mixtures. Hydrocolloids also differ in their sensitivity to acid and to heat, in their solubility, and in their gel-forming ability.

The basic physical property that hydrocolloids have in common is their ability to absorb relatively large amounts of water. This absorption results in high viscosity, which sometimes is an end in itself. Sometimes the increased viscosity contributes to other desired effects of hydrocolloids in food formulation, such as control of crystallization and prevention of the settling-out of solids. Other functions in food include reduction of syneresis (separation of liquid from a gel or very viscous mixture), reduction of evaporation rate, protection of emulsions (possibly through an effect on viscosity), formation of gel structure, and contributions to special textural effects, such as gumminess. A study of the labels on a large variety of processed and formulated foods gives insight into the widespread use of hydrocolloids.

The subject of hydrocolloids is pursued further in recent reviews by Kovacs and Kang (1977) and Meer (1977).

ENZYMES

Enzymes are biological catalysts. The reactions they catalyze would occur very slowly in their absence. Many of the chemical changes in living organisms are catalyzed by enzymes. In plant and animal foods, enzyme activity continues after death unless the enzymes are inactivated. Enzymes are used in the food industry to produce specific effects. Lactase treatment of milk (Chapter 13), enzymatic tenderization of meat (Chapter 5), and glucose isomerase treatment of corn sirups (Chapter 13) are among the numerous examples mentioned in this book.

Mechanism of Action

Enzyme activity involves the formation of a complex between the enzyme and the substrate. This complex changes the conformation of the substrate molecule and expedites the reaction by putting a strain on the bond to be broken. This step, activation, greatly reduces the amount of energy required to cause the reaction.

The reaction then proceeds and the enzyme is regenerated, becoming available to complex with another molecule of the substrate. This reuse is the reason that enzymes, like inorganic catalysts, are effective in minute quantities.

Certain enzyme-catalyzed reactions in food do not occur in the intact tissue because of the physical separation of the enzyme and the substrate. Mechanical damage to the tissue can bring the enzyme and the substrate into contact with each other and permit such a reaction to occur.

Nomenclature and Nature of Enzymes

Nomenclature has not been very systematic. Enzymes frequently are identified according to the substrates on which they act. Examples are sucrase, lipase, and chlorophyllase. Indication of the type of reaction catalyzed is also common, as with dehydrogenase and invertase. Some enzyme names identify both the substrate and the reaction catalyzed. Pectinmethylesterase and ascorbic acid oxidase are examples.

Specificity of enzymes is variable. An enzyme might be so specific that it catalyzes only one reaction involving one substrate. Sucrase, or invertase, is an example. Esterases are examples of enzymes having a limited degree of specificity.

Enzymes are proteins. Their inactivation by heat, therefore, is an example of heat denaturation. Some enzymes function only in the presence of nonprotein coenzymes. Ascorbic acid oxidase, for example, requires copper.

Factors Affecting Activity

Enzyme activity is affected by several factors, one of the most important being temperature. For each enzyme there is an optimum temperature.

It is around 35–40°C for most enzymes, though there are notable exceptions. Enzyme activity is low at low temperatures and increases with increasing temperature, approximately doubling with every 10°C increase in temperature up to the optimum temperature. Further increases in temperature result in increasing denaturation until finally inactivation is complete. The optimum temperature depends on the pH of the medium.

In addition to the above indirect effect, pH also has a direct effect on enzyme activity at a given temperature. For every enzyme there is an optimum pH, which is in the vicinity of pH 7 for all but a few enzymes. Each enzyme shows activity within a pH range that is very narrow for some enzymes and relatively broad for others. In the case of a narrow pH range for activity, buffering is important to continued enzyme activity. In any case, pH levels that cause denaturation also cause enzyme inactivation.

Substrate concentration affects enzyme activity. If all other factors are constant, the initial reaction rate increases with the initial substrate concentration. When the initial substrate concentration is increased beyond a certain level, the enzyme becomes the limiting factor. Increasing the substrate concentration further, therefore, has no effect.

Enzyme concentration affects reaction rate in a similar fashion. As the enzyme concentration increases, the initial reaction rate also increases until the substrate concentration becomes limiting.

With a given combination of substrate concentration and enzyme concentration, the reaction rate changes during the course of the reaction. The initial reaction rate is relatively rapid. Later, the rate slows down and ultimately levels off. Product accumulation is responsible for this effect. A recent technological development called *enzyme immobilization* has the potential of nearly eliminating the retarding effect. The enzyme is immobilized on a solid support that can bind the enzyme without masking the active sites that are needed for complexing the substrate. If the substrate is then passed through a column or packed bed of the solid, the reaction proceeds continuously, catalyzed by the enzyme, with the product(s) collected beyond the column or packed bed. Thus the products have little or no effect on the reaction rate. Even if a batch method is used, immobilization of the enzyme on the solid support has the advantage over traditional methods of permitting indefinite reuse of the enzyme because the solid is readily separated from the system. In previous enzyme applications it was necessary to destroy enzymes in order to stop the reactions they catalyze. The history, methods, and applications of this rapidly developing technology are described by Messing (1975).

Among recent references on the general subject of enzymes is the review by Eskin et al. (1971). Several enzymes and the nature of the reactions they catalyze will be mentioned throughout this book.

ACIDITY AND HYDROGEN ION CONCENTRATION (pH)

The acidity of foods or of the medium in which they are cooked frequently has an important effect on certain qualities of the product, such as the color of a cooked green vegetable or the thiamin retention of quick breads. Acidity often is expressed as pH.

Acidity

Acids are substances that donate the hydrogen ion, H^+, and *bases* are substances that furnish the hydroxide ion, OH^-. Acids and bases also may be called *proton donors* and *receptors,* respectively. The concentration of acids and bases often is expressed as molarity or normality. Normality is somewhat more convenient than molarity in considering solutions that react with one another, because it takes into account the number of replaceable hydrogen atoms or hydroxyl groups. A given volume of a normal solution of any acid will react with the same volume of a normal solution of any base.

Although it is convenient for many purposes, normality does not always give all the necessary information about the acidity of a solution. For example, 0.1 N solutions of hydrochloric acid and of acetic acid are quite different. The 0.1 N acetic acid solution, which is equivalent to highly diluted vinegar, tastes much less sour than the hydrochloric acid solution and reacts less strongly with indicators and other substances. The reason for these differences is that sensation of sourness and certain other effects depend on the concentration of hydrogen ions in the solution. Far more hydrogen ions are present in the hydrochloric acid solution than in the acetic acid solution because the hydrochloric acid is almost completely ionized and the acetic acid is only about 1%

ionized. Since both are 0.1 N, however, they react to the same extent with a base in titration. Titration is a measure of total acidity rather than of hydrogen ion concentration.

Hydrogen Ion Concentration (pH)

Hydrogen ion concentration, sometimes called *active acidity* to distinguish it from *total acidity,* is important in a study of food. It usually is expressed as pH so that the values obtained will be simple numbers. If the hydrogen ion concentration is expressed as moles of hydrogen ion per liter, an awkward number results, such as 0.0000001 mole for pure water. This decimal also can be written $1/10^7$ or 10^{-7}. For convenience, the exponent in the expression $1 \times 1/10^7$ is used to designate pH. The p stands for power and the H for hydrogen ion.

The corresponding values for pH and hydrogen ion concentration in Table 2.1 show that very acidic solutions have a high concentration of hydrogen ions and, therefore, a low pH. Note that the most acidic solution contains some hydroxyl ions and the most basic solution contains some hydrogen ions. This is explained by the slight ionization of water to form both hydrogen and hydroxyl ions. Pure water, with equal concentrations of hydrogen and hydroxyl ions, has a neutral reaction (pH 7). A solution at any pH contains ten times as many hydrogen ions as an equal volume of a solution of the next higher pH. Hydrogen ion concentration is 0.01 mole per liter at pH 2 and 0.001 mole per liter at pH 3.

It has been mentioned that although acid solutions of the same normality give the same value on titration, they may be quite different in hydrogen ion concentration. This is illustrated in Table 2.2, in which the pH values of 0.1 N solutions of several acids and bases are shown. Hydrochloric and sulfuric acids, because they are almost completely ionized, have pH values approaching 1 at a normality of 0.1, whereas acetic acid, which is only slightly ionized, has a pH of 2.87 at a concentration of 0.1 N. Similarly, the 0.1 N solution of the highly ionized base sodium hydroxide has a much higher pH than that of a 0.1 N solution of the slightly ionized ammonium hydroxide.

Change of pH in solutions The pH of distilled water usually is below 7 because water absorbs carbon dioxide from the atmosphere. The unstable acid, carbonic acid, is formed when carbon dioxide dissolves in water. The pH of a solution or of water can be changed by the addition of acidic or basic substances. Often the acidic or basic compounds that alter the hydrogen ion concentration of water and of aqueous solutions are salts. Many salts ionize almost completely. A salt of a strong base and a weak acid raises the pH; a salt of a weak base and a strong acid lowers the pH; and a salt of a strong base and a strong acid does not affect pH.

Changes in pH resulting from the addition of acids, bases, or salts are much more pronounced in water than in foods because foods are buffered. Buffered solutions resist changes in hydrogen ion concentration on the addition of acid or base or on dilution. They do this by taking up hydrogen or hydroxyl ions to form slightly ionized substances. Buffered solutions prepared in the laboratory usually contain

Table 2.1 Concentration of hydrogen and hydroxyl ions in solutions of different pH

Reaction	pH	Hydrogen ion concentration[a]	Hydroxyl ion concentration[a]
↑		mole/liter	mole/liter
Acidic	0	10^{0}	10^{-14}
	1	10^{-1}	10^{-13}
	2	10^{-2}	10^{-12}
	3	10^{-3}	10^{-11}
	4	10^{-4}	10^{-10}
	5	10^{-5}	10^{-9}
	6	10^{-6}	10^{-8}
Neutral	7	10^{-7}	10^{-7}
	8	10^{-8}	10^{-6}
	9	10^{-9}	10^{-5}
Basic	10	10^{-10}	10^{-4}
	11	10^{-11}	10^{-3}
	12	10^{-12}	10^{-2}
	13	10^{-13}	10^{-1}
↓	14	10^{-14}	10^{0}

[a]The product of the hydrogen ion concentration and the hydroxyl ion concentration in water and in dilute aqueous solutions is a constant, 1.00×10^{-14} at 25°C. This fact and the knowledge that numbers can be multiplied by adding their exponents explain why the sum of the exponents in these two columns for the same pH is always -14. It will be recalled that $10^{0} = 1$.

Table 2.2 The pH values of 0.1 N solutions of several acids and bases

Acid or base	pH
Hydrochloric acid	1.07
Sulfuric acid	1.23
Acetic acid	2.87
Ammonium hydroxide	11.27
Sodium hydroxide	13.07

mixtures of weak acids, such as acetic acid or phosphoric acid, and their salts, although weak bases and their salts also can be used. The addition of hydrochloric acid to a solution containing equal parts of acetic acid and sodium acetate illustrates buffer action. As hydrochloric acid is added, the proportion of acetic acid increases while that of sodium acetate decreases, and sodium chloride is formed. Because of the slight ionization of acetic acid, there is no great change in pH until the sodium acetate in the original mixture is nearly exhausted. Similarly, the addition of sodium hydroxide to such a buffer solution increases the proportion of sodium acetate, but it changes the pH little until the acetic acid is almost exhausted. Foods are buffered by proteins and by salts.

Determination of pH Indicator solutions and papers treated with indicator solutions can be used for colorimetric estimation of pH. An indicator changes color at a known pH and is a useful way to obtain information quickly without the use of special equipment. However, the accuracy of indicators is limited and the use of indicators with colored materials is difficult; therefore, pH is measured electrometrically in food research. Electrometric determination of pH is possible because hydrogen ions carry a charge. When the reference electrode and a measuring electrode of a pH-meter are put into a solution, an electromotive force between the electrodes is proportional to the hydrogen ion concentration of the solution. Measured as voltage, the electromotive force is converted into pH by the meter. Buffers of known pH are used for standardization of the instrument prior to use. The standard buffer solution and the unknown solution should be at the same temperature.

BROWNING

Browning is a common occurrence in foods. It is undesirable in some cases and desirable in others. The goal in handling food is to inhibit undesirable browning reactions as much as possible and to control desirable browning reactions. Both enzymatically catalyzed and nonenzymatic browning reactions occur in food.

Enzymatic Browning

Enzymatic (or oxidative) *browning* occurs on the cut surfaces of certain fruits, such as apples, peaches, bananas, and pears. This type of browning requires three factors: substrate, which consists of polyphenolic compounds; a polyphenol oxidase (or phenolase), an enzyme that can catalyze the first step in the reaction; and oxygen, a reactant. Both the substrate and the enzyme are present in susceptible fruits, and cutting into the tissue gives them access to one another. Oxygen becomes available when the cut surface is exposed to air. The enzyme catalyzes the oxidation of the polyphenolic substrate to a quinone. Quinones are not dark in color but are readily polymerized to dark-colored compounds.

Preventive steps include inactivation of the enzyme by heat denaturation, the use of acid to inhibit enzyme activity, exclusion of oxygen, and reduction of quinones by ascorbic acid back to polyphenols before they can polymerize. Bisulfites also interfere with browning, possibly through an effect on the enzyme and possibly through reduction of quinones. The occurrence and prevention of enzymatic browning are discussed further in Chapter 6.

Nonenzymatic Browning

Nonenzymatic browning has been studied extensively and much remains to be learned. Many reactions contribute to nonenzymatic browning. It is possible that several mechanisms can contribute to the browning of a specific food. Some nonenzymatic browning is accelerated by high pH and some by low pH. Nitrogenous compounds are involved in some, but not all, nonenzymatic browning. Sugars participate in more than one browning mechanism. Ascorbic acid plays a role in some browning. Apparently, all types of nonenzymatic browning are accelerated by heat.

Carbonyl-amine browning *Carbonyl-amine browning* also has been called the *Maillard reaction.* Like enzymatic browning, it actually involves more than one reaction, but the total process is more complex than the process of enzymatic browning. The initial step is the condensation of the carbonyl group of a reducing sugar and a free amino group of a protein or a free amino acid. The condensation product is a Schiff base, which undergoes cyclization to a substituted glycosylamine. The glycosylamine in turn undergoes Amadori rearrangement to form a ketose derivative from the aldose. The brown pigments may arise thereafter through a series of fragmentation, condensation, and polymerization reactions. Different pathways are possible, each resulting in brown pigments called *melanoidins.*

The factors affecting carbonyl-amine browning include pH, temperature, and moisture content, as well as the sugars and amino acids available. Carbonyl-amine browning increases with increasing pH. It also increases as temperature increases. Browning rate generally is higher at low-to-intermediate moisture levels than at very high or very low levels. Pentoses are more reactive than hexoses. Sucrose, being a nonreducing sugar, does not participate. Lysine is a particularly reactive amino acid. Both desirable and undersirable cases of carbonyl-amine browning will be mentioned in various chapters.

Sugar browning Also called *caramelization,* sugar browning differs from carbonyl-amine browning in several respects. It does not require nitrogen. It can involve any kind of sugar, including the nonreducing sugar sucrose. It is favored by both high and low pH levels. It requires very high temperatures, whereas carbonyl-amine browning, although accelerated by high temperatures, can occur at room temperature and lower if other conditions are favorable. When a sugar is heated to temperatures above its melting point (for example, above 186°C for sucrose), dehydration reactions occur, resulting in the formation of furfural derivatives which undergo a series of reactions ending with polymerization to brown-colored compounds. Acid and alkali apparently contribute to different reaction steps.

Ascorbic acid browning Ascorbic acid plays a role in the browning of citrus juices and concentrates. Oxidation of ascorbic acid is involved. Nitrogen is not; nor, apparently, is sugar. Over the pH range characteristic of citrus products, the lowest pH levels are associated with the greatest susceptibility to browning.

Prevention of nonenzymatic browning Since there probably is considerable overlap among nonenzymatic browning reactions, control of browning in a given food system is subject to some trial and error. Certainly low temperature storage is helpful in any case. Bisulfites also seem to inhibit a variety of browning reactions. To the extent that variation in moisture and pH levels is feasible, these factors are best controlled according to the experimentally established conditions for stability of a specific food.

Among recent reviews of browning reactions is that by Eskin et al. (1971).

SUGGESTED EXERCISES

1. Find sucrose and sodium chloride in the organic and inorganic sections respectively of the tables of physical constants in the *Handbook of Chemistry and Physics* (CRC, 1977). Note the magnitude of the values for solubility in water in each case. Which is more soluble? Comment on the effect of temperature on the solubility of sucrose and of sodium chloride. Note the molecular weights of the two compounds. What is the significance of their molecular weights with regard to their effects on the boiling point of water?
2. In separate pans of boiling distilled water, make stepwise additions of sucrose and of sodium chloride. (The increments need to be larger for sucrose than for sodium chloride.) After each addition, stir to dissolve the solute, bring the solution back to a boil, and record the boiling point. Continue making additions and recording boiling points until you have demonstrated: (a) a difference in the amounts of the solutes required to bring about a given increase in the boiling point of a given amount of water (What controls will be necessary?); (b) a difference in the solubility of sucrose and sodium chloride; and (c) the relationship between solubility and effect on boiling point.

3. Sometimes comparisons are made in laboratory classes of the crystallization of sucrose and of glucose in a simple mixture such as a fondant. The subject matter of this chapter should have made it clear that heating both solutions to the same boiling point will not produce solutions of equal concentration. (A mole of sucrose is 342 g and a mole of glucose is 180 g.) Calculate the boiling point of a glucose solution that is theoretically equal in concentration (on a weight percentage basis) to a sucrose solution that boils at 114°C.
4. Mix equal parts of a commercial liquid pectin and distilled water. To 10-ml portions in 50-ml beakers, make the following additions: (a) 1.5 ml 0.1 N HCl, (b) 10 g sucrose, (c) 1.5 ml 0.1 N HCl and 10 g sucrose, and (d) 3 ml 70% ethanol. Prepare a second series of samples and heat to the boiling point. Cool to room temperature. Observe and explain the results.
5. Study labels of processed and formulated foods and make a list of the hydrocolloids included and the foods containing them. State the probable function(s) of the gum(s) in each food.
6. Measure the pH of tap water and distilled water. Boil and cool each type of water and again measure pH. Explain the results.
7. Study the buffering action of foods by finding the effects of acid and base on pH as determined by a colorimetric or an electrometric method. Find the pH of boiled, cooled distilled water. Then add to 10 ml of the water enough 0.1 N HCl to lower the pH by two units. Measure the amount of acid required. Next find the pH of a mixture containing half milk and half boiled distilled water before and after adding to 10 ml of the mixture the same amount of acid as was used with the water. If the pH change is less in the presence of milk than of water, the buffering effects of salts and protein are responsible.
8. Repeat 7, using enough 0.1 N NaOH to raise the pH by two units.
9. Heat solutions containing three sugars (sucrose, glucose, and any pentose) and three amino acids (glutamic, glycine, and lysine) in all nine combinations. Compare as to extent of browning.

REFERENCES

Aurand, L. W. and Woods, A. E. 1973. "Food Chemistry." Avi Publ. Co., Westport, CT.

CRC. 1977. "Handbook of Chemistry and Physics," 58th ed., Ed. Weast, R.C. CRC Press, Inc., Cleveland.

Eisenberg, D. and Kauzmann, W. 1969. "The Structure and Properties of Water." Oxford University Press, New York.

Eskin, N. A. M., Henderson, H. M. and Townsend, R. J. 1971. "Biochemistry of Foods." Academic Press, New York.

Kovacs, P. and Kang, K. S. 1977. Xanthan gum. In "Food Colloids," Ed. Graham, H. D. Avi Publ. Co., Westport, CT.

Meer, W. A. 1977. Plant hydrocolloids. In "Food Colloids," Ed. Graham, H. D. Avi Publ. Co., Westport, CT.

Messing, R. A. 1975. "Immobilized Enzymes for Industrial Reactors." Academic Press, New York.

Whitaker, J. R. 1977. Denaturation and renaturation of proteins. In "Food Proteins," Eds. Whitaker, J. R. and Tannenbaum, S. R. Avi Publ. Co., Westport, CT.

Whitney, R. McL. 1977. Chemistry of colloid substances: general principles. In "Food Colloids," Ed. Graham, H. D. Avi Publ. Co., Westport, CT.

I. FOOD SCIENCE TODAY

3. Eggs

Eggs serve important roles in many food products because of their functional properties. The structure and composition of eggs in relation to their functional properties will be considered in this chapter. The three functional properties of eggs that are of importance with respect to food preparation are coagulation, emulsification, and foaming.

STRUCTURE AND COMPOSITION

A diagram of an egg is shown in Figure 3.1. The yolk contains alternate light and dark layers of yolk material surrounded by a colorless vitelline membrane. The yolk is held in place within the egg by means of the *chalazae,* opaque fibrous structures that extend into the white on either end of the egg and are continuous with the chalaziferous layer immediately surrounding the yolk. Just outside the chalaziferous layer is a layer of thin white, surrounded by the thick white and an outer layer of thin white. The thick white constitutes over half of the total fresh egg white. The shell is granular in structure and sufficiently porous to permit respiration of a developing embryo. The shell is covered with a thin outer layer of organic matter. Within the shell are two membranes that separate to form the air cell at the large end of the egg. The air cell forms as the egg cools and the contents contract following laying.

All the nutrients needed by the developing chick are supplied by the egg. The composition of the egg with respect to components important to its role in food processing and preparation is given in Table 3.1. The white, frequently referred to as *albumen,* is comprised largely of protein and water with only a trace of fat. The carbohydrate, though small in amount, can cause problems in the drying of egg whites unless removed. Reducing sugars, primarily glucose, react with proteins in the carbonyl-amine reaction to produce undesirable brown discoloration in dried egg whites as well as the whites of hard-cooked eggs (Baker and Darfler, 1969).

The protein of egg white is a complex mixture. Egg albumen proteins were discussed in detail by Vadehra and Nath (1973). Ovalbumin, a phosphoglycoprotein, constitutes 54% of the protein in egg white. In its denatured state, ovalbumin is an important structural component of baked products. Conalbumin constitutes 13% of the proteins and is characterized by its ability to bind divalent and trivalent metallic ions such as iron. Several globulins are present including lysozyme, which

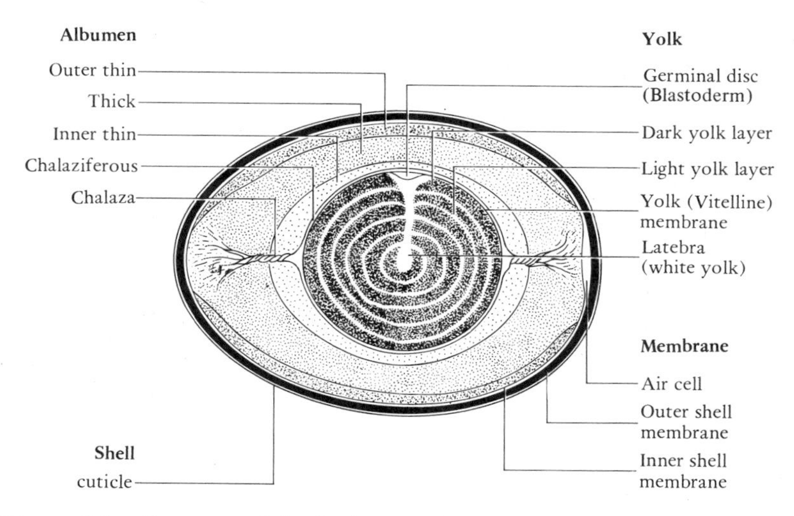

Figure 3.1. Structure of the hen's egg. (AMS, 1975.)

Table 3.1 Relative weights and composition of egg

	Weight[a] g	Water %	Protein %	Fat %	Carbohydrate %
Whole	50	73.7	13.0	11.6	1.0
White	33	87.6	10.9	trace	0.9
Yolk	17	51.1	15.9	30.6	0.6

[a]Large egg.
SOURCE: Adams, 1975.

exhibits bacteriolytic activity. The globulins, as a group, are important for the foaming properties of egg white. This has been shown in studies of the functional properties of duck egg white proteins. The addition of globulins to duck egg whites, which are inherently lacking in this group, results in improvement of the volume of angel food cakes made with the whites (MacDonald et al., 1955). The globulins represent about 8% of the proteins in egg white. Ovomucoid, a heat-resistant glycoprotein, constitutes about 11% of the protein portion of egg white. Other proteins are present in smaller quantities. Ovomucin, another glycoprotein, is thought to contribute to the thickness of the white. Thick white contains four times as much ovomucin as thin white (Feeney et al., 1952). Ovomucin and lysozyme interact to form a complex that may have a role in the thinning of egg white during storage (Powrie, 1973). Other proteins present in small amounts include ovoinhibitor, a proteolytic enzyme inhibitor; an apoprotein that binds riboflavin as flavoprotein; and avidin, which binds biotin, making it biologically unavailable. Avidin is rendered inactive by heating.

Yolk is more concentrated than white, containing less water, more protein, and a considerable portion of fat, as shown in Table 3.1. Yolk also contains several proteins, but has not been studied as extensively as the white. Two lipoproteins, or proteins loosely bound to lipid, have been indentified as lipovitellin and lipovitellenin. Egg yolk also contains phosvitin, a phosphoprotein, and a group of water-soluble proteins, the livetins (Powrie, 1973). The lipids of egg yolk include phospholipids such as lecithin, glycerides, and sterols, primarily cholesterol.

The color of the yolk depends on the presence of the carotenoids, primarily the xanthophylls. It varies from a pale yellow to a bright red depending on the amount and type of pigment present in the diet of the hen (Carlson and Halverson, 1964; Alam et al., 1968).

EGG QUALITY

Evaluation of Egg Quality

Commercial evaluation Eggs are sold commercially according to weight and grade, factors that are independent of each other. The weight of eggs for the purpose of size classification is expressed in ounces per dozen: a minimum of 15 ounces per dozen for Peewee eggs, 21 ounces per dozen for Medium, 24 ounces per dozen for Large, 27 ounces per dozen for Extra Large, and 30 ounces per dozen for Jumbo eggs. This variation in size illustrates the desirability of weighing broken-out egg when doing experimental work. When recipes specify a certain number of eggs, it is assumed that Medium or Large eggs will be used. For commercial sorting of eggs into size categories, a range including the given weights is used. The weights for a Large whole egg, the white, and the yolk were given in Table 3.1. For Medium eggs, the values are 45, 29, and 15 g respectively (Adams, 1975). These weights may be used to determine how much frozen bulk egg should be used in a formula.

Commercial quality evaluation and grading of eggs involve procedures outlined by the United States Department of Agriculture (USDA) (AMS, 1975). The quality of an egg is determined by evaluating a number of factors. The final score given to an egg can be no higher than the lowest score given to the egg for any one quality factor. Therefore, one factor alone may determine the quality score. Quality factors include both exterior and interior factors. External factors include shell characteristics: shape and texture, soundness, cleanliness, and color. Egg quality is not affected by shell color in spite of the opinion of consumers regarding the relative quality of brown versus white eggs. Shell color is used only for sorting eggs into groups of like color for sale and is not considered in USDA standards of quality or grades. Interior quality is evaluated commercially by *candling,* where the condition of the egg is judged as it is held up to a light in a dark room and rotated. The candling light reveals the condition of the shell, the size of the air cell, and the size, distinctness, color, and mobility of the yolk. Abnormalities indicative of inedible yolk also are evident. These include blood rings, embryonic development, mold growth, discoloration, and other signs of spoilage. As the grade of the egg decreases, the size of the air cell increases, the white becomes thinner, and the yolk becomes more distinct in candling and appears flattened and enlarged when the egg is broken from the shell. The appearance of graded eggs that have been broken from the shell is shown in Figure 3.2.

It cannot be assumed that every egg in a carton of a certain grade will meet all the requirements of that grade, because grade specifications allow individual eggs in a lot to be of lower quality than indicated. However, 80% of the eggs must fall within the specified grade at the destination (AMS, 1975). This tolerance is intended to cover differences in the efficiency of graders, normal changes in the product under favorable conditions during reasonable periods between grading and subsequent inspection, and reasonable variations in interpretations by graders.

The relationship of the candled quality of eggs to their performance in food preparation was studied by Dawson and coworkers (1956). Eggs of lower grades were obtained by holding fresh eggs under varying conditions. There was a direct relationship between the candled quality of the eggs and the volume, texture, tenderness, and acceptability of angel food cakes made from them. However, there was little or no relationship between the candled quality of the eggs and the firmness of baked custards as indicated by penetration values or standing height after unmolding. It was evident from this study that the whites of AA or A quality are better for food preparation requiring whites than are those grading B or C. B or C grades may be satisfactory for formulas requiring whole eggs. However, eggs grading B or C are not readily available on the retail market.

Evaluation for research purposes For research studies, candling often is supplemented by examination of the quality of the broken-out

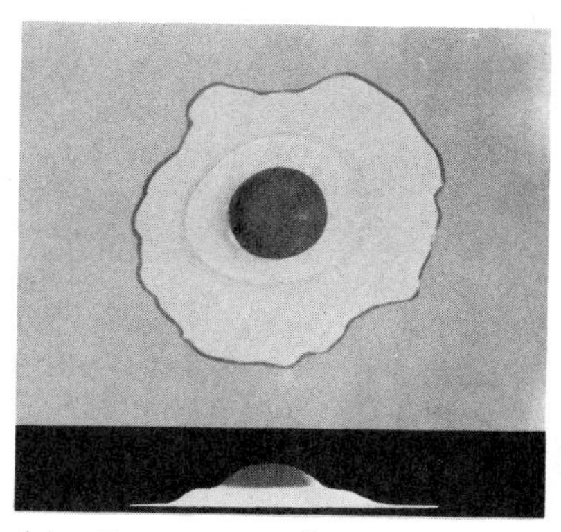

AA. Egg covers small area; much thick white surrounds yolk; has small amount of thin white; yolk round and upstanding.

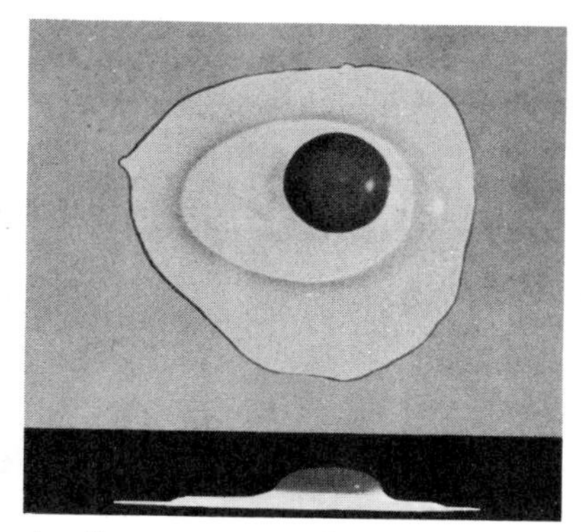

A. Egg covers moderate area; has considerable thick white; medium amount of thin white; yolk round and upstanding.

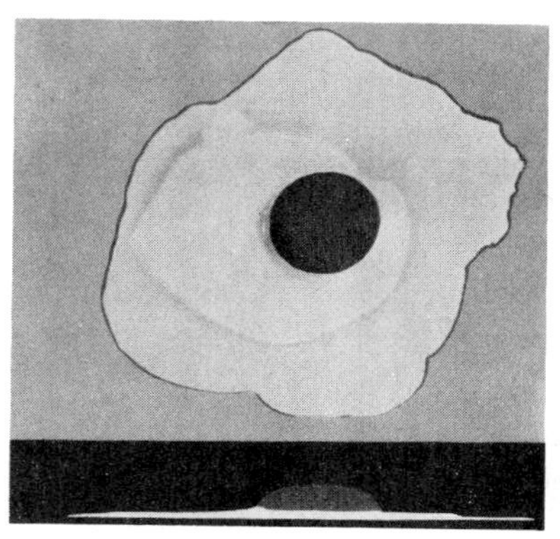

B. Egg covers wide area; has small amount of thick white; much thin white; yolk somewhat flattened and enlarged.

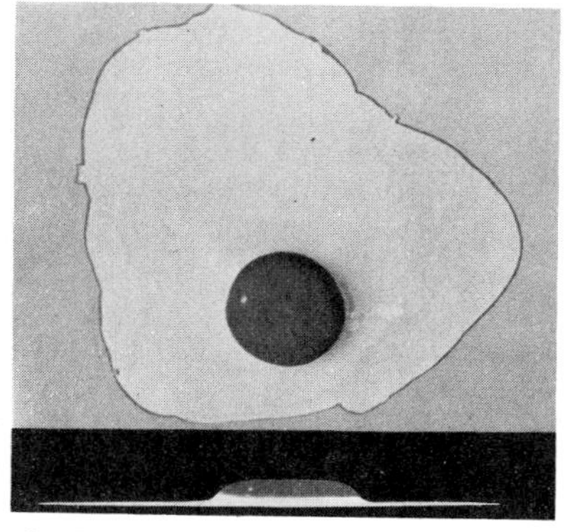

C. Egg covers very wide area; has no thick white; large amount of thin white thinly spread; yolk very flat and enlarged.

Figure 3.2. Broken-out appearance (top and side views) of eggs of varying grades. (Courtesy of Agricultural Marketing Service. U.S. Department of Agriculture.)

egg (Brant et al., 1951; Dawson et al., 1956). Several of the methods depend on measurement of the height of the thick white, which is associated with quality. The height of the albumen can be measured with a commercially available micrometer attached to a tripod[1] (Brant, 1951) or with the probe of a vernier caliper. To find the albumen index, the height of the thick albumen is divided by the width of the albumen. The *Haugh unit* is the most widely used measure of albumen quality. For this unit, 100 times the logarithm of the height of the thick albumen is adjusted to be equivalent to that of a 2-ounce (56.8 g) egg (Haugh, 1937). Tables for direct determination of Haugh units from height and weight measurements are available (AMS, 1975). As albumen quality decreases, Haugh units also decrease. Sauter and coworkers (1953) found relationships between several albumen quality measurements and candled quality. Eisen and Bohren (1963) found low but significant correlations between albumen height and volume of either outer thin white or thick white.

For experimental work it may be desirable to determine the relative proportions of thick, thin, and adhering albumen. The eggs are cracked and allowed to drain for 30 sec, after which the albumen adhering to the shell can be removed. The thin outer white will drain through a sieve and can be collected in a weighed container. The remaining portions of white can be transferred after separation from the yolk to a weighed container. After necessary weighing, the percentage of each fraction can be determined (Reinke et al., 1973).

The condition of the yolk may be evaluated by determining the yolk index, which represents the height of the yolk divided by its width when the yolk is resting on a flat surface (Sauter et al., 1951). As with the albumen index, deterioration of the egg causes a decrease in the yolk index.

A close association between commercial and research methods of grading eggs is shown by the finding that higher candled quality of stored eggs is associated with higher internal quality as indicated by albumen index, yolk index, and Haugh unit (Dawson et al., 1956).

[1]Egg quality micrometers, available from B. C. Ames Co., 131 Lexington St., Waltham, MA 02154, cost $125–$200.

Changes in Egg Quality During Storage

As soon as the egg is laid, changes begin to take place that lower its quality and eventually cause spoilage. These changes can be retarded by proper handling, but they cannot be prevented entirely. During aging, the size of the air cell increases, the yolk enlarges and its membrane weakens, the egg white becomes thinner, the egg becomes more alkaline, and the odor and flavor of the egg deteriorate. These changes and factors affecting the rate of change are discussed below. All of the changes are influenced by temperature. As temperature of storage increases, the rate of quality loss also increases. Thus, cold storage (−1.1–0°C and relatively high humidity) sometimes is used to maintain quality for long periods of time.

The increase in size of the air cell during storage is commercially important because it affects the appearance of eggs in candling. The air cell formed after the egg is laid continues to enlarge as water evaporates from the egg during storage. Moisture loss is retarded by storing the eggs at a relative humidity of 70–80% (Stadelman, 1973). Although of great importance in candling, the size of the air cell in itself is of less importance to the consumer even though it represents a loss of moisture, or a reduced yield and a changed appearance (Hale and Britton, 1974). The influence of air movement on weight loss was evaluated by Maurer (1975a) because refrigerators for home use have fans for air circulation and open containers for storage of eggs. Air speed did not influence weight loss. Eggs stored in open containers lost weight to a greater but not statistically significant extent. Carton type has little influence on quality retention at refrigerator temperatures but is important at room temperature (Mellor et al., 1975).

In addition to the loss of water through the shell, there is movement of water from the white to the yolk. The water content of yolk stored 50 days at 30°C increased from 49 to 50% (Vadehra and Nath, 1973). The transfer results in enlargement and decreased viscosity of the yolk, weakening of the vitelline membrane, and consequent flattening of the yolk when the egg is broken.

The thick egg white becomes thinner during storage. This is reflected in lowered albumen indices. A decrease in ovomucin content occurs (Vadehra and Nath, 1973), perhaps due to reduction of disulfide bonds resulting in depolymerization of ovomucin molecules (Beveridge and

Nakai, 1975). The interaction between ovomucin and lysozyme may contribute to thinning of the egg white, but whether this contribution is due to increased interaction (Hawthorne, 1950) or decreased interaction (Robinson and Monsey, 1972) is not clear.

During storage, the pH of the egg rises with the loss of carbon dioxide from the egg. The pH of white, originally about 7.9, rises to about 9.3 in the first 3 da of storage, changing little thereafter (Brooks and Taylor, 1955). The pH of the yolk, initially about 6.2, increases slowly during prolonged storage. Changes in pH influence the functional properties of egg, which will be discussed later.

Some deterioration in odor and flavor occurs during storage of eggs. Unpleasant flavors are absorbed by eggs if care is not taken to prevent odors in storage areas (MacLeod and Cave, 1977). In addition, characteristic stale odors and flavors develop in eggs during prolonged storage. If eggs are dipped in oil prior to storage, the off flavors that develop may be more intense since volatile flavor components do not

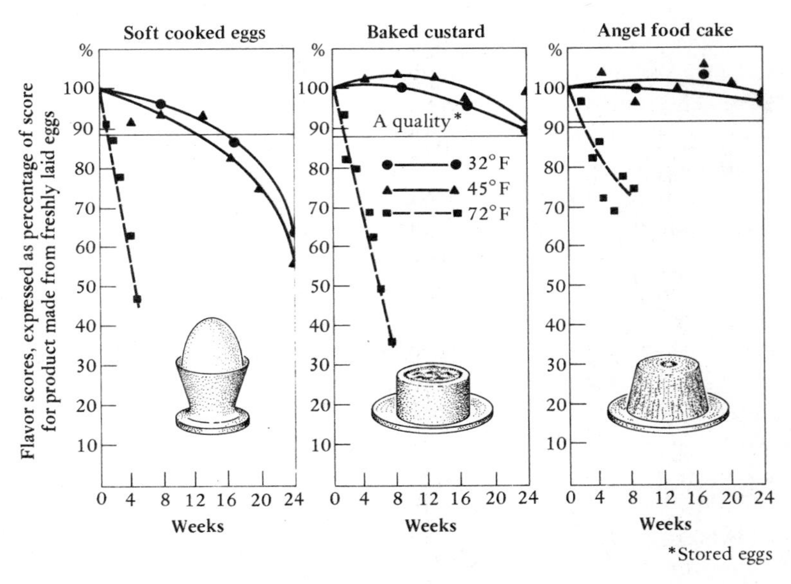

Figure 3.3. Flavor of products made from eggs held at three temperatures for varying lengths of time. (Dawson et al., 1956.)

diffuse from the egg (Stadelman, 1973). The influence of storage temperature on flavor changes in products containing eggs of varying ages was studied by Dawson and coworkers (1956). Results from the study, which are illustrated in Figure 3.3, suggest that changes occurring during storage at refrigerator temperatures are not rapid. Eggs evaluated when soft-cooked maintained flavor equivalent to that of Grade A eggs for 15 wk in cold storage, 13 wk in the refrigerator, or 1 wk at room temperature. As used for baked custards or angel food cake, however, eggs maintained quality equivalent to Grade A for at least 24 wk in cold storage or a refrigerator. In cold storage, Grade A quality can be maintained for as long as 6 mo. Deterioration seems to be most rapid during the first 3 mo of storage (Brooks and Taylor, 1955). Storage of eggs at room temperature may cause a decrease from Grade A to B in 1 wk. The importance of yolk changes in flavor deterioration is evident in Figure 3.3. Scores for angel food cakes remained higher than scores for baked custards and soft-cooked eggs.

In addition to the inevitable changes that occur during the aging of eggs, microbial spoilage sometimes takes place. When the egg is laid, its contents are usually sterile. As the egg cools, however, microorganisms may be drawn in through the porous shell. The chances for microbial invasion are increased if the shell is dirty and is washed, even if a bacteriocide is present in the water (Board, 1973). Even after the microorganisms penetrate the shell, they encounter the natural defenses of the egg, which include the shell membranes, an alkaline pH, the bacteria-dissolving protein lysozyme, and the iron and biotin binding proteins, conalbumin and avidin. Usually heavy contamination overcomes the defensive mechanisms of the egg and causes spoilage during storage.

Treatment of Eggs Prior to Storage

Before they are stored, eggs sometimes are dipped in or sprayed with light mineral oil. The deposited oil film partially closes the pores of the shell, retarding loss of moisture and the rise in pH resulting from loss of carbon dioxide. Thinning of the albumen and changes in candled quality are retarded, but the development of an off flavor is not retarded (Dawson et al., 1956). In general, oiling does not influence the functional properties of eggs with respect to food preparation (Palmer, 1972).

Another process used for improving the keeping quality of eggs in cold storage is *thermostabilization.* Thermostabilization is achieved by holding the eggs in water or oil at about 54°C for about 15 min (Funk, 1950). However, this treatment has not been widely used in industry.

PRINCIPLES OF EGG COOKERY

Eggs are used in many food products because of their functional properties. Eggs may serve more than one function in a product and are therefore difficult to replace.

Coagulation

As with other high-protein foods, the denaturation and coagulation of egg protein are important in any cooking method. *Denaturation* is a rearrangement of the orderly alignment of the molecules in a native protein. Peptide chains of a protein usually are coiled into a globular, compact shape. This arrangement is stabilized with weak bonds between amino acid residues along the peptide chains. During denaturation, some of the forces that hold the chains in coiled and folded positions are disrupted, allowing the peptide chains to uncoil and unfold. New bonds are formed to give a new arrangement to the protein molecules. The solubility of a protein is decreased by denaturation.

Coagulation is a term used to describe this entire process, which results in a loss of solubility or a change from a fluid to a more solid state. The term *gelation,* meaning the formation of a gel, also is used to designate the loss of fluidity of egg white and yolk. The coagulation of eggs is responsible for the thickening effect that eggs have in products such as custards. Egg white begins to thicken as the temperature reaches 62°C, and at 65°C it will not flow. At 70°C the mass is fairly firm. Yolk coagulates at a somewhat higher temperature than white. It begins to thicken at 65°C and loses its fluidity at about 70°C. Coagulation does not occur instantaneously but gradually over a period of time. The reaction proceeds more rapidly as the temperature of heating is increased. Since heat is absorbed during coagulation of egg proteins, the reaction is endothermic. This means that as custards are heated, the temperature will remain the same or even fall as the custard is thickening (Palmer, 1972).

Hard-cooked eggs The length of time required for coagulation of eggs cooked in the shell depends on the heating temperature. When eggs were cooked at unusually low temperatures—85°C for 30 min—the yolk was not firm enough for slicing. The white did not coagulate firmly in 90 min at 72°C (Andross, 1940). The temperature at which eggs are cooked in the shell is varied by starting the eggs in cold water and bringing to a boil, or by starting them in boiling water and turning off the heat. In either method, the proportion of water to eggs and the size of the pan must be controlled for uniform results. Irmiter and coworkers (1970) compared eggs cooked by starting in cold water with eggs placed in boiling water and simmered for 18 min. Both methods produced satisfactory products, but the boiling water start was preferred because there were fewer cracked eggs and the eggs were easier to peel. Difficulty in peeling is one problem associated with hard-cooked eggs. Eggs cooked after 7 to 10 da of storage peel easily with minimum loss of egg white and surface damage (Hale and Britton, 1974; Spencer and Tryhnew, 1973). Ease of peeling is related to the rise in pH that occurs with aging (Reinke et al., 1973). Undesirable changes are associated with the change in pH. Therefore, other methods of increasing ease of peeling fresh eggs have been sought. Hale and Britton (1974) reported that eggs cooked within 24 hr after laying were easier to peel if they were cooled rapidly in liquid nitrogen (30 sec), ice water (1 min), or cool air (25°C for 10 min) and reheated immediately in boiling water for 10 sec.

The age of the egg influences the appearance of hard-cooked eggs in another way. Yolk position is important in institutional and other large food-service operations. The yolks of eggs cooked when only 4 da old were more ideally positioned than the yolks of eggs cooked after 11 days (Grunden et al., 1975).

Sometimes the surface of the yolks of hard-cooked eggs becomes greenish due to the formation of ferrous sulfide during cooking. The iron for the compound comes from the yolk. Tinkler and Soar (1920) believed that the sulfur came from the hydrogen sulfide formed from the egg white upon heating. Later work indicated that the green color can occur when the yolk is heated alone if the pH is high, showing that the yolk can furnish both the sulfur and the iron necessary to produce the green color (Salwin et al., 1953). Baker et al. (1967) reported that as the pH of the yolk increases, more ferrous sulfide is formed. Formation of ferrous sulfide thus is more likely to occur in older eggs. As with other chemical reactions, the formation of the green color is favored by heat, whether due to overcooking or to slow cooling. Extreme discoloration of the yolk was seen in eggs boiled for 20 min and allowed to cool with the cooking water (Maurer, 1975b). Therefore, eggs should be cooked for the minimum length of time and cooled quickly with cold water.

Poached and fried eggs An excellent poached egg can be prepared by dropping an egg having a thick white into boiling water. If the white is thin, it may be impossible to obtain a satisfactory product with any method. The addition of vinegar or salt to the cooking water hastens coagulation, and hence may improve the shape of poached eggs when they have thin whites (Andross, 1940). The shape of a fried egg depends on the viscosity of the white and on the temperature at which it is cooked. If the temperature of the cooking utensil is too low, such as 115°C, the egg will spread excessively, and if the temperature is too high, such as 146°C, the egg will be overcooked. Temperatures of 126 to 137°C are satisfactory (Andross, 1940).

Scrambled eggs The amount of liquid added to the egg determines whether a small, firm mass or a larger, soft mass will be formed. Andross (1940) studied the coagulation and curdling temperatures of scrambled eggs made with 10 to 70 ml of milk per egg. The greatest difference between the coagulation and curdling temperatures occurred when 20 ml of milk were added per egg. Overcooking results in a firm, rubbery mass if little or no liquid is added, or in a curdlike mass that may separate during cooking if much liquid is used.

Stirred and baked custards Coagulation of the protein of egg is responsible for the thickening of a stirred custard and the gel structure of a baked custard. The same proportions of ingredients can be used for both types. One egg (50 g) is sufficient to thicken 250 ml of milk containing 25 g of sugar. Small amounts of salt and vanilla also are added. The effects of different types of milk on the quality and preparation of custards will be discussed in Chapter 4.

Success in making both stirred and baked custards depends on avoiding the application of excessive heat. Excessive heat results in overcoagulation and syneresis of the protein, characterized by curdling in stirred custards and porosity in baked custards.

Stirred custards are likely to curdle if they are cooked above boiling water in a double boiler. Andross (1940) found that custards cooked over rapidly boiling water thickened at 87°C and curdled at 90°C, leaving a narrow margin for safety. However, if the water in the bottom of the double boiler was cold when cooking started and was brought to a boil slowly, the custard thickened at 82°C and curdled at 87°C. Thus, the cold-water start produced custard that thickened at a lower temperature and that could be cooked with a greater margin of safety. A similar result can be achieved by keeping the water at a temperature just below the boiling point for the entire heating period. The consistency of curdled custard can be improved greatly by pouring it into a cool dish immediately after cooking and beating with a rotary beater.

Scalding the milk used in custards shortens the cooking time and also may improve the flavor and texture of the product. Baked custards made from milk preheated to 90°C were better than custards made from milk heated to 50°C. The lower temperature resulted in custards that contained many small holes (Jordan et al., 1954). Even when baked at relatively low oven temperatures, custards are improved by being placed in a pan of hot water for protection from oven heat. An oven temperature of 177°C was preferred to 163 or 204°C because the baking period was unduly long at 163°C and so short at 204°C that the custard could be easily overcooked. In addition, custards baked at 204°C were less desirable in appearance than those baked at the other two temperatures (Jordan et al., 1954). Heating custards in a microwave oven is difficult because the heating period is short and overcooking may occur. Custards can be made successfully in a microwave oven by placing the custard cups in a deep dish containing water. A flat dish containing a shallow layer of water is placed over the deep dish to serve as a lid. This arrangement slows the coagulation process and produces custards as acceptable as those baked in a conventional oven.[2] This arrangement may not be needed in a microwave oven with variable power level.

The temperature at which custard coagulates depends on a number of

[2]Class data, Department of Food Science, Nutrition, and Food Systems Administration, The University of Tennessee, Knoxville.

factors. The temperature can be studied by suspending a thermometer or thermocouple in the mixture during baking. The temperature inside the custard rises rapidly during the early portion of the baking period. Since the coagulation process is an endothermic reaction, the heating curve will become level as gelation starts. As with stirred custards, a slower rate of heating will lower the endpoint temperature. When coagulation is complete, the temperature rises again and a knife inserted in the custard comes out "clean" (Longree et al., 1961). Depending on the formula and the rate of heating, the endpoint temperature for baked custards is about 86–92°C. As the endpoint temperature within the range of 86–92°C increased, the custards became firmer and showed less syneresis on standing (Miller et al., 1959). Increasing the proportion of egg lowers the coagulation temperature and increases the firmness of the custard. Because egg white coagulates at a lower temperature than yolk, baked custards made with whites become firm at a lower temperature than those made with yolks.

Sucrose influences the baking time or endpoint temperature and quality of baked custards. Seideman and coworkers (1963) found that sucrose raised the temperature necessary for coagulation of the egg white, especially at a pH above 8.5. Wang and coworkers (1974) studied the effect of sucrose on the quality of custards baked to an internal temperature in the range of 84–86°C. Increasing sucrose levels increased baking times and crust tenderness. Tenderness of gels was shown to depend more on percentage of protein than on the presence of sucrose.

Salts are necessary for gelation of an egg custard mixture. Substitution of distilled water for milk prevents gelation. Adding salts prior to heating will result in normal gel formation. If salts are added after heating, a curdled custard will be formed. As the valence of the cations present increases, the amount of salt required decreases (Lowe, 1955). For example, 1.2, 0.3, and 0.09 g of NaCl, $CaCl_2$, and $FeCl_3$ were required for maximum gelation of a mixture of 48 g egg, 244 g milk, and 25 g sugar. Milk supplies the necessary salts in custards made without NaCl.

Emulsification

Egg yolk, which is itself an emulsion, is a good emulsifying agent for fats or oils and water. It performs this function in many foods, including shortened cakes and mayonnaise. The emulsifying qualities of eggs often are studied in mayonnaise because it is a relatively simple system.

Foaming

The foaming of egg whites is important in many foods, because it makes the products light in texture and contributes leavening. Egg white foam is a colloidal suspension consisting of bubbles of air surrounded by albumen that has undergone some denaturation at the liquid-air interfaces. This denaturation, which is caused by the drying and stretching of the albumen during beating, makes some of the ovomucin component of the albumen insoluble, thus stiffening and stabilizing the foam. The globulins are important in foam formation because they contribute to viscosity and lower the surface tension. During denaturation the protein molecules unfold, and their polypeptide chains lie with their long axes parallel to the surface. Overbeating incorporates too much air, resulting in denaturation of too much ovomucin so that the protein films become thin and less elastic. Elasticity is needed, especially in foams that are to be baked, so that the incorporated air can expand without breaking the cell walls before the ovalbumin is coagulated by the heat of the oven.

Procedures for evaluation of egg white foams A decision as to when to stop beating a foam is made on the basis of the appearance of the foam in the bowl when the beater is withdrawn and when the bowl is tilted. The characteristics of egg whites beaten to various stages are described in Table 3.2. Whip time may be defined as the time required to reach one of these stages (Seideman et al., 1963; Baldwin et al., 1967).

Evaluation of egg white foams also includes measurement of foam volume and foam stability. The volume of the foam usually is not measured directly because of the difficulty of packing it into a measuring device. A rough comparative measure is made by inserting a ruler into a foam that is still in the bowl in which it was beaten (Carney, 1938). An alternative method is to measure the depth of the foam in the bowl and then determine the volume of the bowl to that depth (Baldwin et al., 1968). An indirect measure of foam volume is made by determining the specific gravity of the foam. Directions for this measurement are given in Chapter 16. The greater the total volume of the

Table 3.2 Characteristics and uses of egg whites beaten to various stages

Stage	Description	Uses
I Slightly beaten (foamy)	Frothy, or slightly foamy; large air bubbles; transparent; flows easily	Clarifying, "coating," emulsifying, and thickening
II Stiff foam (wet peak)	Frothy quality disappearing; air cells smaller; whiter; flows if bowl is tipped; very shiny, glossy, and moist in appearance; if left to stand, liquid separates out readily; when beater is withdrawn, white follows, forming rounded peaks	Soft meringues; angel cake, if made by meringue method
III Stiff	No longer foamy; air cells very small and very white; may slip slightly if bowl is tipped; still glossy, smooth, and moist in appearance	Cakes, tortes, omelets, soufflés, marshmallows, hard meringues, ice creams, and sherbets
IV	Very white but dull; small flakes or curds beginning to show; rigid and almost brittle; particles will be thrown off especially by a whisk beater; if left to stand, liquid separates at bottom slowly	Shirred eggs

SOURCE: Niles, 1937, U.S. Egg and Poultry Magazine, Vol. 43, No. 6, June, 1937.

foam, the less will be its weight per unit volume and the lower its specific gravity.

The stability of the foam often is measured by transferring the foam to a funnel and allowing it it stand for a given period of time. The liquid is allowed to drain into a graduated cylinder for volume determinations or into a weighed container for subsequent weighing. Either measurement may be reported on the basis of the rate of liquid loss.

Factors affecting foaming The time required for foam formation, the foam volume, and the foam stability are affected by many factors. These factors include method of beating, time and temperature of beating, characteristics of the egg white, pH, and the presence of other substances such as water, fat, salt, sucrose, and egg yolk.

Method of beating In early studies of egg white foams, twin-beater electric mixers were used (Barmore, 1934; Henry and Barbour, 1933). Bailey (1935) found that the volume of foam obtained with a wire beater moving with a hypocycloidal action in a stationary bowl was greater than that obtained with stationary twin beaters and a moving bowl. The reason for these differences may be that the stationary twin beaters produce a stiff foam that becomes immobilized around the beaters. As a result, the blades rotate in a central cavity. Hand beaters prove less effective than electric mixers in foam formation (Barmore, 1934). A beater with a hypocycloidal action was used most frequently in more recent studies (Sauter and Montoure, 1972, 1975).

The blades of a blender are located so that they cut the fibers of the thick white rather than incorporate air to form a foam. Pre-blending until the ovomucin fibers are 300 μm in length reduces the time subsequently required for foam formation, and cakes baked from the foams have greater volume (Forsythe and Bergquist, 1951). Homogenizing, a more severe treatment, also reduces the whip time, but the foams produce cakes of reduced volume (Forsythe and Bergquist, 1951).

Beating time and temperature As the time of beating egg white is increased, volume of the foam first increases, then decreases. The time required depends on the type and speed of the beater. Time is shorter for higher speeds. Like volume, the stability of egg white foam also increases then decreases as beating time increases (Henry and Barbour,

1933). Maximum stability is reached before maximum volume, however. With prolonged beating, the bubbles become smaller, but more unstable, and break on standing to form a coarse foam (Barmore, 1934). Thus, increased volume of egg white foam is often accompanied by decreased stability (McKellar and Stadelman, 1955). Egg white that is at room temperature can be beaten more readily than that at refrigerator temperature, possibly because of lowered surface tension at the higher temperature. St. John and Flor (1930–31) found that chilled egg white required longer beating or produced less volume than that beaten when at room temperature (21°C), whereas egg white warmed to 30°C produced foams of increased volume but decreased stability. Barmore (1934) found little difference in stability when white was brought to temperatures ranging from 20° to 30°C and beaten with a twin-bladed electric mixer.

Characteristics of the egg white When the thick and the thin white are separated, the thin white can be beaten more readily than the thick white. Thin white produced greater volume than thick white when beaten with either a hand beater (St. John and Flor, 1930–31) or an electric mixer having twin beaters (Henry and Barbour, 1933). When a mixer with hypocycloidal action was used, the foam from the thick white had greater volume and greater stability than the foam from the thin white (Bailey, 1935). Foam volume was less for high lysozyme egg white than for egg white containing less lysozyme. The negative correlation between foam volume and lysozyme content was evident for eggs up to 4 wk old and could have been due to inadequate whip time for the thick egg white or high lysozyme egg white (Sauter and Montoure, 1972). Barmore (1936) reported that if thick white was beaten long enough, it produced angel food cakes of the same quality as did the thin white. Because thick white becomes thinner when eggs are stored, white from stored eggs may beat more quickly than white from fresh eggs.

pH The pH of the egg white is important to foam formation. Cream of tartar frequently is added to lower the pH because it was found to be more effective in increasing foam stability than either acetic acid or citric acid. Lowering the pH changes the protein concentration at the liquid-foam interface. The cream of tartar is best added during the first portion of the beating period (Barmore, 1934). The addition of acid to egg white also makes the foam more stable to heat (Slosberg et al., 1948). As foam stability is increased, shrinkage of angel-food cake is decreased. The increased stability makes it possible for heat to penetrate the product and coagulate the egg-white protein before the air cells collapse (Baldwin, 1973).

The whipping properties of duck egg whites are improved by the addition of lemon juice (Rhodes et al., 1960). The improvement apparently is due to a stabilizing effect on the ovomucin. This in turn reduces the time required for whipping. Acidified duck egg whites produce angel food cakes as acceptable as those produced with chicken egg whites.

Water and fat The finding that water added to egg white in amounts up to 40% by volume increases the volume of the foam but decreases its stability was reported by Henry and Barbour (1933). The same investigators reported that the addition of cottonseed oil reduced foam volume but did not affect stability unless the amount of oil was 0.5% or greater.

Sodium chloride The addition of salt to egg white or whole eggs before beating reduces the stability of the foam. Amounts of salt up to 1.5 g per 66 g egg white decreased volume and stability and increased whipping time for reconstituted dried egg whites (Sechler et al., 1959). The addition of salt to fresh egg whites also lessened the stability of the foam unless heating time was increased from 6 to 9 min (Hanning, 1945). With whole eggs, the addition of salt before beating resulted in a foam of small volume that would not form peaks. Sponge cakes made from such a foam were smaller in volume and less tender than when salt was omitted or was sifted with the flour (Briant and Willman, 1956).

Sucrose The effects of sugar on the formation of egg white foam are especially important in making meringues, angel food cakes, and sponge cakes. Sugar retards the denaturation of the egg white proteins. Therefore, the addition of 50% sugar more than doubled the time required for foam formation and produced a foam that was less stiff, more plastic, and more stable than one containing no sucrose (Hanning, 1945). In a study of methods of adding sugar to egg whites, it was found that soft meringues were better if the whites were beaten until they held a soft peak than if they were either not beaten at all or beaten to the stiff-but-not-dry stage before the addition of sugar (Gillis and Fitch, 1956).

Egg yolk The presence of very small amounts of egg yolk decreases the volume of egg white foam (St. John and Flor, 1930–31). Cunningham and Cotterill (1972) postulated that the lipid fraction of the yolk prevents the normal functioning of the globular proteins in foam formation. Lipovitellin did not reduce the volume of cakes prepared with lysozyme-free egg white, whereas it did reduce the volume of cakes made with normal egg white. This finding suggests the formation of a complex that reduces the volume. The problem of yolk contamination is significant in the commercial processing of eggs because it is impossible to avoid. Therefore, several techniques for improving functionality of egg whites contaminated with yolk have been proposed. The addition of 2% freeze-dried egg white to whites used in the preparation of angel food cake maintained volume and overall acceptability. The freeze-dried egg white probably supplied enough ovoglobulin, ovomucin, and/or lysozyme-ovomucin complex to replace the proteins that were bound to the yolk lipoprotein (Sauter and Montoure, 1975). Proteolytic enzymes such as papain, protease, bromelin, and trypsin reduce whipping time and stability and increase volume of foams. Although cake volume was improved and maintained, cakes made from whites containing enzymes were gummy and had coarse textures (Grunden et al., 1974). The complex formed between egg white and a yolk fraction may dissociate with heat as heating improved the foaming properties of yolk containing egg whites (Cunningham and Cotterill, 1964; Baldwin et al., 1967). The effect of heat was not evident when heat was applied to spray-dried yolk containing egg whites (Cotterill et al., 1965).

Products depending on the foaming properties of egg white Factors affecting foaming and foam characteristics have been studied in relation to specific products such as meringues, angel food cakes, and sponge cakes.

Meringues Soft meringues had more tender crusts when the beating was started at a lower speed and completed at a higher speed (Briant et al., 1954). They were better when 31.25 g of sugar, instead of 25 or 37.5 g of sugar, were used per egg white. Meringues made with 31.25 g of sugar exhibited slightly greater leakage than meringues made with 25 g of sugar. However, the superior appearance and greater ease of cutting meringues made with the highest amount of sugar more than offset the undesirable increase in leakage (Gillis and Fitch, 1956). The term *leakage* refers to the liquid that collects on the filling under the meringue on the pie. Spreading the meringue on hot rather than cold filling results in less leakage (Hester and Personius, 1949) and less tendency for the meringue to slip on the filling (Felt et al., 1956). A wide range of oven temperatures, from 163 to 218°C, can be used if baking times are adjusted so that meringues of similar color are obtained for each temperature. As baking temperature increased from 163° to 191° to 218°C, the internal temperature reached in equally browned meringues decreased. Leakage was less at the lower temperatures, but there was less stickiness and greater tenderness at higher temperatures (Hester and Personius, 1949).

The yield of meringue is sometimes increased for institutional use by the addition of water in the form of sugar sirup. The most stable meringues were obtained when the sirup contained a vegetable gum stabilizer and was added to the egg whites when hot (Felt et al., 1956). Upchurch and Baldwin (1968) reported that the addition of guar gum to fresh egg whites reduced leakage in freshly prepared meringues. Similar results were reported by Morgan and coworkers (1970) for spray-dried egg white meringues with carrageenan. There was no difference in leakage between pies made with and without a stabilizer after holding at refrigerator temperatures for 20 to 22 hr. Stored meringues containing carrageenan were more tender than those that did not contain the stabilizer.

Meringues heated by microwaves are more compact than meringues heated conventionally. Browning does not occur (Baldwin et al., 1968).

Hard meringues usually contain 50 g of sugar per egg white or twice as much as soft meringues. The sugar contributes to the sweet flavor. A relatively low oven temperature (100–120°C) is used for hard meringues. Baking times are 1 to 2 hr followed by holding in the oven after it is turned off to complete the drying process.

Angel food cakes The foaming power of egg white is important in making angel food cakes. Therefore, the factors affecting foaming discussed previously are significant. The high proportion of thick albumen in high-grade eggs contributes to superior palatability and volume of angel cakes, but may require increased beating times (Dawson et al., 1956). Better cakes were produced from whites beaten at room temperature than from those at lower temperatures.

The stage to which egg whites are whipped influences the quality of the finished product. Underbeating or incorporation of too little air to stretch the cell membranes to their potential capacity during baking results in a small cake with thick cell walls. However, overbeating also results in low volume due to the rupture of the overextended cell walls during the baking process (Palmer, 1972). The egg whites can be beaten to a soft peak and sugar beaten in, or they can be beaten until stiff but not dry and the sugar folded into the foam. A small amount of sugar usually is sifted with the flour to facilitate blending. The flour is folded gently into the foam to avoid loss of air from the beaten egg whites.

The proportion and type of ingredients in angel cakes are important factors in cake quality. If the amount of flour is increased, the cake becomes tougher and drier, whereas additional amounts of sugar make the cake more tender until, with an excessive amount, it falls. An excessive amount of sugar prevents adequate coagulation of the egg white proteins. Palmer (1972) suggested that 0.2–0.4 g of flour and 1–1.25 g of sugar can be used per gram of egg white. The larger amount of sugar is used with the larger amount of flour. Sugar must be decreased in cakes baked at higher altitudes (Lorenz, 1975). Cake flour is more satisfactory for angel cakes than is all-purpose flour. Flour supplements the egg protein structure, and so all-purpose flour, which contributes more gluten than cake flour, contributes to shrinkage of the structure during baking and cooling (Palmer, 1972).

In addition to egg white, flour, and sugar, cream of tartar is added to egg whites for angel food cake to make the cake whiter, finer in grain, and more tender than it would be without the acid. This effect can be attributed to the effect of the cream of tartar on the pH of the cake because similar results were obtained with the substitution of citric, malic, or tartaric acid if the pH of the cake was the same (Grewe and Child, 1930). The acid probably stabilizes the foam by coagulating some of the protein that surrounds the air cells of the foam. Stabilization prevents the foam from collapsing to produce thick cell walls before the temperature of coagulation is reached in the oven (Barmore, 1934). Several explanations for increased whiteness with the use of cream of tartar have been proposed. Palmer (1972) suggested that it is effective because flavone pigments contributed by the flour are colorless at neutral or acidic pH but yellow at a higher pH. An angel food cake made with cream of tartar may appear to be whiter than one made without because of a difference in the character of light reflected from the finer-grained cake made with the cream of tartar (Francis and Clydesdale, 1975).

Low oven temperatures of 149–177°C were recommended by early workers for baking angel food cake because it was thought that higher oven temperatures would toughen the protein of the egg whites. It has been found, however, that the maximum temperature attained in cakes prepared from the same formulas does not vary with oven temperature if baking times are adjusted so that the same crust color is obtained in all cakes (Barmore, 1936). When angel food cakes were baked at 177, 191, 204, 218, and 232°C, the volume and tenderness increased up to 218°C. At 232°C, cakes were smaller and very moist. High palatability scores for cakes baked at 204 and 218°C suggested that one of these oven temperatures should be used (Miller and Vail, 1943).

Sponge cakes Sponge cakes can be made from the yolk and whites beaten separately, from unseparated eggs, or from yolks only. A simplified method in which unseparated eggs are used was published (Briant and Willman, 1956). The eggs and lemon juice are beaten in an electric mixer until soft peaks are formed and the sugar is added while the beating continues. The flour and salt are added with gentle mixing. Cakes made by this method compared favorably to those made by the usual method of using separated eggs. The success of the method depends on thorough beating of the eggs.

PROCESSED EGGS AND THEIR PERFORMANCE IN FOOD PREPARATION

Surplus supplies of eggs may be used in the production of dried and frozen eggs and egg products. The production and use of these products in foods have been studied extensively. Research has led to the development of new processing techniques and the adoption of regulations to ensure the safety of processed egg products. Federal law regulates the processing of eggs. To ensure destruction of potentially pathogenic microorganisms, particularly *Salmonella,* all liquid, frozen, and dried whole eggs, yolks, and whites must be pasteurized or otherwise treated (Goddard, 1966). Pasteurization temperatures range from 56 to 63°C, depending on the part(s) of the egg and additive present.

Dehydration

Although dried egg products usually are not available to consumers at the retail level, they are widely used in mixes and in quantity food production. Several techniques are used for drying whole eggs, egg yolks, or egg whites. Spray-drying is the most commonly used method. The eggs are preheated to 60°C, then sprayed into a drying chamber through which air between 121 and 149°C is passing. The powder separates from the air and is continuously removed from the drying chamber. Prior to drying, egg white is treated with glucose oxidase or fermented with yeast for removal of glucose to prevent carbonyl-amine browning during storage of the dried product. A modification of the spray-drying process involves the incorporation of carbon dioxide, air, or nitrogen into the liquid egg before it is sprayed into the drying chamber. This process is called *foam–spray-drying.* Drying also may be accomplished by passing thin films of egg on belts through a drying chamber. Freeze-drying is used to remove water from frozen eggs in a vacuum chamber. Only a few freeze-dried egg products are available on the retail market because the process is very expensive.

The effects of drying on the functional properties of eggs have been of interest. Bergquist (1973) suggested that the heat coagulation properties of eggs are not affected by drying. Scrambled eggs from spray-dried, foam–spray-dried, and freeze-dried eggs did not differ with respect to tenderness and syneresis (Janek and Downs, 1969). However, foam–spray-dried scrambled eggs received lower flavor scores. Baked custards from eggs dried by the same three methods also have been evaluated. A series of studies on the use of these eggs in gels from yolk or albumen and in custards suggested that drying may adversely affect the white. The reason for this conclusion was that gel strength and custard strength were reduced in products containing, respectively, foam–spray-dried egg white and whole egg, but was not reduced in gels containing the processed yolks (Zabik, 1968; Wolfe and Zabik, 1968).

The foaming properties of egg whites are not altered by drying.

However, the foaming potential of yolks and whole eggs is reduced (Joslin and Proctor, 1954). Good to very good sensory ratings were given to angel food cakes made with foam–spray-dried, spray-dried, and freeze-dried egg whites. Spray-dried eggs produced the smallest volume and foam–spray-dried the largest (Franks et al., 1969). In the same study, cakes made from frozen egg whites were given the highest scores for flavor, suggesting that drying affects the flavor of the egg.

Although dried desugared egg white is quite stable during storage, dried whole egg and yolk deteriorate rapidly in flavor, color, solubility, and cooking quality when stored at room temperature. Refrigerator storage extends the shelf life to about 1 yr (Bergquist, 1973).

Freezing

Freezing effectively preserves the quality of eggs. Whites may be frozen without any addition, but 5% sugar or 1.2% salt usually is added to yolks before freezing so that they will be free from gummy or lumpy particles when they thaw. The increased viscosity of frozen yolks after thawing most likely is due to aggregation of the lipoproteins of the yolk to form a network capable of entrapping large quantities of water (Hasiak et al., 1972). Sugar or salt may be added to whole eggs before freezing to minimize the gelling effect. Home freezing of eggs is discussed in Chapter 7. Frozen eggs are thawed before use, but they should not be held long because of possible spoilage and, in the case of home-frozen eggs, the possible growth of microorganisms.

Whole eggs frozen with or without sugar or yolks frozen with sugar or corn sirup have been found entirely satisfactory for plain cakes and custards (Jordan et al., 1952). Angel food cakes made from frozen whites compared favorably with cakes made from fresh whites except for a slightly smaller volume (Clinger et al., 1951). In a study of the influence of pasteurization and the use of the additives sodium chloride and fructose on the gelation of egg yolk, the quality of mayonnaise and sponge cakes made from the yolks was evaluated (Jaax and Travnicek, 1968). Sodium chloride and fructose reduced gelation of the yolks. Fructose decreased the emulsification capacity of the yolk, whereas sodium chloride increased it. Sponge cakes made with yolks containing the additives were greater in volume and more easily compressed. Sensory evaluation suggested that sodium chloride was less satisfactory than fructose. The use of various additives to improve the quality of scrambled eggs made from frozen eggs has been evaluated (Ijichi et al., 1970). Low levels of salt, sucrose, dextrose, or skimmilk combined with homogenization prevented an increase in viscosity and eliminated the undesirable appearance associated with frozen eggs used for scrambling. Heating of egg yolk to about 45°C after thawing partially reverses the gelation that occurs with freezing (Palmer et al., 1970).

Egg Substitutes

Several commercial egg substitutes have been introduced to the market in response to recommendations for cholesterol-free diets. The chemical composition and nutritional properties of several of these products have been reported (Childs and Ostrander, 1976; Pratt, 1975; Navidi and Kummerow, 1974). Scrambled whole eggs may receive higher flavor scores than "scrambled eggs" prepared from an egg substitute. Aroma also may differ, but color does not. In cakes, volume may be decreased and flavor may be rated lower in those containing the substitutes (Leutzinger et al., 1977).

EXPERIMENTS

I. Custards

The effects of variations in cooking time, cooking temperature, and the amount and part of the egg used on the quality of stirred and baked custards will be illustrated. The role of salts and the effect on baked custards of variations in the amount of sugar will be explored.

A. *Basic formula*

Milk, ml	250
Egg, g	50
Sugar, g	25
Salt, g	0.34
Vanilla, ml	1.2

B. Basic procedure

1. Stirred custards
   a. Scald milk over boiling water before measuring.
   b. Beat egg slightly; add sugar and salt. Add milk very slowly at first, then more rapidly, stirring constantly.
   c. Cook over hot water that is kept at simmering temperature (about 90°C). Stir constantly. Continue cooking until the mixture forms a coating on a metal spoon. Add vanilla.
   d. Remove from heat at once and quickly pour into a small, labeled container. Allow the custard to cool to 50°C before testing.
2. Baked custards
   a. Set an oven at 177°C (350°F). Scald milk over boiling water before measuring. Boil additional water.
   b. Beat egg slightly; add sugar and salt. Add milk very slowly at first, then more rapidly, stirring constantly. Add vanilla.
   c. Pour into matched, labeled custard cups to the top line (about 7/8 full). Place cups in a large, shallow baking pan, set pan on oven rack, and pour in boiling water until it reaches the level of the custard mixture.
   d. Bake at 177°C until a knife inserted in one of the custards comes out clean. (Be consistent in testing techniques.) Record baking time.
   e. Remove from water at once with tongs. Cover with foil. Refrigerate overnight, if possible, or for 30 min if testing must be done on the day of preparation.

C. Experimental variation

1. Variation in amount of egg and part of egg used in stirred or baked custards. Follow basic procedures for preparation.
   a. Basic formula.
   b. Substitute 34 g egg yolk for the whole egg in the basic formula.
   c. Substitute 66 g egg white for the whole egg in the basic formula.
   d. Increase the whole egg in the basic formula to 100 g.

2. Role of salt in baked custards. Follow basic procedure for preparation. Increase egg in basic formula to 100 g. Omit the sugar, salt, and vanilla for all variations.
   a. Basic formula with substitutions and omissions described above.
   b. Substitute distilled, demineralized water for milk.
   c. Substitute 0.1 M NaCl for the milk.
   d. Substitute 0.1 M $CaCl_2$ for the milk.
3. Variation in the amount of sugar in baked custards. Follow basic procedures for preparation.
   a. Basic formula.
   b. Increase sugar in basic formula to 50 g. Start this custard first.
4. Variation in cooking time and temperature.
   a. Stirred custard. Make three times the basic formula as directed under the basic procedure. Divide the mixture equally (310 g per pan) among the tops of three identical double boilers. The water in the lower portion should just touch the upper portion of the double boiler. Cook the custards as follows:
      1) Follow the basic procedure. Record the time required for the mixture to coat a spoon.
      2) Put cold water in the bottom of the double boiler. Bring water to a boil slowly after putting the custard in the top of the double boiler. Record the time required for the mixture to coat a spoon.
      3) Have the water in the bottom of the double boiler boiling rapidly throughout the cooking period. Record the time required for the mixture to coat a spoon.
   b. Baked custard. Make four times the basic formula as directed under the basic procedure for baked custards. Divide the mixture evenly (100 g per cup) among 12 matched custard cups. Set one oven at 177°C (350°F) and one at 260°C (500°F). Bake the custards as follows:
      1) Into the 177°C oven place 3 cups of custard mixture directly on a rack. Place 3 cups in a large, shallow baking pan. Pour boiling water into the

pan. Remove each group when a knife inserted into one of the custards comes out clean. Record the baking time, which may be different for the two groups of cups.

2) Repeat step 1) using the 260°C oven. Begin testing doneness in 10 min.

D. *Objective tests*

It is suggested that as many objective tests be made as the equipment and available time permit.

1. Line-spread for stirred custards. Follow directions in Chapter 16. The custard should be 50°C when tested. Allow the custard to spread for 30 sec before taking readings.
2. Percentage sag of baked custards. Insert the probe of a vernier caliper into the center of the custard and measure the custard height in centimeters. Remove the probe. Loosen the custard from the top of the cup with a spatula and turn out onto a plate. Insert the probe of a vernier caliper through the center of the custard 30 sec after it is removed from the cup. Measure the depth of penetration as before. Calculate percentage sag as follows:

$$\text{Percentage sag} = \frac{\text{height in container} - \text{height out of container}}{\text{height in container}} \times 100$$

3. Penetrometer test on baked custards. After the custard has cooled in the refrigerator (30 min or overnight), remove the crust carefully. Place under the cone (about 35 g) of a penetrometer. Move the cone into place and release for 5 sec. Record the depth of penetration in millimeters.
4. Syneresis of baked custards. Custard that has been used for testing percentage sag or penetration can be used for this test. Invert the custard on a fine wire screen supported on a funnel or into a funnel covered with nylon mesh or cheesecloth and allow to drain for 1 hr. Measure the volume of the liquid in a graduated cylinder.

E. *Sensory test*

Score color, flavor, and texture using appropriate scales. (See the Appendix.)

F. *Questions*

What is the effect of too high a temperature on the quality of stirred and baked custards? What is the practical advantage of cooking over simmering or cold water? How does substituting egg yolk or egg white or increasing the amount of egg affect the characteristics of custards? Relate the differences in characteristics to the differences in the amount of protein supplied by each. How does an increase in sugar affect the baking time of the custards? What is the role of milk in a baked custard?

II. The Stability of Egg White Foams

In this experiment, the effects of various factors on the stability of egg white foams will be studied. The stability of the foams will be determined by measuring the amount of liquid that drains from beaten egg white placed in a funnel. Identical rotary beaters and bowls or identical electric mixers used on high speed should be used for beating the egg whites.

A. *Procedure*

1. Support a funnel (about 125 mm in diameter) with a funnel support or support stand. Place a 25-ml graduated cylinder under the funnel. Place glass wool over the stem but do not stuff it into the stem.
2. Beat 25 g of egg white until stiff enough to hold a peak, but still shiny. Record the length of time required for beating.
3. Transfer all of the egg white foam to the funnel, using a plate scraper. Cover with a watch glass or plastic film. Compare the volume and appearance (gloss and fineness of cell structure) of the foams after they are in the funnels.
4. Record the volume of liquid draining from the foam every 10 min for 1 hr.

B. *Experimental variations*

Follow the basic procedure except for the variable noted.

1. Amount of beating
   a. Beat by the standard method.
   b. Reduce beating time by 25%.
   c. Beat until stiff and dry.
2. Freshness of egg white
   a. Fresh egg white
   b. Deteriorated egg white (allow shell eggs to stand at room temperature for 2 wk)
3. Additional ingredients
   a. Control, none
   b. Salt, 0.78 g (1/8 tsp)
   c. Cream of tartar, 0.19 g (1/16 tsp)
   d. Cream of tartar, 0.38 g (1/8 tsp)
   e. Egg yolk, 6 drops
   f. Water, 15 ml
   g. Sugar, 25 g (added when foam is at soft-peak stage)
   h. Both sugar and water (each added as indicated in f and g)

C. *Results*

Make a graph of the results of each completed experiment. Place drainage time on the horizontal axis (abscissa) and volume of liquid drained from the foam on the vertical axis (ordinate). All of the variables from one experiment can be recorded on one graph if properly labeled.

D. *Questions*

Did salt or cream of tartar stabilize the foam? Why is cream of tartar selected, and how does it affect the foam? How did the sugar influence the stability of the foam? Why? What effect did water have on the foam? What differences were noted when both water and sugar were added? Why did the deteriorated egg give a different volume than fresh white?

III. Soft Meringues

In this experiment, meringue is spread on cold, warm, and hot fillings, and baked at two oven temperatures to show the effects of these variations on the quality of meringue.

A. *Basic formulas*

1. Filling	
Cornstarch, g	36
Sugar, g	150
Water, ml	750
2. Meringue	
Egg white, g	66
Salt, g	0.78 (1/8 tsp)
Sugar, g	62.5

B. *Procedure*

1. Filling
   a. Mix cornstarch and sugar in small pan; add water gradually. Cook over direct heat, stirring constantly until the mixture has boiled for several minutes and is almost clear.
   b. Pour filling into four baking pans or dishes to within 5 mm of the top. Save the remaining filling in the pan.
   c. Put two of the pans in the refrigerator and chill until the filling is about 15°C.
   d. Put the other two pans in a cake pan containing simmering water. The temperature of the filling in these pans should be 55–60°C when the meringue is spread.
   e. Just before the meringue is spread, bring the filling remaining in the pan to a boil and pour into two baking pans or dishes to within 5 mm of the top.
2. Meringue
   a. Set two ovens, one at 163°C (325°F) and the other at 218°C (425°F).
   b. Beat egg whites until they hold a soft peak: then add the sugar gradually while beating. Continue to beat until the meringue is stiff, glossy, and stands in soft peaks

that curve over slightly. Add salt near the end of beating.

c. Spread the meringue on the six pans of filling, two at each temperature. Spread quickly with a spatula, leaving typical peaks, and being sure that the meringue touches the sides of the pan at all points. It will be especially difficult to spread the meringue on the boiling filling.

d. Into the 163°C oven, place one pie for each of the three filling temperatures. Bake 15–20 min. Bake the other three pies at 218°C for 5–7 min. Meringues baked at both temperatures should have the same light brown color when removed from the oven.

C. *Evaluation*

Judge the meringues using scales of appropriate adjectives for general appearance, stickiness, tenderness, surface beading, and leakage from the bottom of the meringue. (See the Appendix.)

D. *Questions*

What are the characteristics of meringues spread on fillings of various temperatures? What effect does oven temperature have on the quality of meringues?

IV. Angel Food Cake

In this experiment, factors affecting the quality of angel food cakes will be studied. They include the level of cream of tartar, the mixing method, and the oven temperature. The work of an experimental series can be divided among a group of students so that one student carries out the same step for each cake in a series.

A. *Basic formula*

Flour, cake, g	23
Sugar, g	63
Egg white, g	61

Salt, g	0.3
Cream of tartar, g	0.9 (1/4 tsp)

B. *Basic procedure*

1. Set oven at 177°C (350°F). Wash a pan approximately 19 × 9.5 × 6 cm thoroughly with detergent to remove any traces of fat.
2. Sift the flour with 1/4 of the sugar.
3. Beat the egg white until foamy. Add the salt and cream of tartar.
4. Beat until egg white is stiff but not shiny.
5. Add remaining sugar in three portions, beating enough to blend (about 10 sec) between additions.
6. Sift 1/4 of the flour-sugar mixture over the meringue and fold with a wire whip just enough to blend. Repeat until all is used.
7. Pour into ungreased pan. Bake at 177°C for about 30 min.
8. Invert on a wire rack until cake reaches room temperature before removing it from the pan. If the cake comes over the top of the pan, suspend the inverted pan between two custard cups.
9. Cut with a serrated knife.

C. *Experimental variation*

1. Effect of mixing method. Use the basic formula for all mixing methods.
   a. Use basic procedure. Count the number of strokes used to incorporate the flour-sugar mixture.
   b. Prepare cake by the basic procedure except beat the egg white until stiff and dry before adding any of the sugar.
   c. Prepare cake by the basic procedure except use twice as many folding strokes to incorporate the flour-sugar mixture as were used in variation 1a.
   d. Prepare cake by the basic procedure except beat egg white until stiff but not dry before adding any of the sugar. The white should be beaten more than for varia-

tion 1a but not as much as for variation 1b. Fold the sugar into the beaten egg white with a wire whip. Add the flour-sugar mixture as in the basic procedure.

2. Effect of cream of tartar.
   a. Basic formula.
   b. Basic formula except omit the cream of tartar.
   c. Basic formula except double the cream of tartar.
3. Effect of oven temperature. Use basic procedure for mixing.
   a. Prepare six times the basic formula. Weigh 140 g into each of six pans.
   b. Bake at 163, 177, 191, 204, 218, and 232°C (325, 350, 375, 400, 425, and 450°F). Adjust baking times so that the cakes are similar in color.

D. *Objective test*

If the cakes do not extend above the sides of the pan, the volumes of the cakes can be determined by method 2 for seed displacement (see Chapter 16). Because of the delicate nature of the cakes, it is best to pour the seeds on the baked cake before removal from the pan. The volume of seeds required to fill the pan containing the cake is subtracted from the volume required to fill the empty pan.

E. *Evaluation*

Evaluate volume, cell size, crust color, crumb color, moistness, tenderness, flavor, and acceptability of the cakes using scales of appropriate adjectives. (See the Appendix.) Make a note of any abnormalities that may occur, such as spotting or cracking of the crusts.

F. *Questions*

What mixing method was best? What oven temperature? What effect did cream of tartar have on the quality of the cakes?

SUGGESTED EXERCISES

1. Compare the quality of several products made from fresh, frozen, and dried eggs.
2. Suspend a thermometer in either a stirred or baked custard mixture. Cook beyond the stage of doneness, recording the temperature of the custard frequently during cooking. Graph temperature against time and mark the points at which coagulation and curdling are first observed. Is coagulation an endothermic reaction?
3. Calculate the amount of protein supplied by the whole egg in the basic formula. Determine the amount of white and yolk required to give an equal amount of protein. Prepare custards using these quantities. Adjust the water and fat content of the custards so that they are equal. Use water and vegetable oil for the adjustments.
4. Compare the volume or specific gravity and the stability of foams made from fresh, frozen, and reconstituted dried egg whites, allowing the water and dry albumen to stand for various lengths of time. Experiment II can be used as a guide.
5. Compare the volume and/or specific gravity of foams beaten in various ways. Experiment II can be used as a guide.
6. Study the effect, on the quality of soft meringue, of beating eggs until foamy or until stiff but not dry before adding sugar. Experiment III can be used as a guide.
7. Study the effects on meringue shells of variations in ingredients, mixing methods, and baking conditions.
8. Develop a small formula for chiffon cake and study the effects of variations in ingredients and method on the quality of the cake.
9. Compare the quality of several products made with egg substitutes that are available on the market.

REFERENCES

Adams, C. F. 1975. Nutritive value of American foods. Agric. Handbook 456, Agricultural Research Service, USDA, Washington, DC.

AMS. 1975. Egg grading manual. Agric. Handbook 75, Agricultural Marketing Service, USDA, Washington, DC.

Alam, A. U., Creger, C. R. and Couch, J. R. 1968. Petals of Aztec marigold, Tagetes erecta, as a source of pigment for avian species. J. Food Sci. 33: 635.
Andross, M. 1940. Effect of cooking on eggs. Chem. and Ind. 59: 449.
Bailey, M. I. 1935. Foaming of egg white. Ind. Eng. Chem. 27: 973.
Baker, R. C. and Darfler, J. 1969. Discoloration of egg albumen in hard-cooked eggs. Food Technol. 23: 77.
Baker, R. C., Darfler, J. and Lifshitz, A. 1967. Factors affecting the discoloration of hard-cooked egg yolks. Poultry Sci. 46: 664.
Baldwin, R. E. 1973. Functional properties in foods. In "Egg Science and Technology," Eds. Stadelman, W. J. and Cotterill, O. J. Avi Publ. Co., Westport, CT.
Baldwin, R. E., Cotterill, O. J., Thompson, M. M. and Myers, M. 1967. High temperature storage of spray-dried egg white. Poultry Sci. 46: 1421.
Baldwin, R. E., Upchurch, R. and Cotterill, O. J. 1968. Ingredient effects on meringues cooked by microwaves and by baking. Food Technol. 22: 1573.
Barmore, M. A. 1934. The influence of chemical and physical factors on egg-white foams. Tech. Bull. 9, Colorado Agric. Exp. Stn., Fort Collins.
Barmore, M. A. 1936. The influence of various factors, including altitude, in the production of angel food cake. Tech. Bull. 15, Colorado Agric. Exp. Stn., Fort Collins.
Bergquist, D. H. 1973. Egg dehydration. In "Egg Science and Technology," Eds. Stadelman, W. J. and Cotterill, O. J. Avi Publ. Co., Westport, CT.
Beveridge, T. and Nakai, S. 1975. Effects of sulphydryl blocking on the thinning of egg white. J. Food Sci. 40: 864.
Board, R. G. 1973. The microbiology of eggs. In "Egg Science and Technology," Eds. Stadelman, W. J. and Cotterill, O. J. Avi Publ. Co., Westport, CT.
Brant, A. W. 1951. A new height gage for measuring egg quality. Food Technol. 5: 384.
Brant, A. W., Otte, A. W. and Norris, K. H. 1951. Recommended standards for scoring and measuring opened egg quality. Food Technol. 5: 356.
Briant, A. M. and Willman, A. R. 1956. Whole-egg sponge cakes. J. Home Econ. 48: 420.
Briant, A. M., Zaehringer, M. V., Adams, J. L. and Mondy, N. 1954. Variations in quality of cream pies. J. Amer. Dietet. Assoc. 30: 678.
Brooks, J. and Taylor, D. J. 1955. Eggs and egg products. Food Invest. Spec. Rpt. 60, Great Brit. Dept. Sci. Ind. Res., London.
Carlson, C. W. and Halverson, A. W. 1964. Some effects of dietary pigmenters on egg yolks and mayonnaise. Poultry Sci. 43: 654.
Carney, R. I. 1938. Determination of stability of beaten egg albumen. Food Ind. 10: 77.
Childs, M. T. and Ostrander, J. 1976. Egg substitutes: chemical and biological evaluations. J. Amer. Dietet. Assoc. 68: 229.

Clinger, C., Young, A., Prudent, I. and Winter, A. R. 1951. The influence of pasteurization, freezing, and storage on the functional properties of egg white. Food Technol. 5: 166.

Cotterill, O. J., Seideman, W. E. and Funk, E. M. 1965. Improving yolk-contaminated egg white by heat treatments. Poultry Sci. 44: 228.

Cunningham, F. E. and Cotterill, O. J. 1964. Effect of centrifuging yolk-contaminated liquid egg white on functional performance. Poultry Sci. 43: 283.

Cunningham, F. E. and Cotterill, O. J. 1972. Performance of egg white in the presence of yolk fractions. Poultry Sci. 51: 712.

Dawson, E. H., Miller, C. and Redstrom, R. A. 1956. Cooking quality and flavor of eggs as related to candled quality, storage conditions, and other factors. Agric. Inform. Bull. 164, USDA, Washington, DC.

Eisen, E. J. and Bohren, B. B. 1963. Some problems in the evaluation of egg albumen quality. Poultry Sci. 42: 74.

Feeney, R. E., Ducay, E. D., Silva, R. B. and MacConnell, L. R. 1952. Chemistry of shell egg deterioration: the egg white proteins. Poultry Sci. 31: 639.

Felt, S. A., Longrée, K. and Briant, A. M. 1956. Instability of meringued pies. J. Amer. Dietet. Assoc. 32: 710.

Forsythe, R. H. and Bergquist, D. H. 1951. The effect of physical treatments on some properties of egg whites. Poultry Sci. 30: 302.

Francis, F. J. and Clydesdale, F. M. 1975. "Food Colorimetry: Theory and Applications." Avi Publ. Co., Westport, CT.

Franks, O. J., Zabik, M. E. and Funk, K. 1969. Angel cakes using frozen, foam–spray-dried, freeze-dried, and spray-dried albumen. Cereal Chem. 46: 349.

Funk, E. M. 1950. Maintainance of quality in shell eggs by thermostabilization. Res. Bull. 467, Missouri Agric. Exp. Stn., Columbia.

Gillis, J. N. and Fitch, N. K. 1956. Leakage of baked soft-meringue topping. J. Home Econ. 48: 703.

Goddard, J. L. 1966. Amendments of standards for whole egg and yolk products and establishment of standards for egg white products. Fed. Register 31: 4677.

Grewe, E. and Child, A. M. 1930. The effect of acid potassium tartrate as an ingredient in angel cake. Cereal Chem. 7: 245.

Grunden, L. P., Mulnix, E. J., Darfler, J. M. and Baker, R. C. 1975. Yolk position in hard cooked eggs as related to heredity, age and cooking position. Poultry Sci. 54: 546.

Grunden, L. P., Vadehra, D. V. and Baker, R. C. 1974. Effects of proteolytic enzymes on the functionality of chicken egg albumen. J. Food Sci. 39: 841.

Hale, K. K. Jr. and Britton, W. M. 1974. Peeling hard cooked eggs by rapid cooling and heating. Poultry Sci. 53: 1069.

Hanning, F. M. 1945. Effect of sugar or salt upon denaturation produced by beating and upon the ease of formation and the stability of egg white foams. Iowa State College J. Sci. 20: 10.

Hasiak, R. J., Vadehra, D. V., Baker, R. C. and Hood, L. 1972. Effect of certain physical and chemical treatments on the microstructure of egg yolk. J. Food Sci. 37: 913.

Haugh, R. R. 1937. The Haugh unit for measuring egg quality. U. S. Egg and Poultry Magazine 43: 552.

Hawthorne, J. R. 1950. The action of egg white lysozyme on ovomucoid and ovomucin. Biochim. Biophys. Acta. 6: 28.

Henry, W. C. and Barbour, A. D. 1933. Beating properties of egg white. Ind. Eng. Chem. 25: 1054.

Hester, E. E. and Personius, C. J. 1949. Factors affecting the beading and the leakage of soft meringues. Food Technol. 3: 236.

Ijichi, K., Palmer, H. H. and Lineweaver, H. 1970. Frozen whole eggs for scrambling. J. Food Sci. 35: 695.

Irmiter, T. F., Dawson, L. E. and Reagen, J. G. 1970. Methods of preparing hard cooked eggs. Poultry Sci. 49: 1232.

Jaax, S. and Travnicek, D. 1968. The effect of pasteurization, selected additives and freezing rate on the gelation of frozen-defrosted egg yolk. Poultry Sci. 47: 1013.

Janek, W. A. and Downs, D. M. 1969. Scrambled eggs prepared from three types of dried whole eggs. J. Amer. Dietet. Assoc. 55: 578.

Jordan, R., Luginbill, R. N., Dawson, L. E. and Echterling, C. J. 1952. The effect of selected pretreatments upon the culinary qualities of eggs frozen and stored in a home-type freezer. I. Plain cakes and baked custards. Food Res. 17: 1.

Jordan, R., Wegner, E. S. and Hollender, H. A. 1954. Nonhomogenized vs. homogenized milk in baked custards. J. Amer. Dietet. Assoc. 30: 1126.

Joslin, R. P. and Proctor, B. E. 1954. Some factors affecting the whipping characteristics of dried whole egg powders. Food Technol. 8: 150.

Leutzinger, R. L., Baldwin, R. E. and Cotterill, O. J. 1977. A research note—sensory attributes of commercial egg substitutes. J. Food Sci. 42: 1124.

Longrée, K., Jooste, M. and White, J. C. 1961. Time-temperature relationships of custards made with whole egg solids. III. Baked in large batches. J. Amer. Dietet. Assoc. 38: 147.

Lorenz, K. 1975. High altitude food preparation and processing. CRC Critical Reviews in Food Technol. 5: 403.

Lowe, B. 1955. "Experimental Cookery." John Wiley and Sons, Inc., New York.

MacDonald, L. R., Feeney, R. E., Hanson, H. L., Campbell, A. and Sugihara, T. F. 1955. The functional properties of the egg white proteins. Food Technol. 9: 49.

MacLeod, A. J. and Cave, S. J. 1977. A research note—absorption of a taint aroma by eggs. J. Food Sci. 42: 539.

Maurer, A. J. 1975a. Refrigerated egg storage at two air movement rates. Poultry Sci. 54: 409.

Maurer, A. J. 1975b. Hard-cooking and pickling eggs as teaching aids. Poultry Sci. 54: 1019.

McKellar, D. M. B. and Stadelman, W. J. 1955. A method for measuring volume and drainage of egg white foams. Poultry Sci. 34: 455.

Mellor, D. B., Gardner, F. A. and Campos, E. J. 1975. Effect of type of package and storage temperature on interior quality of shell treated shell eggs. Poultry Sci. 54: 742.

Miller, G. A., Jones, E. M. and Aldrich, P. J. 1959. A comparison of the gelation properties and palatability of shell eggs, frozen whole eggs, and whole egg solids in standard baked custard. Food Res. 24: 584.

Miller, E. L. and Vail, G. E. 1943. Angel food cakes made from fresh and frozen egg whites. Cereal Chem. 20: 528.

Morgan, K. J., Funk, K. and Zabik, M. E. 1970. Comparison of frozen, foam–spray-dried, freeze-dried, and spray-dried eggs. 7. Soft meringues prepared with a Carrageenan stabilizer. J. Food Sci. 35: 699.

Navidi, M. K. and Kummerow, F. A. 1974. Nutritional value of Egg Beaters ® compared with "farm fresh eggs." Pediatrics 53: 565.

Niles, K. B. 1937. Egg whites on parade. U.S. Egg and Poultry Magazine 43: 337.

Palmer, H. H. 1972. Eggs. In "Food Theory and Applications," Eds. Paul, P. C. and Palmer, H. H. John Wiley and Sons, Inc., New York.

Palmer, H. H., Ijichi, K. and Riff, H. 1970. Partial reversal of gelation in thawed egg proteins. J. Food Sci. 35: 403.

Powrie, W. D. 1973. Chemistry of eggs and egg products. In "Egg Science and Technology," Eds. Stadelman, W. J. and Cotterill, O. J. Avi Publ. Co., Westport, CT.

Pratt, D. E. 1975. Lipid analysis of a frozen egg substitute. J. Amer. Dietet. Assoc. 66: 31.

Reinke, W. C., Spencer, J. V. and Tryhnew, L. J. 1973. The effect of storage upon the chemical, physical and functional properties of chicken eggs. Poultry Sci. 52: 692.

Rhodes, M. B., Adams, J. L., Bennett, N. and Feeney, R. E. 1960. Properties and food uses of duck eggs. Poultry Sci. 39: 1473.

Robinson, D. S. and Monsey, J. B. 1972. Changes in the composition of ovomucin during liquefaction of thick egg white: the effect of ionic strength and magnesium salts. J. Sci. Food Agric. 23: 893.

Salwin, H., Bloch, I. and Mitchell, J. H. Jr. 1953. Dehydrated stabilized egg. Importance and determination of pH. Food Technol. 7: 447.

Sauter, E. A., Harns, J. V., Stadelman, W. J. and McLaren, B. A. 1953. Relationship of candled quality of eggs to other quality measurements. Poultry Sci. 32: 850.

Sauter, E. A. and Montroure, J. E. 1972. The relationship of lysozyme content of egg white to volume and stability of foams. J. Food Sci. 37: 918.

Sauter, E. A. and Montoure, J. E. 1975. Effects of adding 2% freeze-dried egg white to batters of angel food cakes made from white containing egg yolk. J. Food Sci. 40: 869.

Sauter, E. A., Stadelman, W. J., Harns, V. and McLaren, B. A. 1951. Methods for measuring yolk index. Poultry Sci. 30: 629.

Sechler, C., Maharg, L. G. and Mangel, M. 1959. The effect of household table salt on the whipping quality of egg white solids. Food Res. 25: 198.

Seideman, W. E., Cotterill, O. J. and Funk, E. M. 1963. Factors affecting heat coagulation of egg white. Poultry Sci. 42: 406.

Slosberg, H. M., Hanson, H. L., Stewart, G. F. and Lowe, B. 1948. Factors influencing the effects of heat treatment on the leavening power of egg white. Poultry Sci. 27: 294.

Spencer, J. V. and Tryhnew, L. J. 1973. Effect of storage on peeling quality and flavor of hard-cooked shell eggs. Poultry Sci. 52: 654.

Stadelman, W. J. 1973. Quality preservation of shell eggs. In "Egg Science and Technology," Eds. Stadelman, W. J. and Cotterill, O. J. Avi Publ. Co., Westport, CT.

St. John, J. L. and Flor, I. H. 1930–31. A study of whipping and coagulation of eggs of varying quality. Poultry Sci. 10: 71.

Tinkler, C. K. and Soar, M. C. 1920. The formation of ferrous sulphide in eggs during cooking. Biochem. J. 14: 114.

Upchurch, R. and Baldwin, R. E. 1968. Guar gum and triacetin in meringues and a meringue product cooked by microwaves. Food Technol. 22: 1309.

Vadehra, D. V. and Nath, K. R. 1973. Eggs as a source of protein. CRC Critical Reviews in Food Technol. 4: 193.

Wang, A. C., Funk, K. and Zabik, M. E. 1974. Effect of sucrose on the quality characteristics of baked custards. Poultry Sci. 53: 807.

Wolfe, N. J. and Zabik, M. E. 1968. Comparison of frozen, foam–spray-dried, freeze-dried, and spray-dried eggs. 3. Baked custards prepared from eggs with corn syrup solids. Food Technol. 22: 1470.

Zabik, M. E. 1968. Comparison of frozen, foam–spray-dried, freeze-dried and spray-dried eggs. 2. Gels made with milk and albumen or yolk containing corn syrup. Food Technol. 22: 1465.

4. Milk and Milk Products

In this chapter, physical properties, components, flavor, and functional properties of milk and of products made from cow's milk will be considered. The United States Public Health Service defined milk as "the lacteal secretion, practically free of colostrum, obtained by the complete milking of one or more healthy cows, which contains not less than 8 1/4% of milk-solids-not-fat and not less than 3 1/4% milk fat" (USPHS, 1965).

PHYSICAL PROPERTIES

Physical State

Milk is a complex physicochemical system whose various constituents differ widely in molecular size and solubility. The smallest molecules, those of salts, lactose, and water-soluble vitamins, are in true solution. The proteins, including the enzymes, are in colloidal state because of the large size of their molecules. The fat in nonhomogenized milk is present as globules of larger-than-colloidal size. The size of the fat globules varies with the individual cow, its breed, and the stage of lactation (Mulder and Walstra, 1974). Reported diameters range from less than 0.1 μm to as great as 20 μm (Mulder and Walstra, 1974). Globules number from 1.5 to 3 $\times$ 10^9 per ml of milk (Trout, 1950). During homogenization, these fat globules are reduced in size to less than 1 μm and are increased greatly in number by being forced under high pressures through a small orifice. The resulting globule size depends on the pressure and the type of homogenizer used (Mulder and Walstra, 1974). The difference in fat globule size is illustrated in Figure 4.1.

Each fat globule has a membrane composed of lipid components (phospholipids, cholesterol, and glycerides) and protein. The membranes prevent coalescence of the fat globules. It generally is accepted that the membrane is divided into several layers. However, various investigators have proposed that there are two to six layers of lipoproteins and/or phospholipids (Mulder and Walstra, 1974).

pH

Milk is slightly acid, usually having a pH between 6.5 and 6.7 at 25°C. It is well buffered by proteins and salts, especially the phosphates. The pH of milk is temperature dependent. When milk is heated, its pH

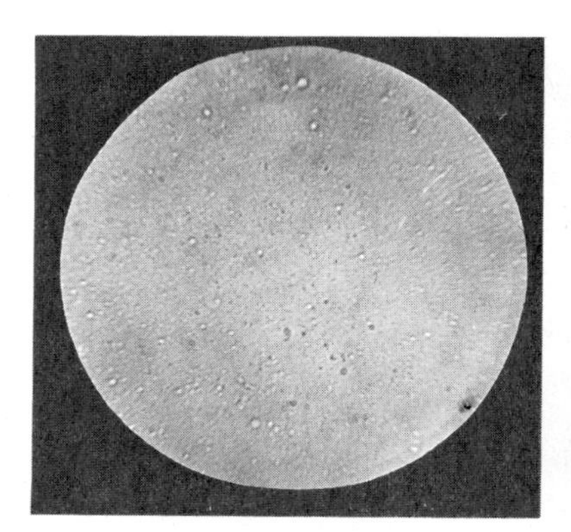
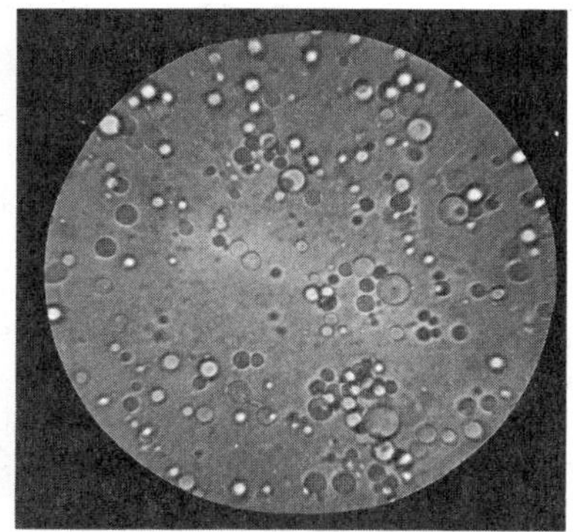

Figure 4.1. Microscopic appearance of the fat globules in homogenized (left) and nonhomogenized milk. (Courtesy of Michigan Agricultural Experiment Station.)

increases as dissolved carbon dioxide is released and then decreases because hydrogen ions are liberated when calcium phosphate precipitates (Pyne, 1962). A balance between these two opposing factors prevents large changes in pH during heating.

Other Physical Properties

Skimmilk and whole milk are essentially Newtonian fluids, which are defined in Chapter 16. Therefore, the viscosity of a given sample depends only on temperature, whereas the viscosity of the non-Newtonian creams, concentrated milks, and butter depends also on shear rate. The quantity of dispersed solids influences the viscosity. Thus whole milk is more viscous than skimmilk, which is more viscous than whey (Jenness et al., 1974).

The freezing point of milk is slightly lower than the freezing point of water because of the presence of lactose and soluble salts. Reported values range from −0.530 to −0.570°C (Jenness et al., 1974). Determination of the freezing point is used for detection of milk to which water has been added.

The presence of milkfat, milk proteins, free fatty acids, and phospholipids lowers the surface tension of the water present in the milk. At 20°C the surface tension of milk is 50 dyne cm^{-1}, whereas water at that temperature has a surface tension of 72.75 dyne cm^{-1} (Jenness et al., 1974).

COMPONENTS OF MILK

The Public Health Service definition includes suggested minimal levels for two components of milk. Composition varies more than is indicated by the definition. State standards for milk-solids-non-fat vary from 3.0 to 8.5%, and standards for milkfat vary from 3.0 to 3.8% (USPHS, 1971). The composition of milk varies with the feed and the physiological condition of the cow. Therefore, tabular values for the various components of milk represent average values rather than absolute values.

Water

As shown in Table 4.1, water is the major component of milk, representing 87% of the total. All other components are suspended or dissolved in this medium. A small amount of the water is bound to the protein of the milk, and some is hydrated to the lactose and salts (Johnson, 1974).

Protein

Some knowledge and understanding of the proteins of milk are essential to a study of the changes in milk that take place during the preparation of products containing milk, the manufacturing of cheese, and homogenization. Basically, milk proteins can be separated into two major fractions, the caseins and the whey, or serum, proteins.

Table 4.1 Composition of certain dairy products per 100 g

Food	Water %	Protein g	Total lipid g	Carbohydrate g	Calcium mg	Vitamin A value I.U.
Fluid milk						
Whole	87.99	3.29	3.34	4.66	119	126
Lowfat, 2%	89.21	3.33	1.92	4.80	122	205
Condensed milk						
Sweetened	27.16	7.91	8.70	54.40	284	328
Evaporated milk						
Whole	74.04	6.81	7.56	10.04	261	243
Skim	79.40	7.55	0.20	11.35	290	392
Dried milk products						
Whole	2.47	26.32	26.71	38.42	912	922
Nonfat	3.16	36.16	0.77	51.98	1257	36
Whey, sweet	3.19	12.93	1.07	74.46	796	44
Half and half	80.57	2.96	11.50	4.30	105	434
Cream, heavy whipping	57.71	2.05	37.00	2.79	65	1470
Cheese						
Cheddar	36.75	24.90	33.14	1.28	721	1059
Cottage, creamed	78.96	12.49	4.51	2.68	60	163

SOURCE: Posati and Orr, 1976.

Caseins The *caseins* are a group of phosphoproteins, accounting for 80% of the total proteins in milk. Acidification of raw skimmilk to pH 4.6 at 20°C will coagulate this fraction (Whitney et al., 1976). Such coagulation is recognized as curdling in a stirred milk, or gelling in a quiescent milk that is acidified gradually by bacterial action. In the latter case, the coagulated casein first forms a continuous mass that breaks into curds when agitated. The precipitated casein can be rendered soluble by the addition of acid or of alkali, either of which shifts the pH away from the isoelectric point. Casein is prepared for research purposes by precipitating it with acid and resolubilizing it in acid several times. Although casein usually is spoken of as a single protein, it really is a mixture of proteins, as shown by electrophoresis. In this method, the distances that proteins move in an electrical field are studied. The casein proteins include α_s-caseins, β-caseins, κ-caseins, and γ-caseins. At the pH of milk, about 6.6, casein is present as a colloidal phosphate complex, calcium caseinate, dispersed as particles called *micelles.* Reflection of light by the micelles is responsible for the white color of milk. Micelles or aggregates of α_s-, β-, and κ-casein are described by Slattery (1976). The percentages of the major caseins in the micelles are α_s, 55%; β, 30%; and κ, 15% (Morr, 1975).

κ-casein has a stabilizing effect on the casein micelle, permitting the existence of the colloidal dispersion. The α- and β-caseins are calcium sensitive. Therefore, if the protective effect of κ-casein is destroyed by the enzyme, rennin, casein reacts with calcium to form a coagulum. Rennin usually is obtained from the fourth stomach of a dairy calf. Rennin coagulation is the preferred system for coagulation of casein for cheese production. As previously discussed, acid also is used for coagulation. Acid can be added as such or formed in the milk by natural souring, during which lactose is changed to lactic acid by the action of bacteria. Although casein usually is not coagulated by boiling, it may be coagulated if the milk is slightly acid or if very high temperatures are used. Thus, fresh milk can be boiled without curdling, but milk that is slightly sour or cooked under pressure is likely to curdle.

Whey proteins After the caseins are precipitated from milk, the whey proteins or milk serum proteins remain. Whey proteins are denatured with heating above 60°C (Morr, 1975). Heating also causes aggregation of the denatured whey proteins (Morr, 1969). Milk protein terminology has changed over time. Early literature contains reference to two components of the whey fraction. Rowland (1938) described the heat-labile portion insoluble in saturated magnesium sulfate solution as the *lactoglobulin fraction,* and the heat-labile portion soluble in this reagent as the *lactalbumin fraction.* These terms, differentiating between two major fractions of whey proteins, are seldom used by dairy chemists today and should not be confused with the names of specific whey proteins as defined by Whitney et al. (1976). Proteins in the whey fraction include β-lactoglobulins, α-lactalbumins, bovine serum albumin, the immunoglobulins, and components of the proteose-peptone fraction (Whitney et al., 1976). β-lactoglobulin is the major whey protein, representing 50% of the whey proteins (Farrell and Thompson, 1974). The α-lactalbumins are the next most abundant proteins, constituting 25% of the whey proteins. Unlike the β-lactoglobulins, this group does not contain any free sulfhydryl groups. Further information on the properties of bovine serum albumin, a protein identical to the albumin of bovine blood serum; the immunoglobulins, the antigenic fraction of milk; and the proteose-peptone fraction, an incompletely characterized fraction, is given by Farrell and Thompson (1974), and by Whitney and other members of the American Dairy Science Association Nomenclature Committee (Whitney et al., 1976).

Enzymes Milk contains a number of enzymes, most of which are inactivated by pasteurization. If the enzyme alkaline phosphomonoesterase (alkaline phosphatase) is not present in pasteurized milk, the process is considered adequate. This test is called the *phosphatase test,* and is an effective index of adequacy of pasteurization because the heat necessary to destroy this enzyme will also destroy pathogenic microorganisms. Other enzymes found in cow's milk include lipase, esterases, phosphatases, xanthine oxidase, lactoperoxidase, protease, amylase, catalase, aldolase, ribonuclease, lysozyme, carbonic anhydrase, and others (Johnson, 1974). Lipase sometimes causes hydrolytic rancidity in dairy products made from milk that has not been heated enough to inactivate this enzyme.

Lipids

Fat is the most variable component of milk. State law regulates the fat content of milk; the required amounts vary from 3.0 to 3.8% (USPHS,

1971). The fat content of fluid milk sold may exceed the minimum amount shown in Table 4.1. The milk from Guernsey and Jersey cows contains more fat than the milk from Holstein, Ayrshire, and Brown Swiss cows (Wilcox et al., 1971). Fat content decreases in May, June, and July and with the aging of the cow, but feed does not appreciably affect fat content (Johnson, 1974).

Milkfat contains the glycerides of a wide variety of fatty acids, both saturated and unsaturated; it is distinguished from other food fats by its content of short-chain saturated fatty acids such as butyric, caproic, caprylic, and capric acids. These short-chain fatty acids are important to the flavor of products made from milk and in off flavors that may develop in milk.

In addition to the glycerides, milk contains several other lipids in varying amounts, including phospholipids, sterols, carotenoids, and fat-soluble vitamins. Factors including type and amount of feed, stage of lactation, and species of animal or breed of cow affect the proportion of these components in milk (Kurtz, 1974).

Carbohydrate

Lactose, the major carbohydrate of milk, is found in cow's milk at levels ranging from 4.4 to 5.2% (Nickerson, 1974). This level of sugar does not make the milk unduly sweet because lactose is less sweet than sucrose. Upon digestion, lactose yields glucose and galactose. In addition to lactose, milk contains small amounts of glucose, galactose, and other saccharides (Johnson, 1974).

Lactose is found as one of two crystalline forms, alpha-hydrate and anhydrous-beta, or as lactose glass, a mixture of α- and β-lactose. Although the lactose of milk is in true solution, its low solubility creates some manufacturing and preparation problems, since lactose crystals are objectionably gritty in texture. Lactose crystallizes from sweetened condensed milk because the small amount of water remaining after evaporation is not sufficient to keep the lactose and the added sucrose in solution. Manufacturers attempt to promote the formation of many nuclei for crystal formation so that crystals formed will be very small and imperceptible (Nickerson, 1974).

In contrast, evaporated milk contains enough water to dissolve all of the lactose. Lactose may crystallize from ice cream since much of the water is frozen and hence is not available to hold the lactose in solution.

As with sweetened condensed milk, the formation of very small crystals is promoted by agitation or seeding with a supersaturated solution of lactose (Nickerson, 1974). Crystallization of lactose also is responsible for lumping and caking of dried milk during storage. Lactose is the source of lactic acid, formed by bacterial action as milk sours. When milk is coagulated, lactose is present in the whey, from which it can be prepared commercially. For this reason, cheese that is prepared from the curd is low in carbohydrates. Some whey cheeses such as Mysost or Primost are high in lactose and therefore sweet in taste.

Salts

The salts of milk, which constitute less than 1% of the milk, include the chlorides, phosphates, and citrates of potassium, sodium, calcium, and magnesium. The heat stability, the rennin coagulation of milk, and the feathering of cream in coffee are influenced by the salts of milk (Johnson, 1974).

Trace Elements and Vitamins

Practically all of the minerals in the soil from which the cow obtains her feed are present in milk, some of them only in trace amounts. Cobalt, copper, and iodine may be low due to deficiencies in soil content. Copper is significant to the sensory quality of milk because it exerts a catalytic effect on the development of oxidized flavor. Other trace elements include iron, magnesium, molybdenum, nickel, and zinc.

Milk contains many vitamins required by humans, some of them in abundance, and some in smaller quantities. Riboflavin is present in significant quantities. This water-soluble compound is responsible for the slight yellowish or greenish tint of skimmilk. The fat-soluble vitamin A precursor, carotene, is responsible for the yellowish color of milkfat.

ALTERATIONS IN MILK AND MILK PRODUCTS BY PROCESSING TREATMENTS

Prior to the consumption of milk as fluid milk or as a product prepared from fluid milk, milk is subjected to one or more treatments that influence the characteristics of the product. Treatment may include one or more heat treatments, coagulation, and/or dehydration and may influence flavor, color, and functional properties.

Heat

Pasteurization Processing affects the flavor of milk, usually because of heating. In fluid milk, a change in flavor is caused by the heat necessary to pasteurize the milk. Pasteurization is used to destroy pathogenic organisms. In addition, enzymes that may cause the development of off flavors may be inactivated.

Some of the most common off flavors in milk are rancid or oxidized flavors. Rancidity resulting from activity of lipase is not usually a problem because the enzyme is destroyed by pasteurization. Probably because of the large surface of its fat globules, homogenized milk becomes rancid within a few hours if not adequately pasteurized or if mixed with small amounts of raw milk. A different type of off flavor, oxidized flavor, is accelerated by traces of copper; this finding has caused a virtual elimination of copper-containing equipment from dairies.

The flavor of fluid milk is changed by the heat of pasteurization. The use of a high temperature for a short time, such as 72°C or higher for 15 sec, changes the flavor less than the holding method of at least 62°C for 30 min. The flavor of milk pasteurized at 78.9°C was preferred over milk pasteurized at higher or lower temperatures (Deane et al., 1967). Boiling, of course, changes the flavor of milk more than does pasteurization. Hutton and Patton (1952) reported that sulfhydryl groups of β-lactoglobulin, which give rise to hydrogen sulfide with denaturation, are responsible for the cooked flavor of milk.

Evaporation and canning Evaporated milk is concentrated to slightly more than double the solids content of fluid whole milk. Then it is homogenized, sealed into cans, and sterilized. The characteristic cooked taste of evaporated milk is caused by the high temperatures required in canning. The milk is sterilized at 117°C for 15 min or 127°C for 2 min (Johnson, 1974). Methyl sulfide, a component that is responsible for a "cowy" flavor in fresh milk (Patton et al., 1956), has been found at elevated levels in evaporated milk, suggesting that it plays a role in the cooked flavor (Patton, 1958).

Flavor deterioration in concentrated milk in the form of cooked, scorched, and stale flavors was greater at 20 and 37°C than at 4°C when concentrated milk was stored for eight mo (Loney et al., 1968). Off colors may be developed in evaporated milk stored at high temperatures for long periods of time. Carbonyl-amine browning and lactose caramelization may be responsible (Hall and Hedrick, 1971).

Sweetened condensed milk also contains a little more than double the milk solids content of fluid whole milk; in addition, sucrose is added after evaporation, resulting in a total carbohydrate concentration of approximately 56%. The flavor of sweetened condensed milk cannot be compared with that of other milks because of its high sugar content. Less change in flavor due to heat can be expected for this milk than for evaporated milk because the added sucrose serves as a preservative, eliminating the need for sterilization. Prior to evaporation, milk to be sweetened is preheated to inactivate enzymes, reduce microbial counts to low numbers, and provide sufficient viscosity to prevent sandiness of the finished product (Hall and Hedrick, 1971).

Drying Several methods are used for the drying of whole milk, skimmilk, and other milk products and are described in detail by Hall and Hedrick (1971). Dried whole milk deteriorates more rapidly during storage than nonfat dry milk. Oxidation of the milkfat results in a tallowy flavor, and the carbonyl-amine reaction is responsible for the stale flavor that develops. Dried whole milk has not been used to a great extent in this country because of this storage problem. Deterioration of dried whole milk may be delayed by reducing its moisture content, preheating, storing at low temperatures, adding small amounts of an antioxidant, or packaging with nitrogen in a sealed container (Hall and Hedrick, 1971).

Nonfat dry milk can be stored at 21°C for up to 1.5 yr. The dispersibility of nonfat dry milk is improved by *agglomeration,* a process that involves rewetting and redrying. Instantized milk powder produced in this process has a light, granular texture and is dispersed easily (Neff and Morris, 1968).

Cheese Production

The production of cheese was used more than 4,000 yr ago as a way of preserving milk for later use. Cheese is made by coagulating milk, cream, skimmilk, buttermilk, or a combination of these and then draining part or most of the whey. Cheese is made most commonly from cow's milk, but the milk of other mammals may be used. Pasteurized milk is used in most cases. The curd, which has been separated from the whey, may be allowed to ripen by the action of enzymes of microorganisms to produce a natural cheese.

Thus, cheese may range from unchanged curd to products with altered moisture content, composition, and flavor (Kon, 1972). Four hundred varieties of cheese are described in a USDA handbook (USDA, 1969).

Coagulation Milk used for cheese production may be clotted with rennin or acid or both. A bacterial culture is added for acid production. Formerly, the enzyme *rennin* was used almost exclusively for the clotting of milk in the production of cheese. Rennin is used as a crude extract, rennet. Rennin acts most effectively at a pH of 6.7 (Bingham, 1975) and a temperature of about 40°C. Enzymatic coagulation of milk involves two phases. Rennin is involved in the first phase, in which a specific bond of κ-casein is cleaved to form insoluble para-κ-casein and a soluble peptide (Bingham, 1975). In the second phase, a clot is formed by the paracasein and calcium (Cheryan et al., 1975). Coagulation time decreases as the pH of the second phase is decreased from 6.7 to 5.6 and as the temperature increases (Cheryan et al., (1975). Heating milk above 65°C and then cooling it prior to treatment with rennin reduces clotting rate and curd firmness. A heat-induced interaction between κ-casein and β-lactoglobulin may delay the action of rennin on κ-casein (Sawyer, 1969).

Because the availability of rennin has decreased, many investigators have studied proteolytic enzymes from other animal and plant sources for use in cheese making. However, in a review of enzymes from plant sources, Ernstrom (1974) suggested that cheese made from curds formed with these enzymes is pasty and bitter. Similar results have been reported for bacterial enzymes, but more encouraging results have been obtained with some fungal enzymes. Mixtures of bovine pepsin and rennin have been used successfully (Emmons et al., 1971; Nelson, 1975).

Curd treatment Following coagulation, the curd is cut to allow loss of the whey. This process is controlled by cutting the curd into various size pieces. The curd is heated at low temperatures to hasten loss of whey and produce a more compact texture. Salts, acids, and bacterial cultures are added at various steps in this process. The curd then is pressed to make the cheese firmer and more compact. Variation in treatments results in the production of various kinds of cheese (Wong, 1974).

The above treatments are important in the control of moisture. Cheeses may be classified on the basis of their moisture content. Hard cheeses, such as Cheddar, contain 30–40% moisture; semi-soft cheeses, such as Muenster, contain 50–75% moisture (Wong, 1974).

For the production of unripened cheeses such as cottage cheese and cream cheese, lactic acid starter bacteria are used for coagulation. The curd then is processed to give the characteristic product.

Curing Cheese may be cured for varying lengths of time and at varying temperatures and humidities in order to allow the product to ripen. The chemical and physical changes associated with ripening result in a change from a bland, tough, rubbery mass to a full-flavored, soft, mellow product. The role of salt is extremely important to the control of ripening. Omission of salt speeds the ripening process, resulting in a pasty texture and the development of an unnatural, bitter, fruity, or flat flavor in Cheddar cheese (Thakur et al., 1975). During ripening, the carbohydrate, fat, and protein are altered. Lactose is rapidly converted to lactic acid, which helps to inhibit the growth of undesirable microorganisms. Lipolytic enzymes liberate fatty acids, which contribute to the flavor. The typical flavor of Cheddar cheese is not developed when skimmilk is used, suggesting an important role for fat in flavor (Ohren and Tuckey, 1969). Proteolytic enzymes such as rennin are responsible for the formation of nitrogenous products of intermediate size, such as peptones and peptides. Enzymes of the microorganisms act on these and other substances to form products like amino acids, amines, fatty acids, esters, aldehydes, alcohols, and ketones. A complex mixture of substances most likely produces the characteristic flavor of cheese. Slight changes in the proportions of the flavor components, caused by variations in the curing process, may alter the flavor of the cheese. A strong shift in the components may lead to off flavors.

Processed cheeses The uniformity of the flavor of processed cheese is due to its preparation by blending of natural cheeses of varying qualities to achieve the desired flavor and texture. First the cheese is ground, then it is mixed with the aid of heat and an emulsifying agent. Acid, cream, water, salt, coloring, and flavoring may be added, but the moisture content may not be more than 1% greater than that of the cheese from which it is made. In the total process, the enzymes are inactivated and the bacteria are destroyed, preventing changes in flavor during storage. Processed cheese food resembles processed cheese, but owing to the inclusion of dairy ingredients such as cream, skimmilk, whey, or lactalbumin it contains more moisture and less fat than processed cheese. Processed cheese spread is even softer than cheese food because it may contain additional moisture. Stabilizing agents such as gums, gelatin, and algin may be added to the spreads.

Fermentation

The introduction of lactic acid bacteria into skim milk, whole milk, or slightly concentrated milk results in the production of cultured dairy products such as buttermilk, yogurt, sour cream, and sweet acidophilus milk. Buttermilk is skimmed or partially skimmed milk that is fermented with a mixed culture of lactic acid bacteria. Sour cream is produced from cream containing 18% fat. Nonfat dry milk solids may be added to increase viscosity. The consistency of yogurt, produced by the action of *Streptococcus thermophilus* and *Lactobacillus bulgaricus,* varies from a liquid resembling a stirred custard to a gel. Control of pH in yogurt production is important. If the pH is less than 5.5, the yogurt is judged to be too sour (Kroger, 1976). Sweet acidophilus milk contains an active culture of *Lactobacillus acidophilus.* The culture is added to cold lowfat or skimmilk and the mixture is kept cold. Therefore, relatively little fermentation occurs to produce lactic acid as in buttermilk or acidophilus milk, which has a high acid content and tart taste. Sweet acidophilus milk is similar in composition to skimmilk except that it contains less lactose and more lactic acid (Hargrove and Alford, 1974).

USE OF MILK AND MILK PRODUCTS IN FOOD PREPARATION

Heating Milk

Effects on proteins As with commercial processing, use of milk in food preparation may be preceded by or otherwise involve heating. The serum proteins are denatured with heating. The implications of this in flavor development, heat stabilization, and rennin coagulation were discussed previously. In addition, the volume of loaf bread made with heated milk is greater than the volume of bread made with unheated milk. At least one component of the serum protein fraction of milk is thought to be responsible for the relatively low volume of bread made with unheated milk (Jenness, 1954). Insufficient heating of milk before spray-drying depresses loaf volume. Holding at 85°C for 20 min is adequate (Hall and Hedrick, 1971). The coagulated material that is deposited on the bottom of the container in which milk is heated is part of the denatured serum protein fraction. The precipitate is likely to scorch when milk is heated directly over a burner.

The casein proteins are not affected by the usual process of heating. However, problems can be encountered when foods containing acids and salts are heated in milk. Thus, the acid from tomatoes may cause cream of tomato soup to curdle, and salt from cured meats will cause coagulation.

Comparisons of use of various forms in cooking The cooking qualities of homogenized milk differ in some ways from those of nonhomogenized milk. When compared with products made of nonhomogenized milk, those made with homogenized milk, like cornstarch pudding (Jordan et al., 1953), white sauce (Towson and Trout, 1946), and cocoa or chocolate beverages (Wegner et al., 1953), are more viscous. Homogenized milk coagulates more easily. Absorption of greater quantities of proteins on the extended fat surfaces may be responsible (Trout, 1950). Thus, homogenized milk is more likely to curdle when used in scalloped potatoes or stirred into hot cooked oatmeal (Hollender and Weckel, 1941). White sauce made of homogenized milk may appear curdled because the added fat does not combine well with that of homogenized milk, causing fat separation and a curdled appearance (Towson and Trout, 1946).

Baked custards made with nonhomogenized milk were superior in appearance, texture, tenderness, sheen, and firmness to those made with homogenized milk (Jordan et al., 1954). In contrast to the thick, light-colored crusts of custards made from homogenized milk, those of custards made with nonhomogenized milk were thin and delicately browned (Jordan et al., 1954), probably because milkfat rose to the top of these custards. Another disadvantage to the use of homogenized milk in making baked custards is that a longer baking period is required than is needed if nonhomogenized milk is used.

Creamed soups and mashed potatoes made with reconstituted evaporated milk were judged superior to those made with fluid milk, whereas practically no differences were noted when the two milks were used in yeast bread, quick bread, cakes, puddings, or baked custard (Atwood and Ehlers, 1933). Hussemann (1951) mentioned the distinctive color and flavor but the satisfactory texture of custards made with evaporated milk. Evaporated milk ranked well in chocolate puddings, perhaps because its distinctive flavor was masked by the chocolate. Curdling occurred in scalloped potatoes more often with evaporated milk than with other forms of milk even though the milks were added as thin white sauces. Such curdling probably is due to the protein of evaporated milk being rendered unstable by the heat of processing and the fact that potatoes have a more acid reaction than milk.

The basic effects of nonfat dry milk in food preparation were studied by Morse et al. (1950a, b). Nonfat dry milk increased the viscosity of pastes of flour, dry milk, and water in proportion to the amount of milk used. This increased viscosity could be offset by reducing the amount of flour. Nonfat dry milk also increased the thickness of white sauces, eggnog, custards, and chocolate pudding made with fluid milk. Although custards made without added dry milk were the most acceptable, those enriched with a moderate amount of nonfat dry milk were quite palatable and had a good flavor. Excessive amounts of dry milk made the consistency of the custards similar to that of pudding, but resulted in a product that was good when fresh and somewhat sticky after 24 hr.

When nonfat dry milk is added to baked products at enrichment levels, changes in the basic recipe may improve the product. Because added milk increases the tendency of the product to brown during baking, it may be necessary to reduce the amount of sugar and to lower the baking temperature. To compensate for the thickening effect of a high solids content, less flour or more liquid can be used. Fat can be increased to offset any toughening effect of milk solids.

The whey that remains after the removal of the curd in making cheese contains many valuable nutrients. Although it often is used for animal feed, more whey is produced than can be used in animal feed and whey cheeses. As a result, cheese whey has become an environmental pollutant, as well as a wasted resource, and much research effort has been devoted to practical means of concentrating and using it. Dried whey has been used successfully in starch puddings (Hanning et al., 1955), cakes (de Goumois and Hanning, 1953), and baked products (Scanlon, 1974). Its addition to soft drinks has been tested (Holsinger et al., 1973).

Heating Cheese

Cheese cookery may consist simply of melting cheese, as in making grilled cheese sandwiches, or of combining it with other ingredients and cooking, as in making cheese sauce. Low temperatures and short times are essential in preventing excessive fat separation, stringing, matting, and toughening. "Aged natural" cheese and processed cheese are better for cooking than natural cheese that has not been aged properly.

A partial explanation of the good cooking qualities of aged cheese was reported by Personius and coworkers (1944). The cooking quality of Cheddar cheese of normal fat content improved as the products of protein hydrolysis, measured by the protein soluble in dilute salt solution, increased with aging. Cheese of high moisture content was superior in cooking quality to samples of normal or low moisture.

The cooking quality of cheese that was low in fat did not improve with age, even though the amount of soluble protein increased during storage as much as that of normal cheese. When fat was not present, the tendency for mat formation and stringing was greatly exaggerated. The fat of normal samples melted during cooking, thus contributing to the consistency and tenderness of the heated product and preventing the formation of a continuous mass.

The harmful effects of excessive heat in cheese cookery also have been demonstrated. Greater fat separation, longer strings, and more toughening were observed in samples heated over water at 100°C than at 75°C and in samples exposed to direct heat of 177°C for the longest periods.

Whipping of Milk Products

Cream Whipped cream is an air-in-water foam in which air cells are surrounded by a film containing fat droplets stabilized by a film of protein. Partial denaturation of this protein occurs as the cream is whipped. There is some clumping of the fat globules in the cell walls of the foam, and the fat is partly solidified, preventing collapse of the cell walls. When whipped cream is heated, the fat is melted and the foam collapses. If whipped cream is beaten too long, further clumping of the fat globules occurs and butter is formed.

The whipping quality of cream has been found to improve with increased butterfat content up to 30%. Further increases in fat content do not improve the quality of the whip, but do improve the standing-up quality and decrease the time required to whip the cream. Cream with less than 22% fat is not satisfactory for whipping.

Raw cream has been found to whip best. Pasteurization impairs whipping ability slightly, and homogenization damages it further (Babcock, 1922). Cream is pasteurized for public health reasons, but it is not homogenized if it is intended for whipping. Impaired whipping ability may be attributed to the increased dispersion of the fat caused by both pasteurization and homogenization, or to damage to the protein necessary for foam formation. This damage may be caused by the heat used in pasteurization and by adsorption on the greatly increased surface of the fat globules in homogenization. If cream is not to be whipped, homogenization is advantageous because it increases viscosity.

The quality of whipped cream is influenced by a number of factors. The whipping ability of cream improves with age, changing rapidly during the first 48 hr and reaching a maximum in 72 hr. Temperatures below 7°C are best for whipping. Although acid improves whipping ability, its use cannot be recommended because it is effective only in amounts large enough to give the cream a sour taste. Sugar decreases the quality of the whip whether added before or after whipping if enough is added to sweeten thoroughly (Babcock, 1922). The effects of granulated and powdered sugar are similar. The addition of gelatin to cream decreases the amount of drainage from it, but changes the character of the whip.

Evaporated milk Evaporated milk can be whipped successfully if it is thoroughly chilled prior to whipping. Air bubbles incorporated during whipping are trapped in the viscous liquid to form a stable foam. The difference in viscosity partially accounts for the difference in stability of foams formed from milk and the more concentrated evaporated milk.

Nonfat dry milk Nonfat dry milk can be whipped if reconstituted to a higher-than-normal solids content. Homogenization of milk prior to drying improves the quality of foams prepared from reconstituted nonfat dry milk (Tamsma et al., 1969).

EXPERIMENT

Rennin Coagulation of Milk

The effect of several conditions on the formation of a rennin gel will be illustrated.

A. *Basic procedure*

1. Place 5 ml milk in a test tube.
2. Add 3–4 drops of the enzyme dispersion prepared from one rennin tablet in 10 ml distilled water.
3. Place the test tube in a 40°C water bath unless otherwise noted.
4. Record the length of time required for coagulation to occur.

B. *Experimental variation*

1. Effect of temperature
   a. Basic procedure.
   b. Place test tube of milk in a beaker of cracked ice. After the milk is thoroughly chilled, add 3–4 drops of rennin solution. Keep the tube in the ice bath and note whether coagulation occurs.
2. Effect of preheating milk
   a. Basic procedure.
   b. Use milk that has been boiled and cooled to 40°C. Follow basic procedure.

C. *Evaluation*

Note differences in coagulation time and size and texture of the curd.

SUGGESTED EXERCISES

1. Enrich some foods with nonfat dry milk at different levels. Use as high a level of dry milk as will make a palatable product. Formulas and procedures for products in the experiments of Chapters 3, 11, and 12 can be used.
2. Study curdling of tomato soup made in several ways. In one series of experiments, add tomato juice to white sauces; in the other, add thickened juice to milk. In each series, combine when both liquids are hot, when both liquids are cold, when the tomato juice is hot and the milk cold, and vice versa. Heat and judge each sample immediately.
3. Prepare baked custards or white sauce using various forms of milk such as fluid whole milk, 2% milk, skimmilk, evaporated whole milk or skimmilk, and reconstituted nonfat dry milk. Evaluate using the objective methods suggested in Chapter 16. The egg custard formula in Chapter 3 may be used if desired.
4. Study the softness of the curd formed by coagulating several forms of milk with rennin. Evaluate the firmness of the curd subjectively and objectively by one of the methods suggested for baked custard in Chapter 3.
5. Add varying amounts of disodium phosphate or potassium citrate to samples of natural Cheddar cheese. Levels of 1, 3, and 5% of the weight of the cheese might be used for the two salts. Use these samples and one of processed cheese with no additions to make cheese sauces, and judge smoothness, viscosity, and palatability. Determine the pH of the sauces if possible.
6. Use one or more of the milk substitutes that are on the market in a product such as custard or cake.

REFERENCES

Atwood, F. J. and Ehlers, M. S. 1933. The use of evaporated milk in quantity cookery. J. Amer. Dietet. Assoc. 9: 306.

Babcock, C. J. 1922. The whipping quality of cream. Bull. 1075, USDA, Washington, DC.

Bingham, E. W. 1975. Action of rennin on κ-casein. J. Dairy Sci. 58: 13.

Cheryan, M., Van Wyk, P. J., Olson, N. F. and Richardson, T. 1975. Secondary phase and mechanism of enzymatic milk coagulation. J. Dairy Sci. 58: 477.
Deane, D. D., Chelesvig, J. A. and Thomas, W. R. 1967. Pasteurization treatment and consumer acceptance of milk. J. Dairy Sci. 50: 1216.
deGoumois, J. and Hanning, F. 1953. Effects of dried whey and various sugars on the quality of yellow cakes containing 100% sucrose. Cereal Chem. 30: 258.
Emmons, D. B., Petrasovits, A., Gillan, R. H. and Bain, J. M. 1971. Cheddar cheese manufacture with pepsin and rennet. Can. Inst. Food Technol. J. 4: 31.
Ernstrom, C. A. 1974. Milk-clotting enzymes and cheese chemistry. Part 1. Milk-clotting enzymes and their action. In "Fundamentals of Dairy Chemistry," 2nd ed., Eds. Webb, B. H., Johnson, A. H. and Alford, J. A. Avi Publ. Co., Westport, CT.
Farrell, H. M. Jr. and Thompson, M. P. 1974. Physical equilibria: proteins. In "Fundamentals of Dairy Chemistry," 2nd ed., Eds. Webb, B. H., Johnson, A. H. and Alford, J. A. Avi Publ. Co., Westport, CT.
Hall, C. W. and Hedrick, T. I. 1971. "Drying of Milk and Milk Products," 2nd ed. Avi Publ. Co., Westport, CT.
Hanning, F., Bloch, J. deG. and Siemers, L. L. 1955. The quality of starch puddings containing whey and/or nonfat milk solids. J. Home Econ. 47: 107.
Hargrove, R. E. and Alford, J. A. 1974. Composition of milk products. In "Fundamentals of Dairy Chemistry," 2nd ed., Eds. Webb, B. H., Johnson, A. H. and Alford, J. A. Avi Publ. Co., Westport, CT.
Hollender, H. and Weckel, K. G. 1941. Stability of homogenized milk in cookery practice. Food Res. 6: 335.
Holsinger, V. H., Posti, L. P., De Vilbiss, E. D. and Pallansch, M. J. 1973. Fortifying soft drinks with cheese whey proteins. Food Technol. 27(2): 59.
Hussemann, D. L. 1951. Effect of altering milk solids content on the acceptability of certain foods. J. Amer. Dietet. Assoc. 27: 583.
Hutton, J. T. and Patton, S. 1952. The origin of sulfhydryl groups in milk proteins and their contributions to "cooked" flavor. J. Dairy Sci. 35: 699.
Jenness, R. 1954. Milk proteins. Effects of heat treatment on serum proteins. J. Agric. Food Chem. 2: 75.
Jenness, R., Shipe, W. F. Jr. and Sherbon, J. W. 1974. Physical properties of milk. In "Fundamental of Dairy Chemistry," 2nd ed., Eds. Webb, B. H., Johnson, A. H. and Alford, J. A. Avi Publ. Co., Westport, CT.
Johnson, A. H. 1974. The composition of milk. In "Fundamentals of Dairy Chemistry," 2nd ed., Eds. Webb, B. H., Johnson, A. H. and Alford, J. A. Avi Publ. Co., Westport, CT.
Jordan, R., Wegner, E. S. and Hollender, H. A. 1953. The effect of homogenized milk upon the viscosity of cornstarch puddings. Food Res. 18: 649.

Kon, S. K. 1972. Milk and milk products in human nutrition. FAO Nutritional Studies No. 27, Food and Agricultural Organization of the United Nations, Rome.

Kroger, M. 1976. Quality of yogurt. J. Dairy Sci. 59: 344.

Kurtz, F. E. 1974. The lipids of milk: composition and properties. In "Fundamentals of Dairy Chemistry," 2nd ed., Eds. Webb, B. H., Johnson, A. H. and Alford, J. A. Avi. Publ. Co., Westport, CT.

Loney, B. E., Bassette, R. and Claydon, J. J. 1968. Chemical and flavor changes in sterile concentrated milk during storage. J. Dairy Sci. 51: 1770.

Morr, C. V. 1969. Protein aggregation in conventional and ultra-high-temperature heated skimmilk. J. Dairy Sci. 52: 1174.

Morr, C. V. 1975. Chemistry of milk proteins in food processing. J. Dairy Sci. 58: 977.

Morse, L. M., Davis, D. S. and Jack, E. L. 1950a. Use and properties of non-fat dry milk solids in food preparation. I. Effect on viscosity and gel strength. Food Res. 15: 200.

Morse, L. M., Davis, D. S. and Jack, E. L. 1950b. Use and properties of non-fat dry milk solids in food preparation. II. Use in typical foods. Food Res. 15: 216.

Mulder, H. and Walstra, P. 1974. "The Milk Fat Globule. Emulsion Science as Applied to Milk Products and Comparable Foods." Commonwealth Bureaux, Farnham Royal, Bucks, England.

Neff, E. and Morris, H. A. L. 1968. Agglomeration of milk powder and its influence on reconstitution properties. J. Dairy Sci. 51: 330.

Nelson, J. H. 1975. Impact of new milk clotting enzymes on cheese technology. J. Dairy Sci. 58: 1739.

Nickerson, T. A. 1974. Lactose. In "Fundamentals of Dairy Chemistry," 2nd ed., Eds. Webb, B. H., Johnson, A. H. and Alford, J. A. Avi Publ. Co., Westport, CT.

Ohren, J. A. and Tuckey, S. L. 1969. Relation of flavor development in Cheddar cheese to chemical changes in the fat of the cheese. J. Dairy Sci. 52: 598.

Patton, S. 1958. Review of organic chemical effects of heat on milk. J. Agric. Food Chem. 6: 132.

Patton, S., Forss, D. A. and Day, E. A. 1956. Methyl sulfide and the flavor of milk. J. Dairy Sci. 39: 1469.

Personius, C., Boardman, E. and Asherman, A. R. 1944. Some factors affecting the behavior of Cheddar cheese in cooking. Food Res. 9: 304.

Posati, L. P. and Orr, M. L. 1976. Composition of foods. Dairy and egg products. Agric. Handbook 8-1, Agricultural Research Service, USDA, Washington, DC.

Pyne, G. T. 1962. Reviews of the progress of dairy science. Section C. Dairy

chemistry. Some aspects of the physical chemistry of the salts of milk. J. Dairy Res. 29: 101.
Rowland, S. J. 1938. The determination of the nitrogen distribution in milk. J. Dairy Res. 9: 42.
Sawyer, W. H. 1969. Complex between β-lactoglobulin and κ-casein. J. Dairy Sci. 52: 1347.
Scanlon, J. 1974. A new sweet whey-soy protein blend for the baker. Baker's Digest 48(5): 30.
Slattery, C. W. 1976. Review: casein micelle structure; an examination of models. J. Dairy Sci. 59: 1547.
Tamsma, A., Kontson, A. and Pallansch, M. J. 1969. Production of whippable nonfat dried milk by homogenation. J. Dairy Sci. 52: 428.
Thakur, M. K., Kirk, J. R. and Hedrick, T. I. 1975. Changes during ripening of unsalted Cheddar cheese. J. Dairy Sci. 58: 175.
Towson, A. M. and Trout, G. M. 1946. Some cooking qualities of homogenized milk. II. White sauces. Food Res. 11: 261.
Trout, G. M. 1950. "Homogenized Milk: A Review and Guide." Michigan State College Press, East Lansing.
USDA. 1969. Cheese varieties and descriptions. Agric. Handbook 54, USDA, Washington, DC.
USPHS, 1965. Milk ordinance and code. United States Dept. Public Health Publ. 229, USPHS, Washington, DC.
USPHS. 1971. Federal and state standards for the composition of milk products. Handbook 51, USPHS, Washington, DC.
Wegner, E. S., Jordan, R. and Hollender, H. A. 1953. Homogenized and nonhomogenized milk in the preparation of selected food products. J. Home Econ. 45: 589.
Whitney, R. McL., Brunner, J. R., Ebner, K. E., Farrell, H. M. Jr., Josephson, R. V., Morr, C. V. and Swaisgood, H. E. 1976. Nomenclature of the proteins of cow's milk: fourth revision. J. Dairy Sci. 59: 795.
Wilcox, J. C., Gaunt, S. N. and Farthing, B. R. 1971. Genetic interrelationships of milk composition and yield. Bull. 155, Southern Coop. Series, Gainesville, FL.
Wong, N. P. 1974. Milk-clotting enzymes and cheese chemistry. Part II. Cheese chemistry. In "Fundamentals of Dairy Chemistry," 2nd ed., Eds. Webb, B. H., Johnson, A. H. and Alford, J. A. Avi Publ. Co., Westport, CT.

5. Meat, Poultry, and Fish

Meat, poultry, and fish are major components of the American diet. Research regarding red meat, beef in particular, has been extensive, whereas there has been less work on poultry and even less on fish. This chapter is an attempt to summarize a representative sample of the research on these foods. The basic composition and structure of muscle foods will be discussed, as will the effect of various treatments on eating quality.

EATING QUALITY OF MEAT AS DETERMINED BY MUSCLE TISSUE COMPONENTS AND STRUCTURE

The gross composition of muscle tissue from several animals is shown in Table 5.1. Water, protein, fat, carbohydrate, and other components are important to the eating quality of meat, poultry, and fish. The contributions of muscle tissue components will be considered in terms of their contributions to three quality attributes: color, texture, and flavor.

Table 5.1 Composition of selected raw muscle foods[a]

Muscle food	Water %	Protein %	Fat %
Beef			
Chuck	70.3	21.2	7.4
Loin	47.5–55.7	13.0–17.0	27–35
Round	66.6	20.1	12.2
Chicken			
Light meat	63.8	31.6	3.4
Dark meat	64.4	28.0	6.3
Fish			
Flounder	58.1	30.0	8.2
Cod	64.6	25.0	4.7
Scallops	73.1	23.1	1.4
Lamb			
Leg	64.8	17.8	16.2
Loin	57.7	14.0	21.4
Pork			
Loin	57.2	17.1	25.0
Shoulder cuts	59.3	15.4	24.4
Turkey			
Light meat	62.1	32.9	3.9
Dark meat	60.5	30.0	8.2

[a]Carbohydrate listed as 0 for all muscle foods in Adams (1975).
SOURCE: Adams, 1975.

Color

Myoglobin and hemoglobin The color of muscle tissue is due primarily to the heme pigment myoglobin. Variation in color of meat depends on the chemical state of the pigment. Hemoglobin, the blood pigment, also contributes to meat color. Between 20% and 30% of the pigment present may be hemoglobin (Fox, 1966). Both pigments combine reversibly with oxygen, thereby supplying it for metabolic processes. Hemoglobin transports oxygen in the blood stream, whereas myoglobin holds it in the tissues. In muscle such as heart, which is used constantly and has a high oxygen demand, myoglobin is present in relatively high amounts and contributes to intense color. Muscles of older animals contain increased amounts of myoglobin. Thus, veal is lighter in color than beef. There also is a species difference; for example, pork, which is marketed at a later age than veal, contains approximately the same amount of myoglobin as veal.

Myoglobin is a conjugated protein. The iron-porphyrin compound heme is combined with a simple protein of the globin class to form the pigment. Hemoglobin is also a conjugated protein containing heme and globin. Although the heme portion of these two molecules is identical, the globins are different, and the hemoglobin molecule contains four heme protein units whereas myoglobin has only one. The structure of heme is shown in Figure 5.1. The iron atom of heme is bound covalently to four nitrogens, each in a pyrrole ring of the porphyrin structure. In myoglobin the iron atom also is bound to a nitrogen of a histidyl residue in the globin moiety.

A sixth position is open for the formation of complexes with several compounds (Bodwell and McClain, 1971). Variations at the sixth position are in part responsible for differences in color of meat. An abbreviated formula for myoglobin is shown in Figure 5.1. The uppermost vertical line in the abbreviated formula shows where the element sharing the sixth pair of electrons will be positioned.

Oxymyoglobin In living tissue, myoglobin, which is purplish red, exists in equilibrium with its bright red oxygenated form, oxymyoglobin, as shown in Figure 5.2. After the death of the animal, the oxygen in the tissues is used quickly and the pigment exists almost entirely in the purplish reduced form. In the interior of the meat this pigment is stable at low temperatures for long periods of time since there is no free oxygen present. On cutting, the color of the surface of the meat becomes bright red because oxygen from the air combines with myoglobin to form oxyglobin. (Note: iron is still reduced.) It remains red for a limited time if sufficient oxygen is present, a condition that is met in most prepackaged meats by the use of packaging film that is permeable to oxygen. If oxygen is excluded from the cut meat, as by vacuum packaging, the color of the meat may turn purple again because the pigment is deoxygenated to myoglobin. Such packaging extends shelf life of prepackaged beef because it suppresses microbiological growth and retards possible spoilage (Gôvindarajan, 1973).

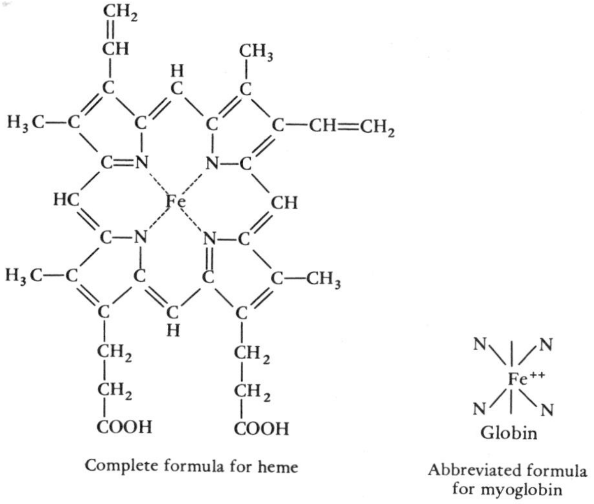

Complete formula for heme

Abbreviated formula for myoglobin

Figure 5.1. Complete formula for heme and abbreviated formula for myoglobin.

Metmyoglobin The major discoloration problem of fresh meat is the formation of the brownish red pigment metmyoglobin. As shown in Figure 5.2, this discoloration is caused by oxidation of the iron of myoglobin from the ferrous to the ferric state. As long as there are reducing substances present in the meat, any metmyoglobin that is formed is changed back to myoglobin; but when the reducing power of the muscle is lost, the color of the meat becomes brownish. The

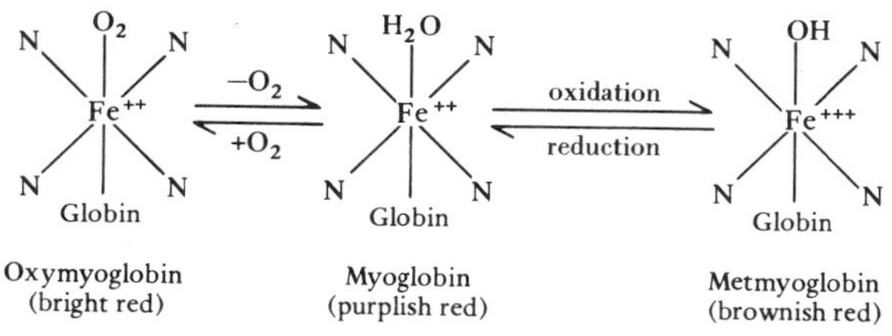

$-O_2$ / $+O_2$ — oxidation / reduction

Oxymyoglobin (bright red) — Myoglobin (purplish red) — Metmyoglobin (brownish red)

Figure 5.2. Pigment changes in red meat.

palatability of such meat may, however, still be satisfactory after cooking (Mangel, 1951). The formation of metmyoglobin is accelerated by high temperatures and microorganisms that use oxygen. It also is increased in the presence of fluorescent and incandescent light of 250 ft-c intensity as used in meat departments of some retail stores (Satterlee and Hansmeyer, 1974). Surface discoloration also depends on salt concentration and the presence of various metals (Watts, 1954; Giddings, 1977). After the formation of metmyoglobin, further oxidative changes in meat pigment, caused by enzymes and bacteria, produce a series of brown, green, and faded-looking compounds (Watts, 1954).

Changes during cooking The red pigment of meat cooked until rare is the same oxymyoglobin that is present in raw meat (Bernofsky et al., 1959). With further heating, the globin of oxymyoglobin is denatured and the iron is oxidized from the ferrous to the ferric state. Probably this denatured, oxidized form of the pigment is in all well-done meats, such as well-done beef, pork, or chicken (Tappel, 1957). Other reactions also contribute to cooked meat color.

Changes during curing During curing, the pigment of meat changes to a form that remains red during cooking. The traditional curing process involved the preservation of meat by the addition of salt, sodium or potassium nitrate, and sugar. After it was learned that the nitrate functions in the pigment conversion only after being reduced to nitrite, both nitrate and nitrite were added. Recent public health considerations have resulted in a decreased addition of both nitrites and nitrates. The use of nitrate probably will be eliminated completely because of the difficulty in controlling residual nitrite levels when nitrate is the source of nitrite.

Reactions involving myoglobin and occurring during curing are shown in Figure 5.3. In the older methods of curing, which are still being used to cure solid pieces of meat like hams, salts diffuse into the tissues from brine and from dry salt mixtures applied to the outside of the meat; in newer methods, a curing solution is pumped into the tissues through arteries. The pigment characteristic of cured meat is formed by the combining of nitric oxide with myoglobin to form an unstable, red pigment, nitric oxide myoglobin. When the meat is heated, as it often is during the curing process, the more stable, pink

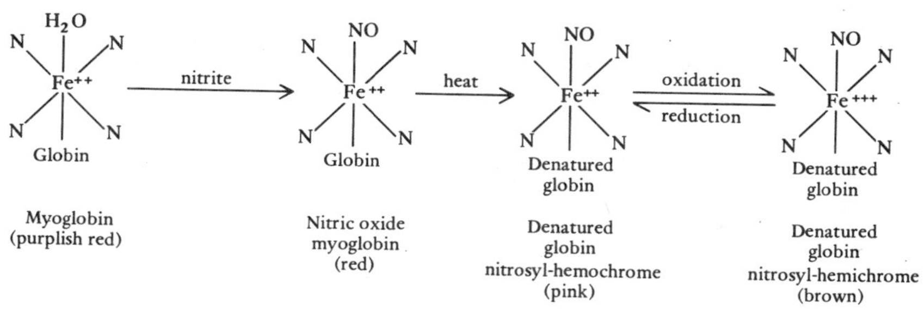

Figure 5.3. Pigment changes in cured meat.

pigment, nitrosyl hemochrome, is formed. This pigment is responsible for the color of ham, bacon, and corned beef as well as that of many table-ready meats. Its color is stable during cooking, but becomes brown when exposed to light and air because the iron of the pigment is oxidized from the ferrous to the ferric state, thus changing the pigment to denatured globin nitrosyl-hemichrome. Further oxidation produces a variety of decomposition products, some of which are green in color. Decomposition may include removal of the iron atom and breakage of the porphyrin ring (Bard and Townsend, 1971). Because the pigments of cured meats fade rapidly in transparent packages unless oxygen is excluded, they are often vacuum packaged.

Browning Surface browning of meat during heating is attributed to three components. Fats decompose and react with the products of carbohydrate and protein decomposition (Bratzler, 1971). One reaction that occurs is carbonyl-amine browning.

Texture

Texture of meat is considered by most to be synonymous with tenderness, or the ease of chewing a piece of meat. However, texture is more complex than that. Cover et al. (1962a, b, c) recognized that tenderness is a multicomponent sensation. Panelists participating in the studies conducted by Cover's group scored juiciness, softness, muscle fiber, and connective tissue components of tenderness. Thus, understanding of tenderness or, more correctly, texture, requires an understanding of water and water binding capacity, muscle fiber proteins and ultrastructure, connective tissue proteins and organization, and fat content and distribution.

Water Water is quantitatively the most important muscle tissue constituent, as shown in Table 5.1. Since it does represent about 75% of the weight of the tissue it is important to the texture of meat as well as the color and flavor.

Water-holding capacity A very small portion of the water in muscle tissue is bound very closely to the proteins of muscle tissue. All but 4–10 g of the 300–360 g of water per 100 g of protein in muscle tissue exist as free molecules within the muscle fibers and the associated connective tissue (Wismer-Pedersen, 1971). Most of this so-called free water is located within muscle fibers, with smaller amounts located within connective tissue and the sarcoplasm. Mechanical immobilization of water within muscle fibers is possible because of the three-dimensional structure of the muscle fibers. Thus disruptions of the spaces within this structure will reduce water-holding capacity (WHC). Treatments that can reduce WHC include grinding, proteolysis, freezing, salting, pH alteration, and heating.

Juiciness Water-holding capacity is thought to be related to the juiciness of meat as evaluated by sensory panelists. Two phases of juiciness perception can be identified: the initial release of juices with early chews and the continued impression of juiciness with further release of serum and increased flow of saliva resulting from the stimulation by fat in the meat (Bratzler, 1971).

Muscle fibers The fibers of muscle tissue are long, slender multinucleated cells ranging in length from a few millimeters up to 34 cm and averaging 60 μm in diameter (Bailey, 1972). The fibers are comprised of smaller structures, myofibrils. Approximately 2000 myofibrils, 1.0 μm in diameter, are found in each average-sized fiber. Molecules of the proteins actin, tropomyosin, and troponin are arranged in an orderly manner as illustrated in Figure 5.4. Microscopically, muscle tissue samples appear to have a pattern of cross striations as shown in Figure 5.5. The light and dark banding pattern is produced by the orderly

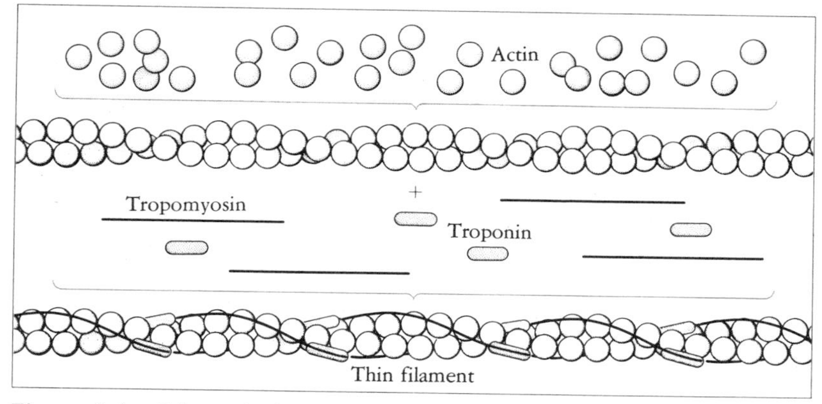

Figure 5.4. Schematic diagram showing assembly of the thin filament from actin, tropomyosin, and troponin, and the molecular architecture of the assembled thin filament. (From "The Cooperative Action of Muscle Proteins," by John M. Murray and Annemarie Weber. Copyright © February, 1974 by Scientific American, Inc. All rights reserved.)

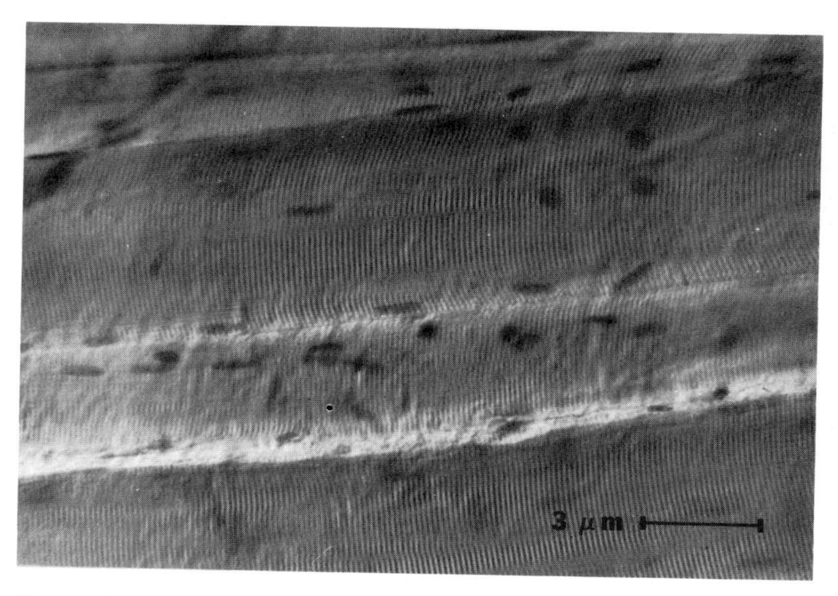

Figure 5.5. Light micrograph of raw beef semitendinosus tissue showing cross striations and nuclei.

arrangement of the thin actin filaments and the thicker myosin filaments. Tropomyosin is located along the length of the thin filaments and probably is located, along with α-actin, in the Z-disc, a structure that apparently holds the thin filaments together at the Z-line. The Z-lines are visible in Figure 5.6. Actin and myosin combine to form actomyosin as they slide past each other in the process of muscle contraction. The nucleotide adenosine triphosphate (ATP) supplies the energy for the contraction process.

The size of the muscle fibers and the effect of heat on them are related to the tenderness of cooked meat. Slender muscle fibers appear to be associated with tenderness, since as fiber diameter increases, tenderness decreases (Hiner et al., 1953). The diameter of the fiber increases as the animal increases in age.

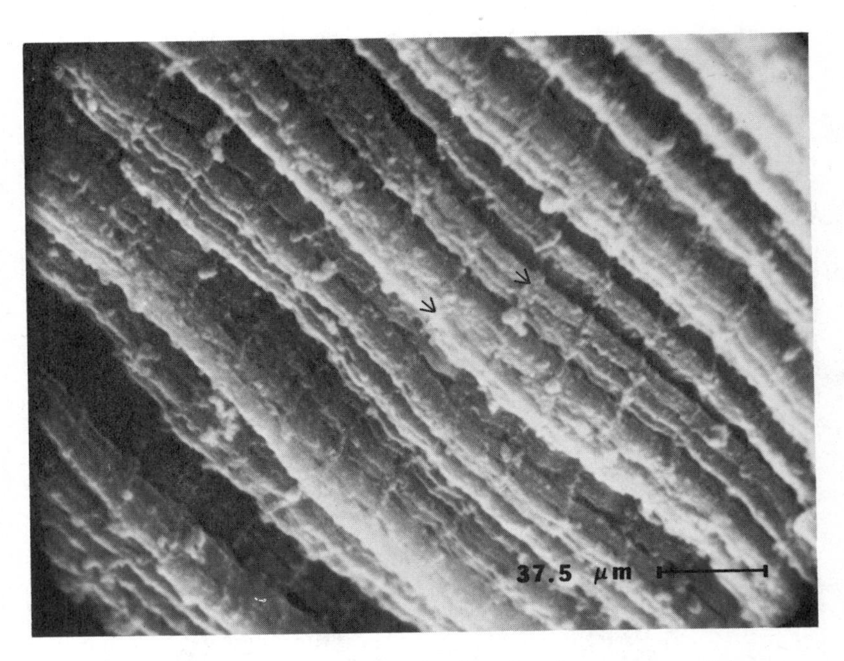

Figure 5.6. SEM micrograph of raw beef semitendinosus tissue. Z-line structures are indicated by arrows. (Courtesy of Tennessee Agricultural Experiment Station, University of Tennessee, Knoxville.)

Sarcomere length has been measured as an indicator of muscle fiber characteristics. Shortened sarcomeres are associated with toughness (Marsh and Leet, 1966). Divisions between sarcomeres are clearly visible in the three-dimensional photograph from a scanning electron microscope shown in Figure 5.6. Electron microscopy is a recently employed tool in the study of meat texture as described by Jones (1977) and Hearne and coworkers (1978).

It generally is accepted that muscle fibers become tougher as meat is heated. Direct evidence that cooking may toughen muscle fibers has been obtained by heating isolated fibers (Wang et al., 1956). Probably not all cooking methods toughen muscle fibers, however. The degree of toughening depends on the time and temperature of heating as well as the cooking method.

As in cooking eggs and other protein foods, heating of meat causes denaturation and coagulation of the proteins with an unfolding of peptide chains and the formation of new salt and hydrogen bonds. At temperatures below 40°C denaturation is mild, but it becomes greater as the temperature increases and is almost complete at 65°C (Hamm and Deatherage, 1960). Coagulation tightens the network of protein structure of meat, and increases the pH by decreasing the acidic but not the basic groups of the protein. The change in pH during heating is very small.

Connective tissue Muscle fibers are supported in the body by connective tissue. Each fiber is enclosed in the cell membrane, or *sarcolemma,* and groups of fibers are organized into bundles by an extremely thin network of connective tissue called the *endomysium.* A bundle is surrounded by a sheet of connective tissue called the *perimysium.* A group of such bundles, surrounded by an outer layer of connective tissue called the *epimysium,* forms the muscle. The relationship of the muscle fibers to these three types of connective tissue is shown in Figure 5.7. This harnessing of the muscle by connective tissue and its attachment by means of a tendon to bone can be seen clearly in a chicken leg or in the heel of a round of beef.

Collagen *Collagen* is one of two fibrous connective tissue proteins. It is the principal component of tendons, which attach muscle to bone, and the connective tissue within and around muscles. Collagen frequently is described as a "coiled coil," suggesting an entwinement of coiled fibers.

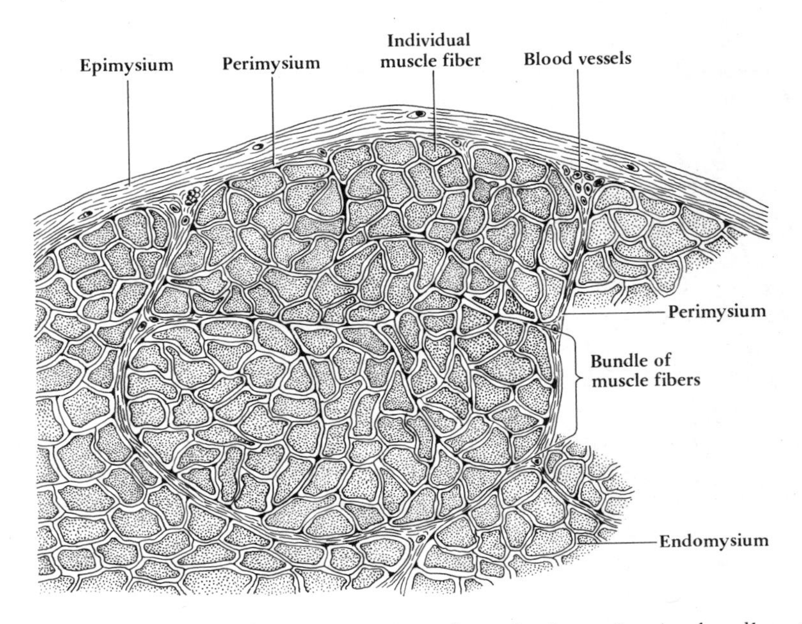

Figure 5.7. Diagram of a cross section of muscle tissue showing bundles of individual fibers.

Approximately 13% of the amino acid residues in collagen are hydroxyproline. So unique is this hydroxyproline content of collagen that determinations of hydroxyproline are used for estimating the amount of collagen in a meat sample. Collagen fibers contract to about 1/3 of their original length when heated to a specific temperature, the thermal shrink temperature. When heated to higher temperatures in the presence of water, collagen may be converted to gelatin.

Elastin *Elastin,* the second fibrous connective tissue protein, is important in ligaments, which perform various functions such as connecting two bones or cartilages. Elastin fibers stretch, but return to their original shape when pulling forces are removed. Changes in elastin with heating may occur but are probably not significant because the elastin content of muscles is very low.

Reticulin fibers In addition to collagen and elastin fibers, muscle tissue contains *reticulin fibers,* which may be similar in composition to collagen but also contain some carbohydrates and bound lipids. Changes in reticular fibers with heating have not been studied extensively.

Ground substance Collagen and elastin are embedded in an amorphous material called *ground substance.* This material contains mucopolysaccharides (nitrogen-containing polymers of glucuronic acid) in combination with proteins.

Influence on texture Because connective tissue is an important factor in the tenderness of cooked meat, the content of both collagen and elastin in various cuts of meat has been studied. The amounts of collagenous and elastin connective tissue can be rated histologically after staining. A significant relationship was found between an objective tenderness rating of cooked muscles and the histological rating of raw muscles (Ramsbottom et al., 1945; Ramsbottom and Strandine, 1948). For the most part, muscles with large amounts of connective tissue were less tender. Muscles that are used constantly may contain more collagen and larger, more numerous, elastin fibers than do muscles used less frequently (Hiner et al., 1955). Although the amount of collagen in muscle tissue does not increase with chronological age, the number and strength of cross-bonds between peptide chains does increase (Hill, 1966). The increased bonding associated with aging decreases the amount of collagen that may be solubilized during heating and contributes to decreased tenderness (Herring et al., 1967; Hill, 1966; Goll et al., 1964).

The tenderness of isolated connective tissue increases with heating in water at 65°C for 16 min or at higher temperatures for shorter periods of time (Winegarden et al., 1952). In steak, softening occurs at an even lower temperature of 61°C, that of rare beef, as shown by loss from the meat of about 1/4 of the collagen during broiling (Irvin and Cover, 1959). Connective tissue becomes shorter and thicker during heating, a change that may account for the plump appearance of some cooked meat, in which the length has decreased and the width increased. Clearly, any changes occurring in connective tissue, primarily collagen, during cooking are for the better, since the softening of connective tissue contributes to tenderness of meat.

Fat As an animal is fattened, the fat is deposited first around certain organs and under the skin, and later as marbling within the muscles. Intramuscular fat cells are located in the spaces in the perimysium, most frequently along small blood vessels. Both quantity and distribution of marbling are important factors in the grading system used by the Department of Agriculture (USDA, 1975). However, research has failed to show a definite relationship between fat content and the tenderness of raw or cooked meat (Ramsbottom et al., 1945; Cover and Hostetler, 1960).

Fat is sometimes yellowish due to the presence of the pigment carotene. The color is more pronounced in dairy than in beef animals, in older animals, and in those consuming feed high in carotene.

The characteristics of fatty tissue vary with species because fatty acids in the triglycerides (neutral lipids) vary in degree of saturation and in chain length. Most of the fatty acids are monounsaturated and saturated. As the degree of saturation increases, hardness of the fat increases. Thus, lamb fat is harder than beef fat. Details of the fatty acid composition have been reported by many investigators (Duncan and Garton, 1967; Stinson et al., 1967; Terrell et al., 1967).

In addition to the neutral triglycerides in fatty tissue, animal tissue also contains small quantities of phospholipids. As membrane components, the phospholipids are of structural and functional importance to muscle tissue. The phospholipids are less saturated than the neutral lipids and probably affect flavor and keeping quality of meat and meat products (Dugan, 1971).

As a cut of meat containing both fatty and lean tissues is cooked, surface fat melts and penetrates the lean meat, thus increasing the fat content of the lean (Thille et al., 1932). A microscopic study of the distribution of fat in meat during cooking suggested that fat released from the cells by heating is dispersed finely in locations where collagen has been hydrolyzed during cooking. (Wang et al., 1954) The effect of this melted and dispersed fat on tenderness has not been demonstrated.

The fat content of meat may influence its apparent juiciness, as shown by two experiments. As the percentage of fat in fluid pressed from a beef rib roast with a hydraulic press increased, judges' scores for quantity and quality of the juice in the meat increased, although less press fluid was removed from the meat (Gaddis et al., 1950). Fat, because of its flavor, seemed to stimulate the flow of saliva, giving the impression of juiciness. The impression of moistness remained because the fat coated the mouth, which is probably why many investigators have failed to find a significant relationship between the amount of press fluid and judges' scores for quantity of juice (Gaddis et al., 1950). An entirely different approach to the relationship of juiciness to fat content was made by Siemers and Hanning (1953), who mixed lean beef and suet in various proportions. As more suet was used in the mixture, less juice was lost during cooking. Judges' scores for juiciness increased as the amount of juice lost decreased. Thus, meat with high fat content appears to be more juicy than meat with a lower fat content.

Carbohydrate Although present in very small quantities in meat, the carbohydrate components are important to the texture and other eating qualities of meat. Conversion of glycogen stores to lactic acid during postmortem aging determines the final pH of the meat and therefore influences water-holding capacity, firmness, and color. Mucopolysaccharides are components of the ground substance and thus are important in structure and texture.

Flavor

The odor and taste of cooked meat vary with preslaughter factors as well as time, temperature, and method of cooking. Hornstein (1971) suggested, in a review of meat flavor, that the carbonyl-amine reaction between reducing sugars and amino acids is important to the development of meat flavors. Specific components responsible for meat flavor were reviewed by Chang and Peterson (1977). These investigators suggest that flavor is a complex characteristic arising from the presence of lactones; acyclic sulfur-containing compounds; and aromatic and nonaromatic heterocyclic compounds containing sulfur, nitrogen, and oxygen.

MUSCLES AND MEAT QUALITY

Muscles that make up a retail cut of meat may differ in tenderness. Muscle locations within the carcass are shown diagrammatically by Tucker et al. (1952). A few of the muscles most often used in meat cookery research are shown in Figures 5.8 and 5.9. Some muscles

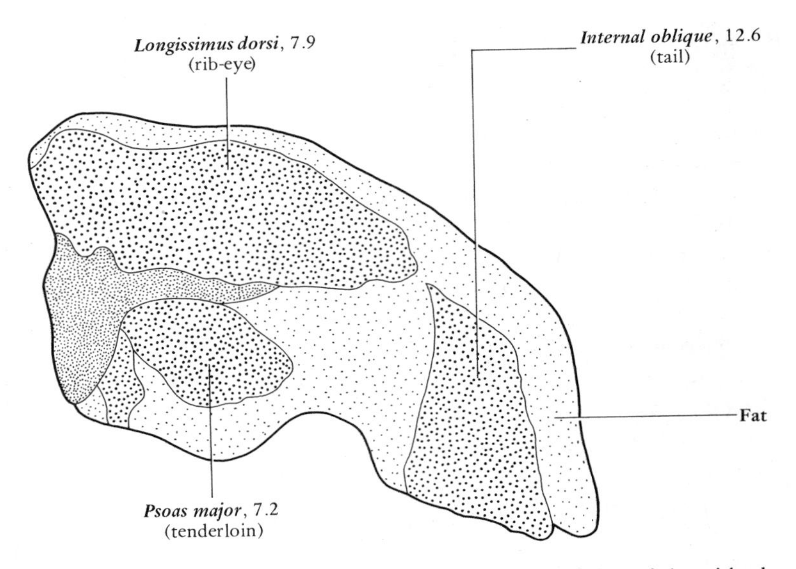

Figure 5.8. The names of the major muscles of beef short loin with shear readings, in pounds, of the cooked meat. (Reprinted from Food Research, Vol. 13, No. 4, pp. 315–330, 1948, Copyright © by Institute of Food Technologists.)

appear in several cuts. For example, longissimus extends the length of the spine from the neck to the pelvis, thus appearing in the chuck, rib, short loin, and sirloin wholesale cuts, whereas psoas major begins in the short loin and extends through the sirloin. The biceps femoris extends from the sirloin into the round and makes up most of the lean meat of the rump.

Shear values for some of the muscles after heating in lard at 121°C (Ramsbottom and Strandine, 1948) also are shown in Figures 5.8 and 5.9. *Shear value,* which is discussed in more detail in Chapter 16, is the force necessary to shear a standard-size sample of meat against the dull blade of a triangular opening. It is measured in pounds or grams. Low readings suggest that the sample was taken from a tender piece of meat. The shear test used by Ramsbottom and Strandine (1948) indicated that the short loin contains the most tender muscle of beef, psoas major, and another, longissimus, that is nearly as tender. The toughest muscle, not shown in these figures, was found to be the rhomboideus, the muscle just outside the longissimus in chuck. As indicated in Figure 5.9, no striking differences were found in the tenderness of the principal muscles of the round. Many of the retail cuts were found to contain muscles of widely varying tenderness. For example, the rump, the chuck, and the shank contain some muscles that are among the most tender of the animal, and others that are quite tough.

For research in meat cookery, a large uniform muscle is preferred so that adequate samples will be available for the judges and the objective tests. The cut that is used frequently for roasting is the standing rib roast. Because of its relative uniformity, the semitendinosus muscle also is used frequently for meat studies. Paired cuts or identical cuts from the right and left sides of the same animal are used in comparisons of two treatments if possible. A second choice for research material is

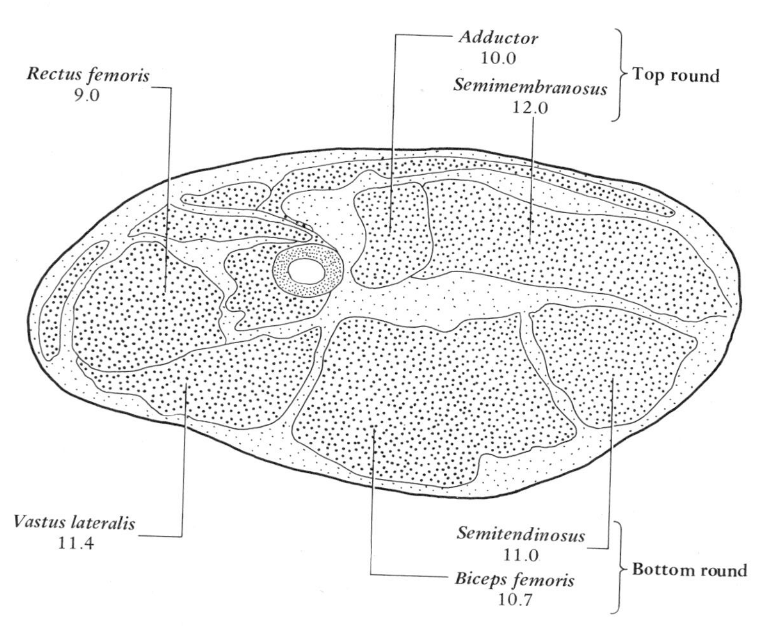

Figure 5.9. The names of some important muscles of beef round with the shear readings, in pounds, of the cooked meat. (Reprinted from Food Research, Vol. 13, No. 4, pp. 315–330, 1948, Copyright © by Institute of Food Technologists.)

adjacent cuts from the same animal. Variation must, however, be expected among muscles of the same cut (Paul et al., 1970) as well as among different parts of the same muscle (Ramsbottom et al., 1945). There are, of course, wide variations in the meat of different animals.

POSTMORTEM AGING

Processing of meat prior to preparation for consumption involves slaughter and a period of postmortem aging. Several changes that influence the eating quality of meat occur in that period.

Rigor Mortis

A few hours after the slaughter of the animal, the carcass becomes rigid, a condition known as *rigor mortis.* This development is accompanied by the disappearance of adenosine triphosphate from the tissues and the interaction of actin and myosin filaments as they slide past one another to form actomyosin (Huxley, 1969). Tension is created in the muscle fibers by the formation of actomyosin. Histological examination (Paul et al., 1952) of the fibers reveals dense nodes of contraction alternated with areas of extreme stretch that make some of the fibers look like an accordian, as shown in Figure 5.10. A day or two after the onset of rigor, the muscles become soft again, the fibers become straighter, and breaks appear in the muscle fibers. These breaks are either sharp fractures or disintegrated areas that appear granulated under the microscope. The breaks occur when the actin filaments separate from their connections to the Z-line (Paul, 1972). The degradation of the muscle fibers, which continues throughout the aging period, may be caused by hydrolysis of the muscle fiber proteins by catheptic enzymes, but further investigation in this area is needed.

Decreased pH

The pH of muscle tissue decreases after slaughter. The pH of living tissue is usually in the range of 7.0–7.2. With depletion of the supply of oxygen to the tissues after slaughter, lactic acid, which is produced from glycogen, accumulates, resulting in a decrease in pH.

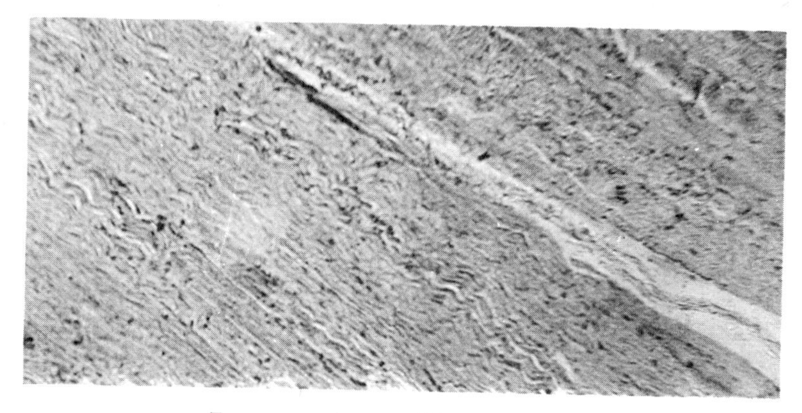

Zero storage (note straight unbroken fibers)

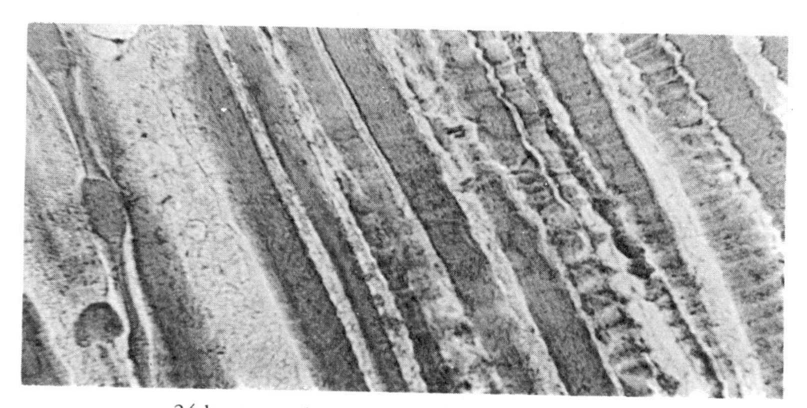

24-hr storage (note contracted appearance of fibers)

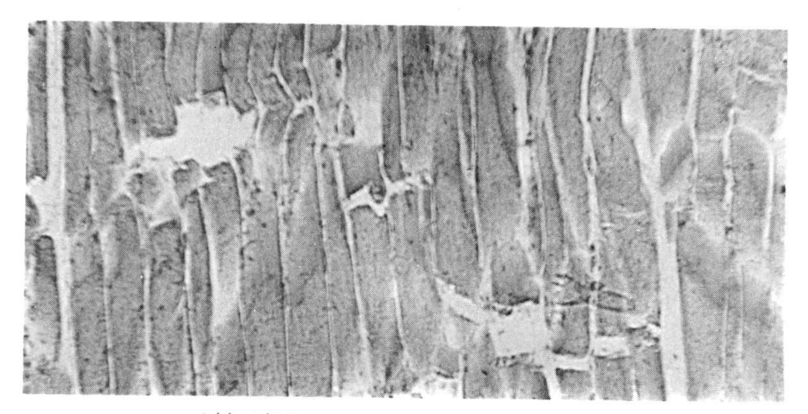

144–149-hr storage (note breaks in fibers)

Figure 5.10. Histological appearance of raw beef after various periods of cold storage. (Reprinted from Food Research, Vol. 17, No. 6, pp. 504–510, 1952, Copyright © by Institute of Food Technologists.)

The ultimate pH of meat depends on the supply of oxygen in the muscle tissue when the animal is slaughtered. If the supply of glycogen is normal, the pH of beef drops to about 5.5. If the glycogen content is low due to starvation or exercise before slaughter, however, less lactic acid will be formed, and the pH may be 6.6, higher than normal (Hall et al., 1944). Beef with an abnormally high pH is brownish red to purplish black in color and sticky or gummy in texture. Beef of this type is referred to as *dark-cutting beef.* Dark, firm, and dry pork is similar in that the ultimate pH is higher than normal. In PSE (pale, soft, and exudative) pork, the pH rapidly decreases to an abnormally low level of 5.1–5.4 (Briskey, 1964). Certain porcine breeds are particularly susceptible to the development of the PSE condition. The problem is minimized by avoidance of antemortem stress and by rapid chilling postmortem.

Changes in Quality

At first, quality decreases, as shown by the toughness of meat, including poultry, that is cooked while in rigor. Because of this, all meats are best aged until rigor has passed before they are cooked or frozen. After rigor has passed, meat gradually becomes more tender. When beef was aged for 2 wk, the tenderness and flavor of the fat and lean of broiled rib-eye improved, but an additional 2-wk aging period usually was detrimental to flavor and had little additional tenderizing influence (Doty and Pierce, 1961). Altered flavor during aging may result from increased amounts of free sugars, amino acids, and hypoxanthine (Lawrie, 1968). Tenderness of meat can be increased more rapidly at temperatures above the usual 1–2°C if the growth of microorganisms on the meat is retarded by the use of carbon dioxide, ozone, or ultraviolet light in the cooler (Urbain, 1971).

MEAT COOKERY

Research on meat cookery was started earlier and has been pursued more vigorously than that on many other foods. The literature on meat preparation studies is complex and voluminous and therefore difficult to summarize. Thus, the discussion that follows must be considered an overview. Further reports on the effects of heat on the quality of meat as well as other aspects of meat cookery that are discussed can be found easily in the food science literature.

Many theories on the effects of heat on the eating quality of meat have been proposed and examined. Some of the theories developed and applied early in this century have been re-examined and revised. One of these theories is that moist heat is required to cook tough meat because added moisture is required to break down the large amount of collagen that it contains. This change in collagen actually takes place in meat cooked by dry-heat methods as well. Another theory is that high temperatures are to be avoided because they toughen meat. This statement requires qualification because steaks may be tender when broiled at about 288°C, while an oven temperature of 204°C has been considered too high for roasting meat and 100°C too high for braising meat. In general, it is accepted that muscle fibers harden or become less tender with heating, and connective tissue softens with heating. Thus an ideal cooking method will minimize hardening but maximize softening of the connective tissue. Appropriate methods vary with the cut of meat. The hypothesis that very tender cuts of meat containing little collagen decrease in tenderness during heating whereas less tender cuts become more tender has been tested many times. Time and temperature of heating also must be considered. Discussion of several methods of preparation will include a review of literature on the effects of various methods of heating on the eating quality of meat. The discussion will primarily include studies on beef, since it has been studied most frequently. Studies on lamb and pork will be cited when available and pertinent. In addition, methods used in meat research will be described.

Roasting

Roasting, or more specifically *oven roasting,* is the cooking of meat uncovered in dry heat. This term is to be distinguished from *pot roasting,* which is a form of braising, a moist-heat method. For roasting, meat is placed on a flat open pan, with a rack if the bony structure is not sufficient to keep the meat out of the drip as the meat is heated in an oven. The meat is not basted and no water is added. Such dry-heat roasting results in meat that is quite different from that produced by heating in a covered pan, because covering the pan holds in the moisture given off from the meat or from added liquid, resulting in moist-heat cookery. The method of heat transfer in various meat cookery methods differs. Therefore, the characteristics of the product heated by the different methods will vary. In roasting, heat energy is transferred through the air by convection and the meat is heated by conduction.

Early experimental work on roasting involved the use of *searing.* The meat was placed in an open pan in a hot oven (265°C) for about 20 min, after which the temperature was reduced to 125°C. It was thought that the crust formed on the outside of the meat would hold in the juices. However, meat roasted at a constant oven temperature shrank less during heating than meat that was seared (Alexander and Clark, 1939; Cline et al., 1930). Searing improved the external appearance of the roasts but did not increase palatability. Roasting at a constant oven temperature is more convenient. Investigators, therefore, became involved in studies to determine the optimum oven temperature.

Oven temperature Cline and coworkers (1930) compared beef ribs roasted until rare at oven temperatures of 110, 125, 163, 218, and 260°C. Meat roasted at the lower oven temperatures was more tender, juicy, flavorful, and uniformly done throughout, required longer heating times, and showed less cooking loss than meat heated at the higher temperatures. Of the temperatures tested, 125 and 163°C were selected, the former temperature being slightly superior. A temperature of 163°C usually is recommended for roasting of all meats.

Endpoint temperature Endpoint temperature is a second consideration in the selection of a preparation method. Suggested temperatures for general use are 60, 71, and 77°C for rare, medium, and well-done beef respectively. As the internal temperature of tender cuts of meat increases during roasting, the meat becomes less tender (Cover and Hostetler, 1960) and less juicy, while cooking losses increase (Bayne et al., 1973). Control of endpoint temperature is important in experimental work. Thermometers or thermocouples should be placed in the geometric center of the piece of meat. Location should be checked periodically because thermometers or thermocouples may be displaced as meat expands and contracts during heating. The rate of temperature rise varies at different depths of a cut of meat (Funk et al., 1966).

The internal temperature of meat continues to rise during the first few minutes after it is removed from the oven. The rise is greater if the oven temperature is relatively high and if the meat is rare, because the difference in temperatures between the outside and the center of the roast is relatively large when the roast is removed from the oven. The rise in internal temperature is due to the transfer of heat from the warmer outside of the roast to its cooler center and is greater in a large roast than in a smaller one.

Cooking losses In meat research, cooking losses usually are measured. If the meat is roasted, these losses consist of drip that remains in the pan and evaporative loss, which is the water that evaporates from the meat and the drip during cooking. The evaporative loss is found by subtracting the weight of the cooked roast and drip from the weight of the raw roast. Because these losses go on continuously during cooking, it is not surprising to find that they increase as the endpoint temperature of the meat or the oven temperature is increased (Bayne et al., 1973). This is of considerable economic importance in commercial food service, because with higher cooking losses, fewer servings will be obtained per pound of raw meat. For example, Vail and O'Neill (1937) obtained 62 slices from a rib roast heated at 141°C but only 51 from a matching roast heated at 232°C to the same internal temperature.

Oven roasting of less tender cuts of meat Roasting at low temperatures (66–121°C) has been studied extensively as a method of increasing the tenderness of less tender cuts of meat. Cover (1943) found that well-done chuck and round roasted at the unusually low oven temperature of 80°C were more tender than chuck and round roasted at 125°C. The development of tenderness seemed to accompany a long cooking time. If 30 hr or more were required for the meat to lose its pink color, the roasts were always tender. The connective tissue changed to a moist, viscous mass without resistance to the knife or teeth. The texture of the meat was mealy. Although meat roasted at 80°C looked dry and required 21–43 hr to roast, cooking losses were about the same as those for meat roasted at 125°C. Bayne et al. (1969) reported that large and small top round roasts heated at 93°C to an internal temperature of 67°C were more tender than roasts heated at 149°C to an internal temperature of 70°C. Laakkonen et al. (1970) reported that a major

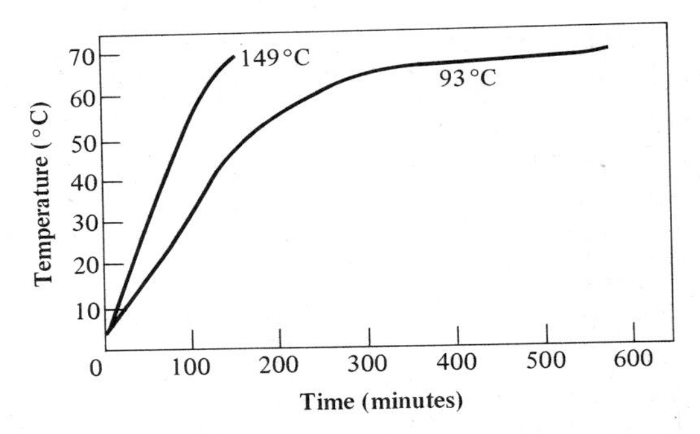

Figure 5.11. Time-temperature curves from oven roasting of top round roasts at 93 and 149°C. (Reprinted from Journal of Food Science, Vol. 40, No. 1, pp. 150–154, 1975, Copyright © by Institute of Food Technologists.)

increase in tenderness occurred as semitendinosus samples were heated from 50 to 60°C. The time in that temperature range during heating at 93°C is almost 3.5 times as long as the time for heating in a 149°C oven, as illustrated in Figure 5.11. Long, slow heating results in greater solubilization of collagen, and shear values decrease as the percentage of collagen solubilized increases (Penfield and Meyer, 1975).

Although the breakdown of collagen is important in cooking cuts of meat with high collagen content, other factors also are important. Extensive collagen solubilization during heating may not be reflected in increased tenderness because hardening of muscle fiber may contribute to decreased tenderness (Paul et al., 1973; Penfield and Meyer, 1975).

Although very low oven temperatures produce tender meat, they may not be practical because not all ovens will maintain them and because the times are very long. Temperatures such as 93°C may be practical, however, especially for the consumer who wishes to start less tender cuts of meat before going to work. For cuts of meat that are sufficiently tender when roasted at ordinary temperatures, unusually low oven temperatures may be undesirable.

Top rounds of Choice beef were found desirable for institutional use when roasted at 149°C (Marshall et al., 1959). Reduction of the oven temperature to 93, 107, or 121°C for these roasts was not recommended because cooking times were long and variable, and because roasts cooked to the well-done stage at 93°C crumbled when sliced (Marshall et al., 1960). The use of lower temperatures for heating sirloin roasts in forced convection ovens was studied by Davenport and Meyer (1970). In a comparison between 93 and 149°C, the lower temperature reduced cooking losses and cost per serving and increased yield and cooking time. Shear values and sensory ratings of tenderness, juiciness, and flavor were not affected. Since the roasts heated at both temperatures were rated "good" to "very good," the benefits of the lower temperature were obvious.

Moist versus dry heat The finding that meat can be made more tender by slow heat penetration in dry-heat cooking was supported by an extensive study done in several colleges and universities and sponsored by the United States Department of Agriculture (Dawson et al., 1959). Palatability scores for tenderness and flavor were in favor of dry-heat cooking when corresponding cuts of beef round were cooked by dry- and moist-heat methods to the same endpoint temperature. The dry-heat methods were roasting and broiling and the moist-heat methods were braising either on the surface of a range or in the oven. The study also indicated that dry-heat methods need not be reserved for meat of the higher grades because beef of lower grades was satisfactory in eating quality when cooked by methods usually reserved for Prime and Choice grades. More recently, Schock and coworkers (1970) reported that semitendinosus roasts heated to 70°C by deep-fat frying, oven roasting, oven braising, and pressure braising did not differ in flavor, tenderness, or overall acceptability. Juiciness differed, but it did not influence the overall acceptability scores as much as flavor and tenderness did.

Braising

Braising is cooking meat in a small amount of liquid or steam in a covered utensil. The rate of heat penetration in braising is more rapid than in roasting because steam conducts heat more rapidly than does air. Braising methods include braising on the range, in an oven, in a pressure cooker, or in a slow cooker, as well as braising in aluminum foil or oven film in an oven.

Browning in a small amount of fat before steaming is an optional

(AHEA, 1975) but desirable step in the braising process because it improves both the color and the flavor of the meat. Browning in a pan preheated to 246°C proved more satisfactory than browning at a lower temperature (Paul and Bean, 1956). When the temperature was too low, the meat tended to stew in liquid that escaped from the meat rather than to brown. The addition of 50 ml of water after browning of a small steak 2.5 cm thick proved to be essential to avoid burning steaks cooked to high internal temperatures and improved the palatability of those heated for shorter periods. The palatability and cooking losses of meat that was oven braised at 121, 149, or 177°C were similar, but 121°C was selected as best because less watching was required to prevent scorching the meat. Bowers and Goertz (1966) reported that eating quality, cooking loss, and time were not affected by the addition of water to skillet-braised pork chops. Brownness was decreased, however. Oven temperature did not affect the quality of pork chops that were oven braised. When meat is braised on the range, a temperature below the boiling point usually is recommended.

On the basis of their research, Cover and Hostetler (1960) reported that a gentle boil is as satisfactory as a simmering temperature, which is often difficult to maintain. In another study, little difference was found in the quality of meat braised in an oven at 149°C or held below the boiling point on the surface of the range (Hood et al., 1955).

Endpoint temperature There is no general agreement concerning what internal temperature is best for braising meat. In one study (Griswold, 1955a), meat braised to 85°C received higher palatability scores than meat braised to 80 or 93°C, although the differences were not significant. In another study (Paul and Bean, 1956), longer cooking up to 98°C plus 30 min generally improved the flavor, but it decreased the juiciness and increased the cooking losses of beef rounds. Dawson et al. (1959) found that braised beef was more juicy when cooked to low rather than to high endpoint temperatures but was about equally tender and flavorful regardless of endpoint temperature. Bowers and Goertz (1966) reported that the cooking losses and juiciness decreased as the internal temperature of both oven- and skillet-braised pork chops increased.

Most consumers probably braise meat until it is tender when tested with a fork rather than to a certain endpoint temperature. This stage has been reproduced in the laboratory by bringing the meat to an internal temperature of 100°C and holding it there for 25 min. Cover (1959) observed that meat is very tender at this point. The connective tissue is soft, and muscle fibers often seem crumbly or friable when cut with a fork or chewed. The texture of such muscle fibers is very different from that of rare meat, which seems soft to the teeth and tongue. Very friable meat offers less resistance to teeth or fork and is therefore very tender. It has the disadvantage, however, of seeming dry and crumbly. Cover (1959) studied some of these effects by having panelists score three sensations related to tenderness: tenderness of connective tissue, softness of muscle fibers, and friability of muscle fibers.

Braising of tender and less tender cuts of meat Tender and less tender cuts of meat do not respond alike to cooking, as shown by Cover (1959). Samples from the loin and bottom round were broiled to 85 or 100°C plus 25 min. Judges gave high scores to the tenderness of the connective tissue of the loin cuts, indicating that there was little possibility for improvement with increased doneness or braising. The loin contained only a small amount of soft connective tissue. By contrast, the biceps femoris of the round had a large amount of firm connective tissue that became more tender in both methods as doneness increased. It was tough when broiled rare, about equally tender when broiled well done and braised to 85°C, but very tender when braised for the longest period. The muscle fibers of bottom round differed from those of the loin, becoming softer and more friable when braised to the higher temperature. The muscle fibers of the loin, which are soft when rare, may be stringy when braised because little connective tissue is present to hold them together. As the cooking time was increased in the Cover study (1959), making the meat more well done, the muscle fibers of the loin seemed to toughen, while those of the round became more friable and therefore more tender.

Pressure cookery Cooking time is reduced by pressure cooking, a modification of braising. In a comparison of pressure braising with deep-fat frying, oven broiling, and oven roasting of semitendinosus muscle samples, Schock et al. (1970) observed that pressure-braised pieces were scored least juicy and had the greatest cooking losses. Tenderness—as evaluated by both sensory panel and objective methods

—flavor, and overall acceptability did not differ with the methods. More recently, Brady and Penfield (1977) reported that microwave-heated pressure-cooked beef semitendinosus roasts were less tender by objective measurement than roasts braised in a conventional oven or a slow cooker. Cooking losses and sensory scores for color, tenderness, flavor, juiciness, and overall acceptability did not differ with the methods.

Slow cookers Appliances designed to allow slow cooking of food for very long periods of time were introduced several years ago. Few studies on the use of these appliances for heating meat by a moist method of cooking have been published. Rump roasts heated in a slow cooker for 10 hr reached an internal temperature of 81°C as compared to internal temperatures of 85°C at a low oven temperature of 107°C for 9 hr and 77°C at an oven temperature of 177°C for 2 1/4 hr. Cooking losses were similar at the lower oven temperature and in the slow cooker, but greater than at the higher oven temperature (Sundberg and Carlin, 1976). Brady and Penfield (1977) reported a trend toward greater losses from semitendinosus roasts heated in slow cookers than from roasts heated in a conventional oven, a microwave oven, or a pressure cooker. Results of the sensory evaluation were discussed previously.

Foil and films Meat can be wrapped tightly in aluminum foil to retain steam as effectively as a covered pan. Hood (1960) studied the effects of aluminum foil wrapping by cooking paired beef roasts to an endpoint temperature of 77°C in an oven at 149°C. Foil-wrapped roasts lost more weight during cooking and were judged to be less juicy, less tender, and of less satisfactory flavor than roasts heated on racks in shallow, uncovered pans.

Bramblett and coworkers (1959) compared two oven temperatures, 63 and 68°C, for heating of foil-wrapped muscles from beef round. The lower temperature produced the more tender, juicy meat. Beef roasts cooked in foil may have a steamed flavor (Blaker et al., 1959). More recently, Baity and coworkers (1969) reported that at a low oven temperature (93°C), beef loaves cooked more rapidly when the foil was tightly sealed than when the loaves were not wrapped or loosely wrapped. At a higher temperature (232°C), cooking times increased with foil wraps. Quality attributes of the loaves were not discussed.

Polyester cooking films and bags are also available for use in the heating of meat. Results of studies of the quality of meat prepared in these films and bags suggest that the quality is comparable to that of meat heated in foil. Beef top round roasts heated in bags to 80°C in ovens at 177 and 205°C took less time but had greater cooking losses than the roasts heated similarly but without bags (Shaffer et al., 1973). Roasts heated in film also have a more well-done appearance (Ferger et al., 1972; Shaffer et al., 1973).

Broiling and Frying

Oven broiling, pan broiling, and frying are quick methods of cooking meat by dry heat. In broiling, radiant heat is applied directly to the meat with minimal heating by conduction of heat from the pan, whereas in pan broiling the meat is put into a hot pan and heated by conduction. No fat is added during pan broiling, and fat that accumulates during cooking is poured off. Frying is cooking in fat, either a small amount or a deep layer.

For all of the above quick methods, characteristic results depend on the use of enough heat to brown the outside of the meat without overcooking the inside. If the temperature is too low, the meat seems to stew in the liquid that cooks out of it, thus becoming gray instead of brown on the outside. The high temperatures commonly used in these methods do not seem to toughen the meat, perhaps because the cooking time is short, or because the meat used is tender.

Broiling Broiling of pork chops has been studied frequently. Oven-broiled chops were more tender, flavorful, and acceptable than deep-fat fried chops (Flynn and Bramblett, 1975). Juiciness was not affected by heating method. Broiling also has been compared with braising. Cooking time was shorter and cooking losses greater with broiling of pork chops than with braising (Weir et al., 1962). Sensory evaluation suggested that the broiled chops were more tender, more juicy, and less flavorful than the braised chops. The thickness of chops and steaks influences the quality of the broiled product. Thick broiled chops (1.5 in. thick) were less juicy than thin chops (0.75 in. thick) (Weir et al., 1962). Holmes et al. (1966) reported similar results.

Because broiled meat has not been compared directly with roasted meat, Batcher and Deary (1975) compared beef round steaks heated by the two methods. They concluded that the steaks prepared by roasting were more tender, more juicy, and less mealy than the broiled steaks. Increased cooking time for roasting may have influenced the increased tenderness.

Deep-fat frying Meat heated by deep-fat frying is heated by transfer of heat from the oil to the meat. Because the temperature of the oil is very high, a crust forms rapidly on the outside of the meat. Deep-fat fried roasts lost more moisture and were scored as more juicy and more well done in appearance than oven-roasted semitendinosus roasts (Schock et al., 1970). Differences in flavor, tenderness, and overall acceptability were not found. Carpenter et al. (1968) reported that beef rib steaks that had been deep-fat fried were more tender than those heated by microwaves or broiled conventionally.

Microwave Heating

Microwave or electronic heating is quite different from the previously described methods. Energy is supplied by short electromagnetic waves. The waves penetrate the food to a depth of about 5 cm and cause polar molecules within the food to oscillate, resulting in conversion of the electromagnetic energy to heat. Although some questions about microwave heating have been answered, more work must be done. Microwave heating is discussed in detail by Van Zante (1973).

Many investigators have reported that cooking losses are greater with microwave heating than with conventional heating of meat (Bowers et al., 1974; Carpenter et al., 1968; Ream et al., 1974). Greater losses most frequently are attributed to greater evaporative losses (Bowers et al., 1974). Denaturation of proteins within the muscle may cause juices to be forced out of the meat (Carpenter et al., 1968). However, Korschgen and coworkers (1976) found that cooking losses were greater for beef longissimus but not for pork longissimus or deboned leg of lamb when heated by microwaves. Law et al. (1967) reported that losses from beef loin and top round steaks were greatest during conventional broiling, intermediate with microwave heating, and lowest with oven roasting. In ground beef patties, in addition to greater moisture loss, more fat was lost from microwave heated patties than from conventionally grilled or broiled patties (Janicki and Appledorf, 1974). Similar results were seen with beef loaves (Ziprin and Carlin, 1976). Van Zante (1968) reported that bone reflects microwaves. Therefore, meat next to a bone cooks more rapidly and loses more moisture. Juiciness scores are lower for microwave heated meat (Ream et al., 1974).

Because of the low surface temperatures associated with microwave heating, a crust may not form on beef patties (Janicki and Appledorf, 1974). Low surface temperatures and the associated lack of surface browning may influence eating quality. Korschgen and coworkers (1976) found that flavor scores for the outside portions of pork and lamb roasts were higher for conventionally heated roasts than for those heated in the microwave oven, whereas scores for internal portions did not differ. Beef roasts heated by microwaves have been described as less flavorful (Ream et al., 1974). Although microwave heating may be detrimental to flavor, microwave reheating may result in better flavor than conventional reheating. Penner and Bowers (1973) reported that pork, conventionally roasted and then reheated in a microwave oven, had a sweeter aroma and less metallic flavor than conventionally reheated meat. Similarly, turkey reheated by microwaves had greater turkey flavor and less stale flavor than conventionally reheated turkey (Cipra and Bowers, 1971). Turkey scored immediately after conventional heating and turkey reheated in a microwave had higher meaty brothy flavor and lower rancidity scores than conventionally reheated turkey (Bowers, 1972).

In addition to juiciness and flavor, tenderness of microwave heated meat has been compared to that of conventionally heated meat. Differences in tenderness seem to depend on the kind of meat, the size of the piece heated, and the conventional method of heating used in the study. In studies comparing conventional roasting to microwave heating, beef arm and rib roasts (Ream et al., 1974) heated by microwaves were shown to be less tender by sensory evaluation. However, Headley and Jacobson (1960) reported that legs of lamb prepared by the two methods did not differ in tenderness.

Oven broiling produced more tender beef rib steaks and pork chops than microwave heating (Carpenter et al., 1968). Cooking time in the microwave oven is much shorter than with conventional methods. Therefore, less tender cuts of meat cannot be expected to become tender (Van Zante, 1973). However, more tender cuts may be prepared acceptably in the microwave oven.

Studies on the use of film bags in microwave cooking and conventional heating have indicated that both wraps and microwave heating increase cooking losses when compared with conventional methods (Ruyack and Paul, 1972). More recently, Armbruster and Haefele (1975) reported that plastic film covers improved appearance, color, mouth feel, moistness, flavor, and overall acceptability of meatloaf and turkey, and appearance, color, and flavor of pork loin. Cooking times were reduced and foods were more uniformly done due to the more even heat distribution achieved with the films.

Tenderizers

The effectiveness of some of the agents used to tenderize meat will be discussed in this section. These include pounding, scoring, cubing, grinding, needling or pinning, and the addition of substances such as enzymes, salt, and vinegar.

Mechanical methods of tenderization Meat can be tenderized by mechanical methods that cut or break the muscle fibers and connective tissues. Pounding but not scoring before braising increased the tenderness of beef round (Griswold, 1955b). Cubing with a commercial device, a more vigorous treatment than scoring or pounding, probably is even more effective. Needling or blade tenderization is a more recently introduced technique for tenderization of meat. Interest in such techniques was stimulated by desires to use grass-fed or short-fed beef to supply a greater portion of the meat consumed in the United States. Needlelike blades approximately 0.5 cm wide are inserted into the meat automatically as it passes through the tenderizer (Glover et al., 1977). Glover and coworkers (1977) reported that sensory scores for tenderness were improved for tenderized round steaks but not for round roasts, chuck roasts, or loin steaks. Roasts cooked in less time lost more drip and were less juicy when tenderized; these effects were not seen with tenderized steaks. Savell et al. (1977) reported that steaks from the semitendinosus, biceps femoris, gluteus media, and longissimus muscles could be tenderized as many as two times in order to improve tenderness without decreasing overall palatability or increasing cooking losses. The tenderizing effect of grinding meat is evident to anyone who has broiled or roasted ground meat that was suitable only for soup or stew prior to grinding.

Enzymes Proteolytic enzymes frequently are used to tenderize meat. The most commonly used enzyme is papain, which is prepared from the green fruit of the papaya plant. Other enzymes that exhibit the necessary proteolytic activity for use in meat tenderizers are bromelin, which is extracted from the pineapple, ficin from figs, trypsin from the pancreas of animals, and fungal proteases (Miyada and Tappel, 1956). Other sources of proteolytic enzymes, such as rhizome from ginger, are being studied (Thompson et al., 1973).

The mechanism of the tenderizing action of enzymes on meat was studied. Tappel et al. (1956) indicated that tenderization cannot be ascribed to one specific component of beef muscle. Tsen and Tappel (1959) described the enzymatic effect as beginning with destruction of the sarcolemma at the surface of the sample, then hydrolysis of actomyosin, and finally the muscle fibers themselves. Because muscle fiber protein accounts for 75% of the edible portion of beef, its hydrolysis is probably the most important mechanism in the enzymatic tenderization of beef. Enzymes also attack the connective tissue fibrous proteins, producing granulation of collagen and segmentation of elastin fibers (Wang et al., 1958). Kang and Rice (1970) investigated the effects of several enzymes on meat components and reported that the extent of enzymatic activity varied with the enzyme as well as the substrate. Thus bromelin, papain, or any of the other proteolytic enzymes that may be used as meat tenderizers affect muscle fibers and connective tissue differently. It is possible that these studies may exaggerate the true situation because much higher levels of the enzymes were used than are common in meat cookery.

The way in which enzymatic tenderizers are used influences their effectiveness. Since papain penetrates only 0.5–2 mm when applied to the surface of beef, a physical method of incorporation such as forking the enzyme into the meat may be more effective than mere surface treatment, whether the enzyme is in liquid or powdered form (Tappel et al., 1956). Directions for the use of some commercial enzyme tenderizers suggest that the meat stand for a time after application of the enzyme and before cooking, usually at room temperature. This is probably unnecessary, as shown by an experiment in which no differences were found in the tenderness of meat cooked immediately after treatment and that of meat held for 1–5 hr. The facts that meat cooked immediately after treatment is more tender than untreated meat and that the action of papain is greatest at temperatures of 60–80°C indicate that the enzyme acts during the cooking process (Tappel et al., 1956). Furthermore, Hinrichs and Whitaker (1962) reported that collagen is not affected by the proteolytic enzymes used in meat tenderizers at low temperatures, but that the enzyme will attack the collagen when it has been denatured by heat.

Commercially, papain may be injected into the veins of an animal about 10 min prior to slaughter (Robinson and Goesner, 1962). The animal's circulatory system distributes the enzyme through its body. The tenderizing action starts when the meat is heated. Aging is reduced to 1–3 da. Antemortem injections of papain have been shown to improve tenderness of cooked steaks (Huffman et al., 1962) and poultry (Fry et al., 1965). Injection of papain also has been used to increase the palatability of bullock meat, which is lower in quality than steer meat (Smith et al., 1973).

The proper application of proteolytic enzymes to thin cuts of meat improves their tenderness (Hay et al., 1953). The misuse of enzymes in meat cookery is possible. Enzymes probably are ineffective with roasts unless injection methods are used because they do not penetrate the meat. Overtenderization is likely to result in meat that has a mushy, crumbly texture similar to that of very tender liver. Decreased juiciness is another possible result of the unnecessary application of enzymes. Overtenderization may result from the use of excessive amounts of enzymes, from the use of enzymes in cooking tender steaks or meat made tender by mechanical means, or from a cooking method in which the enzyme-treated meat is held for long periods in the temperature range favorable for enzyme activity. Such temperatures may exist when meat is being held warm for serving as well as during cooking. The enzyme is inactivated if the temperature is high enough to denature it.

Other means of tenderizing meat Salt may, under certain conditions, have an effect on tenderness. Draudt (1972) suggested that sodium chloride and sodium bicarbonate decreased the severity of the hardening of muscle fibers during heating, thus increasing tenderness.

These salts increase the water-holding capacity of the muscle fiber proteins unless excessive amounts are used. Salt in amounts large enough to precipitate protein is not likely to be used in cookery because of its undesirable taste. Some of the effects of salt on the hydration of meat were described by Hamm (1960) and Deatherage (1963). The amounts of sodium chloride normally used with ground meat probably increase its ability to hold water, as shown by an experiment in which the addition of 1.3% sodium chloride to ground beef decreased the amount of juice lost from it during cooking (Wierbicki et al., 1957a). Larger amounts of juice were lost from meat cooked without salt or with concentrations of salt beyond the limit of taste acceptance (Wierbicki et al., 1957b). The possible effects of water retention on the tenderness of meat are best observed on meat that has not been ground. So these effects can be studied, salt has been incorporated in various ways. Tenderness was improved when freeze-dried steaks were rehydrated in 2% sodium chloride instead of in water (Wang et al., 1958), when a solution of sodium and magnesium chlorides was injected into beef rounds (Wierbicki et al., 1957b), and when steaks were injected with a sodium tripolyphosphate and sodium chloride solution prior to freeze-drying (Hinnergardt et al., 1975).

Another way to increase the hydration, and thus perhaps the tenderness, of meat is to make the meat more acidic or more basic. Decreasing or increasing the pH of meat reduces the loss of moisture on heating, as does the addition of salts (Hamm and Deatherage, 1960). Raising the pH has the disadvantage of darkening the meat. The addition of acid is feasible, as in the German dish called "sauerbraten," for which meat is soaked in vinegar before cooking. In one experiment, however, soaking beef in diluted vinegar for 48 hr failed to increase the tenderness of the meat and lowered its acceptability (Griswold, 1955a, b). The ineffectiveness of this treatment may be due to the only slight reduction in pH achieved with the vinegar. The strong buffering effect of meat necessitates rather large amounts of acid and base for changing its pH. Lind and coworkers (1971) reported that soaking commercial-grade top round steaks in a wine vinegar solution for 48 hr prior to braising increased juiciness and tenderness but decreased odor, taste, and overall acceptability scores. Higher concentrations of vinegar were used in this study than in earlier studies and may account for the differences in the results as to tenderness.

POULTRY

Composition, Structure, and Quality Characteristics

Much of the general information included in the first portion of this chapter is applicable to poultry. Composition of poultry is included in Table 5.1. Poultry tissue structure does not differ from that of other muscle foods, as illustrated by Johnson and Bowers (1976).

Like meat, poultry cooked during rigor mortis is tough. Therefore, chickens (Pool et al., 1959) should be held for at least 4 hr and turkey (Klose et al., 1959) for at least 12 hr before cooking or freezing. Holding inadequately aged poultry at freezer temperatures does not improve its tenderness, but holding it at 1.7°C after frozen storage has as much tenderizing effect as an equal period before freezing (Klose et al., 1959). Removing muscles within 1 to 2 hr after slaughter results in decreased tenderization of the muscles. Injection of polyphosphate into the muscle prevents this toughening effect by slowing the postmortem reduction in pH (Peterson, 1977).

Dark and light meat The dark meat of poultry usually is more juicy but less tender than the light meat. Perhaps the greater juiciness of the dark meat is due to its higher fat content (Hanson et al., 1942). The hemoglobin and myoglobin contents of the dark and light meat of chicken and turkey and the meat of duckling were reported by Saffle (1973).

Flavor Raw chicken has little flavor. Flavor develops during cooking. Cooking increases the concentration of carbonyl compounds (Dimick et al., 1972), which are thought to be major contributors to "chickeny" flavor (Pippen et al., 1958; Minor et al., 1965). Sulfides also are contributors to the flavor (Pippen and Eyring, 1957). The small amount of phospholipids is important in any flavor deterioration that may occur during frozen storage of cooked chicken (Lee and Dawson, 1976).

Cooking Methods

Chicken An oven temperature of 163°C is now recommended for roasting poultry as well as other meats (AHEA, 1975). However, Hoke (1968) suggested that if time is a factor, 191°C is suitable although yields will be lower than at 163°C.

Steaming and simmering were found superior to pressure cooking of hens when palatability scores, shear values, and press fluid values were considered. Less time was required to steam hens than to simmer them at 91°C (Swickard et al., 1954). Older chickens and turkeys cooked by moist heat are most suitable for precooked frozen products, according to Hanson and coworkers (1950). These investigators found that older chickens yield a more flavorful precooked frozen product than younger birds.

Shear values and cooking losses did not vary when chickens were broiled at temperatures varying from 175 to 230°C (Goertz et al., 1964). Microwave heating did not adversely affect the quality of chicken (Phillips et al., 1960). Panelists could distinguish between conventionally and microwave heated chicken, but acceptability scores did not differ.

Turkey The question of roasting turkey covered or uncovered has been studied. Wrapping turkey in aluminum foil is recommended frequently because the oven temperature can be raised to shorten the cooking time without excessive browning. Sensory characteristics of unwrapped turkeys cooked at 163°C were comparable to those of foil-wrapped turkeys heated at 232°C (Lowe et al., 1953). However, foil-wrapped turkeys may have a steamed appearance (Bramblett and Fugate, 1967) and flavor (Blaker et al., 1959). Baity and coworkers (1969) reported that turkeys tightly sealed in foil cooked in less time than unwrapped or loosely wrapped turkeys. Juiciness of turkeys may be influenced more than flavor or tenderness by cooking in foil. Deethardt et al. (1971) reported that juiciness is decreased. The use of cooking bags in oven roasting of turkeys was compared with foil wraps, paper bags, and open pans (Heine et al., 1973). Cooking time was longest with the paper bag and shortest with the film bag. Cooking losses were smaller with foil than with the other methods. Juiciness was greatest in the foil-wrapped and conventionally roasted turkey. Panelists found that flavor did not differ but that samples from the dark muscles of carcasses cooked in paper bags and open pans were more tender than samples cooked by the other methods.

Engler and Bowers (1975), studying the use of slow cookers for turkey preparation, found that tenderness was greater and cooking losses lower for "slow-cooked" turkey breast halves than for halves roasted conventionally. Thus these methods may offer added convenience with quality as acceptable as is obtained by conventional methods.

Research on frozen stuffed turkeys indicated that the practice of covering birds with cheesecloth, basting with fat, and roasting at 163°C results in a cooked product of high quality and attractive appearance (Esselen et al., 1956). Reducing the temperature to 121°C lengthened the roasting time and markedly increased the period during which the meat and stuffing were in the incubation zone for food spoilage microorganisms. In order to ensure the destruction of organisms that may cause food-borne illness, primarily *Staphylococcus aureus,* the stuffing should reach a temperature of not less than 74°C (AHEA, 1975). Procedures for ensuring safety of stuffed turkeys were studied by Woodburn and Ellington (1967) and Bramblett and Fugate (1967).

Endpoint temperatures Satisfactory determinations of endpoint temperatures are difficult with poultry because the muscles are smaller than those of beef and tend to separate during cooking, sometimes leaving the bulb of the thermometer exposed to air or liquid. Because of their large size, however, turkey breast and thigh muscles are fairly satisfactory for this purpose. An endpoint temperature of 95°C in the thigh produced turkeys of optimum doneness (Goertz et al., 1960). However, lower endpoint temperatures (82–85°C) are being recommended currently (AHEA, 1975).

TEXTURED VEGETABLE PROTEINS AND MEAT EXTENDERS

Textured vegetable proteins (TVP) have been used to extend ground meat as a means of reducing cost without reducing nutritional value (Lachance, 1972). Federal school lunch regulations allow for inclusion of up to 30% of TVP in ground meat for that program (USDA, 1971). Addition of TVP influences the quality of ground meat products, as indicated in several studies. The effect on quality depends on the level of TVP added. Textured soy protein (TSP) products have been used as the extender in most studies.

Anderson and Lind (1975) reported that beef patties containing 25%

TVP retained more moisture and less fat than beef patties with comparable initial fat content. Similar results were reported by Seideman et al. (1977). However, they also reported that flavor desirability and overall palatability decreased as the level of TSP increased to 20 and 30%. They concluded that 10% TSP can be added without significantly affecting cooked appearance and palatability of patties. Increased moisture content in extended patties does not increase panel scores for juiciness (Bowers and Engler, 1975). Patties containing 20% TVP were reported to be as acceptable as all-beef patties by Cross et al. (1975). Turkey patties containing 10% TSP were as acceptable as all-turkey patties, but those containing 20% TSP were not (Baldwin et al., 1975).

Reheating of patties containing TSP and all-meat patties reduced flavor differences (Bowers and Engler, 1975). Stale taste and aroma were almost absent in reheated patties containing soy, suggesting that if components responsible for stale taste and aroma were present, other components masked the stale flavors.

The addition of ingredients such as vinegar, pineapple, horseradish, and soy sauce reduces the undesirable effects of soy proteins on the flavor of combination dishes (Baldwin et al., 1975). Cooking method also may influence quality of soy and beef products. Microwave heating of beef-soy loaves resulted in increased loss of moisture and reduced beef flavor (Ziprin and Carlin, 1976).

FISH

Table 5.1 includes information on the components of representative fish. The proteins and muscle structure of fish were described by Connell (1964). Myofibrils appear to be composed of the same proteins as the myofibrils of muscle tissue from warm-blooded animals. Muscle fibers are shorter, however. Fibers, 3 cm long, have their ends embedded in sheets of connective tissue called *myocommata.* When heated, the connective tissue softens and the cells separate, giving a flaky appearance to the fish. Connective tissue is present in very small quantities in fish (Connell, 1964).

Fish may be divided into two groups on the basis of their fat content. All shellfish and many fish with bones contain less than 5% fat in the edible portion. Examples of such lean fish with bones are cod, flounder, haddock, and swordfish (Adams, 1975). Examples of fish containing more than 5% fat are halibut, herring, mackerel, salmon, and shad (Adams, 1975). The fatty fish are desirable for baking and broiling because their fat keeps them from becoming dry. Lean fish may, however, be cooked with these methods if brushed or basted with melted fat, and are less likely to fall apart on boiling or steaming than fatty fish. Both fatty and lean fish can be fried successfully.

Less research on cooking methods has been done for fish than for meat and poultry. Cooking should be stopped as soon as the fish flakes easily when tested with a fork because overcooking results in dryness, crumbliness, and firmness. A more scientific method for testing doneness was suggested by Charley (1952), who used a meat thermometer in salmon steaks 2.5 cm thick. The steaks were baked to an endpoint temperature of 75°C in ovens at 177, 204, 232, and 260°C. Although these temperatures had little effect on palatability scores, 204°C was the most practical of the temperatures tested since there were more spattering and drip at the higher temperatures, while the lowest temperature increased the cooking time and caused a white material to ooze from the fish. Steaks baked at an oven temperature of 204°C to internal temperatures of 70, 75, 80, and 85°C became drier as the endpoint temperature increased but did not differ in tenderness. Salmon steaks cooked to 80 and 85°C were more palatable than those cooked to the lower temperatures. There was no significant difference between the two higher temperatures, except that there was a significant increase in cooking losses at 85°C as compared with 80°C. It will be recalled that salmon is relatively high in fat. In a later study (Charley and Goertz, 1958), larger pieces (900 g) of salmon were baked at the same four oven temperatures as in the earlier study. Total cooking losses and evaporative losses increased as oven temperatures increased.

EXPERIMENTS

The number of experiments that can be done on meat cookery is almost unlimited. One experiment is outlined below to give some basic information on conducting such a study. In addition, several examples of other studies are given in the next section. For the suggested exercises, methods used in the experiment below could be used.

Effect of Oven Temperature on Roast Beef or Pork

This experiment illustrates how variation in oven temperature affects the quality of beef or pork.

A. *Material*

Roasts for the experiment may be "paired"—that is, obtained from corresponding locations on opposites sides of the same animal—or they may be adjacent cuts, which usually are more easily obtained. Beef rib roasts or rump roasts, or pork loin roasts may be selected. If rib or loin roasts are used, each roast should contain at least two ribs. If boneless roasts are used, they should weigh about 900 g each. The weights should be as nearly the same as possible.

B. *Procedures*

1. Set one oven at 163°C (325°F). In planning when to start the meat, allow about 45 min per 450 g for meat with bone or 1 hr per 450 g for meat without bone. Avoid planning the schedule too closely as it is impossible to predict the actual cooking time. In planning when to start the roast to be heated at 232°C (450°F), allow about 20 min per 450 g for meat with or without bone.
2. Weigh the roasting pan, including a rack if a loin or rib roast is not used.
3. Place the meat, fat side up, in the weighed pan. Add neither water nor seasoning. Weigh the pan and roast.
4. Insert a thermometer or thermocouple so that the bulb or point is in the center of the largest muscle.
5. Put the roast in the oven and record the time.
6. Observe and record the temperature of the roast at frequent intervals if a thermometer is used. If possible, record temperatures at 5-min intervals for the first 15 min and after the temperature reaches 50°C. For the remainder of the heating period, 10-min intervals are adequate.
7. Remove the roast from the oven when the internal temperature of beef reaches 60°, 71°, or 77°C, depending on the degree of doneness desired, or when the internal temperature of pork reaches 85°C.

8. Remove the thermocouple or thermometer. Weigh the roast and pan. Remove the roast from the pan. Weigh pan and drip.
9. If a Warner-Bratzler shear is available, take several cores from the largest muscle in the roast for shearing.
10. Cut thin slices for judging. If slices are large, they may be cut in half. Use only the largest muscle and the inner slices for judging. Evaluate flavor, color, tenderness, juiciness, and overall acceptability.

C. *Results*

Record the findings in a table similar to Table 5.2. Do all of the necessary calculations. Graph internal temperatures of the meat against time. Compare the cooking losses and appearance of the drip from the roasts cooked at the two temperatures.

D. *Questions*

How would the number of servings from the two roasts compare? How does the oven temperature affect the evenness of cooking throughout the meat? Were any differences in tenderness or flavor noticed?

SUGGESTED EXERCISES

1. Compare the effect of meat tenderizer on the tenderness of broiled round steak. Use adjacent cuts of meat. Sprinkle a weighed amount of tenderizer onto each side of one steak (2.0 g per side of a 450 g steak). Fork in using 50 strokes per side. Weigh the steak. Treat a second steak in a like manner except omit the tenderizer. Broil each steak for 12 min per side or until the desired degree of doneness is reached. Cook in the same oven at the same time, if possible. After the steaks cook, cool for 10 min. Weigh each one. Calculate cooking losses. If possible, do shear values. For sensory evaluation, cut samples of equal size from the same position in all of the steaks for each of the judges. Ask the judges to record the number of chews that are required before the meat is ready to swallow.

Table 5.2 Cooking data and results for roasts cooked at two oven temperatures

Factor measured	Oven temperature °C 163	232
A. Before cooking		
1. Weight of pan, g	_____	_____
2. Weight of roast, g	_____	_____
3. Weight of pan and roast, g	_____	_____
B. On removal from oven		
1. Weight of pan, roast, and drip, g	_____	_____
2. Weight of pan and drip, g	_____	_____
C. Losses		
1. Loss due to evaporation, g (A3 − B1)	_____	_____
2. Loss due to drip, g (B2 − A1)	_____	_____
3. Total cooking loss, g (C1 + C2)	_____	_____
D. Losses as percentages of weight of uncooked roast		
1. Due to evaporation, % $\left(\frac{100 \times C1}{A2}\right)$	_____	_____
2. Due to drip, % $\left(\frac{100 \times C2}{A2}\right)$	_____	_____
3. Total loss, % (D1 + D2)	_____	_____
E. Cooking time		
1. Time put into oven	_____	_____
2. Time removed from oven	_____	_____
3. Total cooking time, min	_____	_____
4. Cooking time, min per kg	_____	_____

2. Compare dry and moist methods of cooking veal or baby beef.
3. Compare moist and dry methods of cooking various parts of poultry varying in age.
4. Study the effects of proportion of fat, use of fillers, and different methods of cooking on the quality and cooking losses of ground beef cooked as loaves or patties.
5. Compare the quality and cooking losses of simple beef meat loaves baked to 60, 71, 77, 82°C, and the highest temperature possible.

To find the weight of the cooked meat loaf, turn it out on a cooking rack for a few minutes before weighing. Observe the ways in which the meat proteins coagulated in each of the loaves.

6. To study the effects of salt on the retention of water in meat during heating, divide a well-mixed sample of ground beef into three equal portions. To one portion, add no salt; to the second, add 5.5 g of salt per 450 g of meat; and to the third, add 15 g of salt per 450 g of meat. Cook as meat patties or loaves. Bake loaves at 163°C (325°F) to the endpoint temperature found best in exercise 5. Weigh the patties or loaves before and after cooking. Compare cooking losses.
7. Roast meat, poultry, or fish with aluminum foil, with an oven film, and without a wrap.
8. Prepare beef patties containing 0, 10, and 25% textured vegetable protein. Broil equal lengths of time. Compare cooking losses and have a panel of judges compare flavor, juiciness, and tenderness.
9. Compare oven roasting of beef semitendinosus with heating in a microwave oven. One-half of a muscle may be used for each cooking method. Unless a special thermometer is available for use in the microwave oven, the roast must be removed from the oven periodically and a thermometer or thermocouple inserted to see if the roast has reached the desired degree of doneness.
10. Fry or bake fish fillets to various internal temperatures (60, 70, 80, or 90°C). Calculate cooking losses and compare quality of the four samples.
11. Bake fish at several oven temperatures—177, 204, and 260°C (350, 400, and 500°F)—to an internal temperature of 80°C. Compare cooking losses and quality.
12. Compare several heating methods from the list below with respect to their effects on the quality of less tender cuts of meat. If several methods are to be compared, select a large piece of beef, preferably one muscle, from which several uniform roasts can be cut. A muscle from the round that has been boned and tied into a uniform cylinder is satisfactory. Because there will be some variation in the quality of the roasts, selection of a roast for each treatment is best made by chance. This is done easily by putting a card identifying each roast in a container and drawing at random for the experiments. Compare cooking times, rates of heat penetration, cooking losses, and quality.

a. Oven at 93°C (200°F). Roast on a rack in a small baking dish, uncovered, until the temperature reaches 85°C. Allow about 10 hr.
b. Oven at 93°C (200°F) with foil or oven wrap. Wrap meat with heavy duty aluminum foil, place in an oven bag, or wrap with oven film as directed by the manufacturer. Insert meat thermometer or thermocouple through the wrap or film. Put on a rack in a small baking pan and roast uncovered until the temperature of the meat is 85°C. The weights recorded should be without wrap. Allow about 10 hr.
c. Oven at 177°C (350°F). Put meat on a rack in a small baking pan and roast uncovered until the temperature of the meat is 85°C.
d. Simmering water. Fill a 4-liter pan about 2/3 full of water and put on a small unit of the range. Cover the pan and bring the water to a boil. Put the meat in the water, reduce heat, and simmer until the meat reaches an internal temperature of 85°C. Allow about 1 hr.
e. Braising. Set an electric fry pan at 182°C (360°F) and add 13 g of fat. Brown the meat well on all sides and slip a small rack under the meat. Add 60 ml of water and cover tightly. Cook until the temperature of the meat reaches 85°C. Add more water during heating if needed. A small, covered fry pan may be substituted for the electric one. After the meat is browned, cook it slowly on top of the range or in an oven at 177°C (350°F). About 1 hr total cooking time will be required.
f. Electric slow cooker. Add 120 ml of water. Cook on the low setting for the time suggested by the manufacturer. Approximately 10 hr will be required.

REFERENCES

Adams, C. 1975. Nutritive value of American foods. Agric. Handbook 456, Agricultural Research Service, USDA, Washington, DC.

AHEA. 1975. "Handbook of Food Preparation." American Home Economics Assoc., Washington, DC.

Alexander, L. M. and Clark, N. G. 1939. Shrinkage and cooking time of rib roasts of beef of different grades as influenced by style of cutting and method of roasting. Tech. Bull. 676, USDA, Washington, DC.

Anderson, R. H. and Lind, K. D. 1975. Retention of water and fat in cooked patties of beef and of beef extended with textured vegetable protein. Food Technol. 29(2): 44.

Armbruster, G. and Haefele, C. 1975. Quality of foods after cooking in 915 MHz and 2450 MHz microwave appliances using plastic film covers. J. Food Sci. 40: 721.

Bailey, A. J. 1972. The basis of meat texture. J. Sci. Food Agric. 23: 995.

Baity, M. R., Ellington, A. E. and Woodburn, M. 1969. Foil wrap in oven cooking. J. Home Econ. 61: 174.

Baldwin, R. E., Korschgen, B. M., Vandepopuliere, J. M. and Russell, W. D. 1975. Palatability of ground turkey and beef containing soy. Poultry Sci. 54: 1102.

Bard, J. and Townsend, W. E. 1971. Cured meats. Meat curing. In "The Science of Meat and Meat Products," Eds. Price, J. F. and Schweigert, B. S. W. H. Freeman and Co., San Francisco.

Batcher, O. M. and Deary, P. A. 1975. Quality characteristics of broiled and roasted beef steaks. J. Food Sci. 40: 745.

Bayne, B. H., Meyer, B. H. and Cole, J. W. 1969. Response of beef roasts differing in finish, location, and size to two rates of heat application. J. Animal Sci. 29: 283.

Bayne, B. H., Allen, M. B., Large, N. F., Meyer, B. H. and Goertz, G. E. 1973. Sensory and histological characteristics of beef rib cuts heated at two rates to three end point temperatures. Home Econ. Res. J. 2: 29.

Bernofsky, C., Fox, J. B. Jr. and Schweigert, B. S. 1959. Biochemistry of myoglobin. VII. The effect of cooking on myoglobin in beef muscle. Food Res. 24: 339.

Blaker, G. G., Newcomer, J. L. and Stafford, W. D. 1959. Conventional roasting vs. high-temperature foil cookery. J. Amer. Dietet. Assoc. 35: 1255.

Bodwell, C. E. and McClain, P. E. 1971. Chemistry of animal tissues. Proteins. In "The Science of Meat and Meat Products," Eds. Price, J. F. and Schweigert, B. S. W. H. Freeman and Co., San Francisco.

Bowers, J. A. 1972. Eating quality, sulfhydryl content, and TBA values of turkey breast muscle. J. Agric. Food Chem. 20: 706.

Bowers, J. A. and Engler, P. P. 1975. Freshly cooked and cooked, frozen, reheated beef and beef-soy patties. J. Food Sci. 40: 624.

Bowers, J. R. and Goertz, G. E. 1966. Effect of internal temperature on eating quality of pork chops. I. Skillet- and oven-braising. J. Amer. Dietet. Assoc. 48: 116.

Bowers, J. A., Fryer, B. A. and Engler, P. P. 1974. Vitamin B_6 in turkey breast muscle cooked in microwave and conventional ovens. Poultry Sci. 53: 844.

Brady, P. and Penfield, M. P. 1977. A comparison of four methods of heating beef roasts: conventional oven, slow cooker, microwave oven, and pressure cooker. Tennessee Farm and Home Science 101: 15.
Bramblett, V. D. and Fugate, K. W. 1967. Choice of cooking temperature for stuffed turkeys. I. Palatability factors. J. Home Econ. 59: 180.
Bramblett, V. D., Hostetler, R. L., Vail, G. E. and Draudt, H. N. 1959. Qualities of beef as affected by cooking at very low temperatures for long periods of time. Food Technol. 13: 707.
Bratzler, L. J. 1971. Palatability characteristics of meat. Palatability factors and evaluation. In "The Science of Meat and Meat Products," Eds. Price, J. F. and Schweigert, B. S. W. H. Freeman and Co., San Francisco.
Briskey, E. J. 1964. Etiological status and associated studies of pale, soft, exudative porcine musculature. Adv. Food Res. 13: 89.
Carpenter, Z. L., Abraham, H. C. and King, G. T. 1968. Tenderness and cooking loss of beef and pork. I. Relative effects of microwave cooking, deep-fat frying, and oven broiling. J. Amer. Dietet. Assoc. 53: 353.
Chang, S. S. and Peterson, R. J. 1977. Symposium: the basis of quality in muscle foods. Recent developments in the flavor of meat. J. Food Sci. 42: 298.
Charley, H. 1952. Effects of internal temperature and of oven temperature on the cooking losses and palatability of baked salmon steaks. Food Res. 17: 136.
Charley, H. and Goertz, G. E. 1958. The effects of oven temperature on certain characteristics of baked salmon. Food Res. 23: 17.
Cipra, J. S. and Bowers, J. A. 1971. Flavor of microwave and conventionally-reheated turkey. Poultry Sci. 50: 703.
Cline, J. A., Trowbridge, E. A., Foster, M. T. and Fry, H. E. 1930. How certain methods of cooking affect the quality and palatability of beef. Bull. 293, Missouri Agric. Exp. Stn., Columbia.
Connell, J. J. 1964. Fish muscle proteins and some effects on them of processing. In "Symposium on Foods: Proteins and Their Reactions," Eds. Schultz, H. W. and Anglemeir, A. F. Avi Publ. Co., Westport, CT.
Cover, S. 1943. Effect of extremely low rates of heat penetration on tendering of beef. Food Res. 8: 388.
Cover, S. 1959. Scoring for three components of tenderness to characterize differences among beef steaks. Food Res. 24: 564.
Cover, S. and Hostetler, R. L. 1960. An examination of some theories about beef tenderness by using new methods. Bull. 947, Texas Agric. Exp. Stn., College Station.
Cover, S., Ritchey, S. J. and Hostetler, R. L. 1962a. Tenderness of beef. I. The connective-tissue component of tenderness. J. Food Sci. 27: 469.
Cover, S., Ritchey, S. J. and Hostetler, R. L. 1962b. Tenderness of beef. II. Juiciness and the softness components of tenderness. J. Food Sci. 27: 476.

Cover, S., Ritchey, S. J. and Hostetler, R. L. 1962c. Tenderness of beef. III. The muscle fiber components of tenderness. J. Food Sci. 27: 483.

Cross, H. R., Stanfield, M. S., Green, E. C., Heinemeyer, J. M. and Hollick, A. B. 1975. Effect of fat and textured soy protein content on consumer acceptance of ground beef. J. Food Sci. 40: 1331.

Davenport, M. M. and Meyer, B. H. 1970. Forced convection roasting at 200° and 300°F: yield, cost, and acceptability of beef sirloin. J. Amer. Dietet. Assoc. 56: 31.

Dawson, E. H., Linton, G. S., Harkin, A. M. and Miller, C. 1959. Factors influencing the palatability, vitamin content, and yield of cooked beef. Home Econ. Res. Rpt. 9, USDA, Washington, DC.

Deatherage, F. E. 1963. The effect of water and inorganic salts on tenderness. In "Proceedings Meat Tenderness Symposium." Campbell Soup Co., Camden, NJ.

Deethardt, D., Burrill, L. M., Schneider, K. and Carlson, C. W. 1971. Foil-covered versus open-pan procedures for roasting turkey. J. Food Sci. 36: 624.

Dimick, P. S., MacNeil, J. H. and Grunden, L. P. 1972. Poultry product quality. Carbonyl composition and organoleptic evaluation of mechanically deboned poultry meat. J. Food Sci. 37: 544.

Doty, D. M. and Pierce, J. C. 1961. Beef muscle characteristics as related to carcass grade, carcass weight, and degree of aging. Tech. Bull. 1231, USDA, Washington, DC.

Draudt, H. N. 1972. Changes in meat during heating. Proc. 25th Ann. Recip. Meat Conf., p. 243, Natl. Livestock and Meat Board, Chicago.

Dugan, L. R. Jr. 1971. Chemistry of animal tissues. Fats. In "The Science of Meat and Meat Products," Eds. Price, J. F. and Schweigert, B. S. W. H. Freeman and Co., San Francisco.

Duncan, W. R. H. and Garton, G. A. 1967. The fatty acid composition and intramuscular structure of triglycerides derived from different sites in the body of the sheep. J. Sci. Food Agric. 18: 99.

Engler, P. P. and Bowers, J. A. 1975. Eating quality and thiamin retention of turkey breast muscle roasted and "slow-cooked" from frozen and thawed states. Home Econ. Res. J. 4: 27.

Esselen, W. B., Levine, A. S. and Brushway, M. J. 1956. Adequate roasting procedures for frozen stuffed poultry. J. Amer. Dietet. Assoc. 32: 1162.

Ferger, D. C., Harrison, D. L. and Anderson, L. L. 1972. Lamb and beef roasts cooked from the frozen state by dry and moist heat. J. Food Sci. 37: 226.

Flynn, A. W. and Bramblett, V. D. 1975. Effects of frozen storage, cooking method and muscle quality on attributes of pork loins. J. Food Sci. 40: 631.

Fox, J. B. Jr. 1966. The chemistry of meat pigments. J. Agric. Food Chem. 14: 207.

Fry, J. L., Waldroup, P. W., Ahmed, E. M. and Lydick, H. 1965. Enzymatic tenderization of poultry meat. Food Technol. 20: 952.

Funk, K., Aldrich, P. T. and Irmiter, T. F. 1966. Forced convection roasting of loin cuts of beef. J. Amer. Dietet. Assoc. 48: 404.

Gaddis, A. M., Hankins, O. G. and Hiner, R. L. 1950. Relationships between the amount and composition of press fluid, palatability, and other factors of meat. Food Technol. 4: 498.

Giddings. G. G. 1977. Symposium: the basis of quality in muscle foods. The basis of color in muscle foods. J. Food Sci. 42: 288.

Glover, E. E., Forrest, J. C., Johnson, H. R., Bramblett, V. D. and Judge, M. D. 1977. Palatability and cooking characteristics of mechanically tenderized beef. J. Food Sci. 42: 871.

Goertz, G. E., Hooper, A. S. and Harrison, D. L. 1960. Comparison of rate of cooking and doneness of fresh-unfrozen and frozen, defrosted turkey hens. Food Technol. 14: 458.

Goertz, G. E., Meyer, B. H., Weathers, B. and Hooper, A. S. 1964. Effect of cooking temperatures on broiler acceptability. J. Amer. Dietet. Assoc. 45: 526.

Goll, D. E., Hoekstra, W. G. and Bray, R. W. 1964. Age-associated changes in bovine muscle connective tissue. II. Exposure to increasing temperature. J. Food Sci. 29: 615.

Gövindarajan, S. 1973. Fresh meat color. CRC Critical Reviews in Food Technol. 4: 117.

Griswold, R. M. 1955a. The effect of different methods of cooking beef round of Commercial and Prime grades. II. Collagen, fat and nitrogen content. Food Res. 20: 171.

Griswold, R. M. 1955b. The effect of different methods of cooking beef round of Commercial and Prime grades. I. Palatability and shear values. Food Res. 20: 160.

Hall, J. L., Latschar, C. E. and Mackintosh, D. L. 1944. Quality of beef. IV. Characteristics of dark-cutting beef. Survey and preliminary investigation. Tech. Bull. 58, Kansas Agric. Exp. Stn., Manhattan.

Hamm, R. 1960. Biochemistry of meat hydration. Adv. Food Res. 10: 355.

Hamm, R. and Deatherage, F. E. 1960. Changes in hydration, solubility and charges of muscle proteins during heating of meat. Food Res. 25: 587.

Hanson, H. L., Stewart, G. F. and Lowe, B. 1942. Palatability and histological changes occurring in New York dressed broilers held at 1.7°C (35°F). Food Res. 7: 148.

Hanson, H. L., Winegarden, H. M., Horton, M. B. and Lineweaver, H. 1950.

Preparation and storage of frozen cooked poultry and vegetables. Food Technol. 4: 430.

Hay, P. P., Harrison, D. L. and Vail, G. E. 1953. Effects of a meat tenderizer on less tender cuts of beef cooked by four methods. Food Technol. 7: 217.

Headley, M. E. and Jacobson, M. 1960. Electronic and conventional cookery of lamb roasts. J. Amer. Dietet. Assoc. 36: 337.

Hearne, L. E., Penfield, M. P. and Goertz, G. E. 1978. Heating effects on bovine semitendinosus: phase contrast microscopy and scanning electron microscopy. J. Food Sci. 43: 13.

Heine, N., Bowers, J. A. and Johnson, P. G. 1973. Eating quality of half turkey hens cooked by four methods. Home Econ. Res. J. 1: 210.

Herring, H. K., Cassens, R. G. and Briskey, E. J. 1967. Factors affecting collagen solubility in bovine muscles. J. Food Sci. 32: 534.

Hill, F. 1966. The solubility of intramuscular collagen in meat animals of various ages. J. Food Sci. 31: 161.

Hiner, R. L., Anderson, E. E. and Fellers, C. R. 1955. Amount and character of connective tissue as it relates to tenderness in beef muscle. Food Technol. 9: 80.

Hiner, R. L., Hankins, O. G., Sloane, H. S., Fellers, C. R. and Anderson, E. E. 1953. Fiber diameter in relation to tenderness of beef muscle. Food Res. 18: 364.

Hinnergardt, L. C., Drake, S. R. and Kluter, R. A. 1975. Grilled freeze-dried steaks. Effects of mechanical tenderization plus phosphate and salt. J. Food Sci. 40: 621.

Hinrichs, J. R. and Whitaker, J. R. 1962. Enzymatic degradation of collagen. J. Food Sci. 27: 250.

Hoke, I. M. 1968. Evaluation of procedures and temperatures for roasting chickens. J. Home Econ. 60: 661.

Holmes, Z. A., Bowers, J. A. and Goertz, G. E. 1966. Effect of internal temperature on eating quality of pork chops. J. Amer. Dietet. Assoc. 48: 121.

Hood, M. P. 1960. Effect of cooking method and grade on beef roasts. J. Amer. Dietet. Assoc. 37: 363.

Hood, M. P., Thompson, D. W. and Mirone, L. 1955. Effects of cooking methods on low-grade beef. Bull. N.S. 4, Georgia Agric. Exp. Stn., Athens.

Hornstein, I. 1971. Palatability characteristics of meat. Chemistry of meat flavor. In "The Science of Meat and Meat Products," Eds. Price, J. F. and Schweigert, B. S. W. H. Freeman and Co., San Francisco.

Huffman, D. L., Palmer, A. Z., Carpenter, J. W., Hargrove, D. D. and Kroger, M. 1962. Effect of breeding, level of feeding and ante-mortem injection of papain on the tenderness of weanling age calves. (Abstr.) J. Animal Sci. 21: 381.

Huxley, H. E. 1969. The mechanism of muscular contraction. Science 164: 1356.
Irvin, L. and Cover, S. 1959. Effect of a dry heat method of cooking on the collagen content of two beef muscles. Food Technol. 13: 655.
Janicki, L. J. and Appledorf, H. 1974. Effect of broiling, grill frying and microwave cooking on moisture, some lipid components and total fatty acids of ground beef. J. Food Sci. 39: 715.
Johnson, P. G. and Bowers, J. A. 1976. Influence of aging on the electrophoretic and structural characteristics of turkey breast muscle. J. Food Sci. 41: 255.
Jones, S. B. 1977. Ultrastructural characteristics of beef muscle. Food Technol. 31(4): 82.
Kang, C. K. and Rice, E. E. 1970. Degradation of various meat fractions by tenderizing enzymes. J. Food Sci. 35: 563.
Klose, A. A., Pool, M. F., Wiele, M. B., Hanson, H. L. and Lineweaver, H. 1959. Poultry tenderness. I. Influence of processing on tenderness of turkeys. Food Technol. 13: 20.
Korschgen, B. M., Baldwin, R. E. and Snider, S. 1976. Quality factors in beef, pork, and lamb cooked by microwaves. J. Amer. Dietet. Assoc. 69: 635.
Laakkonen, E., Wellington, G. H. and Sherbon, J. W. 1970. Low-temperature, long-time heating of bovine muscle. I. Changes in tenderness, water-binding capacity, pH and amount of water soluble components. J. Food Sci. 35: 175.
Lachance, P. A. 1972. Update: Meat extenders and analogues in child feeding programs. Proc. Meat Ind. Res. Conf. 24: 97.
Law, H. M., Yang, S. P., Mullins, A. M. and Fielder, M. M. 1967. Effect of storage and cooking on qualities of loin and top round steaks. J. Food Sci. 32: 637.
Lawrie, R. A. 1968. Chemical changes in meat due to processing—a review. J. Sci. Food Agric. 19: 233.
Lee, W. T. and Dawson, L. E. 1976. Changes in phospholipids in chicken tissues during cooking in fresh and reused cooking oil, and during frozen storage. J. Food Sci. 41: 598.
Lind, J. M., Griswold, R. M. and Bramblett, V. D. 1971. Tenderizing effect of wine vinegar marinade on beef round. J. Amer. Dietet. Assoc. 58: 133.
Lowe, B., Edgar, M., Schoenleber, F. and Young, J. 1953. Cooking turkey in aluminum foil. Iowa Farm Sci. J. 8: 326.
Mangel, M. 1951. The determination of methemoglobin in beef muscle extracts. Res. Bull. 474, Missouri Univ. Agric. Exp. Stn., Columbia.
Marsh, B. B. and Leet, N. G. 1966. Studies in meat tenderness. III. The effects of cold shortening on tenderness. J. Food Sci. 31: 450.
Marshall, N., Wood, L. and Patton, M. B. 1959. Cooking Choice grade, top round beef roasts. Effect of size and internal temperature. J. Amer. Dietet. Assoc. 35: 569.

Marshall, N., Wood, L. and Patton, M. B. 1960. Cooking Choice grade, top round beef roasts. Effects of internal temperature on yield and cooking time. J. Amer. Dietet. Assoc. 36: 341.

Minor, L. J., Pearson, A. M., Dawson, L. E. and Schweigert, B. S. 1965. Chicken flavor: the identification of some chemical components and the importance of sulfur compounds in the cooked volatile fraction. J. Food Sci. 30: 686.

Miyada, D. S. and Tappel, A. L. 1956. The hydrolysis of beef proteins by various proteolytic enzymes. Food Res. 21: 217.

Murray, J. M. and Weber, A. 1974. The cooperative action of muscle proteins. Scientific American 230(2): 58.

Paul, P. C. 1972. Meat. In "Food Theory and Applications," Eds. Paul, P. C. and Palmer, H. H. John Wiley and Sons, Inc., New York.

Paul, P. C. and Bean, M. 1956. Method for braising beef round steaks. Food Res. 21: 75.

Paul, P., Bratzler, L. J., Farwell, E. D. and Knight, K. 1952. Studies on tenderness of beef. I. Rate of heat penetration. Food Res. 17: 504.

Paul, P. C., Mandigo, R. W. and Arthaud, V. H. 1970. Textural and histological differences among 3 muscles in the same cut of beef. J. Food Sci. 35: 505.

Paul, P. C., McCrae, S. and Hofferber, L. M. 1973. Heat-induced changes in extractability of beef muscle collagen. J. Food Sci. 38: 66.

Penfield, M. P. and Meyer, B. H. 1975. Changes in tenderness and collagen of beef semitendinosus muscle heated at two rates. J. Food Sci. 40: 150.

Penner, K. K. and Bowers, J. A. 1973. Flavor and chemical characteristics of conventionally and microwave reheated pork. J. Food Sci. 38: 553.

Peterson, D. W. 1977. Effect of polyphosphates on tenderness of hot cut chicken breast meat. J. Food Sci. 42: 100.

Phillips, L., Delaney, I. and Mangel, M. 1960. Electronic cooking of chicken. J. Amer. Dietet. Assoc. 37: 462.

Pippen, E. L. and Eyring, E. J. 1957. Characterization of volatile nitrogen and volatile sulfur fractions of cooked chicken and their relation to flavor. Food Technol. 11: 53.

Pippen, E. L., Nonaka, M., Jones, F. T. and Stitt, F. 1958. Volatile carbonyl compounds of cooked chicken. I. Compounds obtained by air entrainment. Food Res. 23: 103.

Pool, M. F., de Fremery, D., Campbell, A. A. and Klose, A. A. 1959. Poultry tenderness. II. Influence of processing on tenderness of chickens. Food Technol. 13: 25.

Ramsbottom, J. M. and Strandine, E. J. 1948. Comparative tenderness and identification of muscles in wholesale beef cuts. Food Res. 13: 315.

Ramsbottom, J., Strandine, E. J. and Koonz, C. H. 1945. Comparative tenderness of representative beef muscles. Food Res. 10: 497.

Ream, E. E., Wilcox, E. B., Taylor, F. G. and Bennett, J. A. 1974. Tenderness of beef roasts. J. Amer. Dietet. Assoc. 65: 155.
Robinson, H. E. and Goesner, P. A. 1962. Enzymatic tenderization of meat. J. Home Econ. 54: 195.
Ruyack, D. F. and Paul, P. C. 1972. Conventional and microwave heating of beef: use of plastic wrap. Home Econ. Res. J. 1: 98.
Saffle, R. L. 1973. Quantitative determination of combined hemoglobin and myoglobin in various poultry meats. J. Food Sci. 38: 968.
Satterlee, L. D. and Hansmeyer, W. 1974. The role of light and surface bacteria in the color stability of prepackaged beef. J. Food Sci. 39: 305.
Savell, J. W., Smith, G. C. and Carpenter, Z. L. 1977. Blade tenderization of four muscles from three weight-grade groups of beef. J. Food Sci. 42: 866.
Schock, D. R., Harrison, D. L. and Anderson, L. L. 1970. Effect of dry and moist heat treatments on selected beef quality factors. J. Food Sci. 35: 195.
Seideman, S. C., Smith, G. C. and Carpenter, Z. L. 1977. Addition of textured soy protein and mechanically deboned beef to ground beef formulations. J. Food Sci. 42: 197.
Shaffer, T. A., Harrison, D. L. and Anderson, L. L. 1973. Effects of end point and oven temperatures on beef roasts cooked in oven film bags and open pans. J. Food Sci. 38: 1205.
Siemers, L. L. and Hanning, F. 1953. A study of certain factors influencing the juiciness of meat. Food Res. 18: 113.
Smith, G. C., West, R. L., Rea, R. H. and Carpenter, Z. L. 1973. A research note—increasing the tenderness of bullock beef by use of antemortem enzyme injection. J. Food Sci. 38: 182.
Stinson, C. G., deMan, J. M. and Bowland, J. P. 1967. Fatty acid composition and glyceride structure of piglet body fat from different sampling sites. J. Amer. Oil Chemists' Soc. 44: 253.
Sundberg, A. D. and Carlin, A. F. 1976. Survival of Clostridium perfringens in rump roasts cooked in an oven at 107° or 177°C or in an electric crockery pot. J. Food Sci. 41: 451.
Swickard, M. T., Harkin, A. M. and Paul, B. J. 1954. Relationship of cooking methods, grades, and frozen storage to quality of cooked mature Leghorn hens. Tech. Bull. 1077, USDA, Washington, DC.
Tappel, A. L. 1957. Reflectance spectral studies of the hematin pigments of cooked beef. Food Res. 22: 404.
Tappel, A. L., Miyada, D. S., Sterling, C. and Maier, V. P. 1956. Meat tenderization. II. Factors affecting the tenderization of beef by papain. Food Res. 21: 375.
Terrell, R. N., Lewis, R. W., Cassens, R. G. and Bray, R. W. 1967. Fatty acid compositions of bovine subcutaneous fat depots determined by gas-liquid chromatography. J. Food Sci. 32: 516.

Thille, M., Williamson, L. J. and Morgan, A. F. 1932. The effect of fat on shrinkage and speed in the roasting of beef. J. Home Econ. 24: 720.

Thompson, E. H., Wolf, I. D. and Allen, C. E. 1973. Ginger rhizome: a new source of proteolytic enzyme. J. Food Sci. 38: 652.

Tsen, C. C. and Tappel, A. L. 1959. Meat tenderization. III. Hydrolysis of actomyosin, actin and collagen by papain. Food Res. 24: 362.

Tucker, H. Q., Voegeli, M. M. and Wellington, G. H. 1952. "A Cross Sectional Muscle Nomenclature of the Beef Carcass." Michigan State College Press, East Lansing.

Urbain, W. M. 1971. Meat preservation. In "The Science of Meat and Meat Products," Eds. Price, J. F. and Schweigert, B. S. W. H. Freeman and Co., San Francisco.

USDA. 1971. Textured vegetable protein products (B-1). Notice 219, Food and Nutrition Service, USDA, Washington, DC.

USDA. 1975. Grades of carcass beef; slaughter cattle. Fed. Register 40: 11535.

Vail, G. E. and O'Neill, L. 1937. Certain factors which affect the palatability and cost of roast beef served in institutions. J. Amer. Dietet. Assoc. 13: 34.

Van Zante. H. J. 1968. Some effects of microwave cooking power upon certain basic food components. Microwave Energy Appl. Newsletter 1(6): 3.

Van Zante, H. 1973. "The Microwave Oven." Houghton Mifflin Co., Boston.

Wang, H., Doty, D. M., Beard, F. J., Pierce, J. C. and Hankins, O. G. 1956. Extensibility of single beef muscle fibers. J. Animal Sci. 15: 97.

Wang, H., Rasch, E. and Bates, V. 1954. Histological observations on fat loci and distribution in cooked beef. Food Res. 19: 314.

Wang, H., Weir, C. E., Birkner, M. L. and Ginger, B. 1958. Studies on enzymatic tenderization of meat. III. Histological and panel analysis of enzyme preparation from three distinct sources. Food Res. 23: 423.

Watts, B. M. 1954. Oxidative rancidity and discoloration in meat. Adv. Food Res. 5: 1.

Weir, C. E., Slover, A., Pohl, C. and Wilson, G. D. 1962. Effect of cooking procedures on the composition and organoleptic properties of pork chops. Food Technol. 16(5): 133.

Wierbicki, E., Kunkle, L. E. and Deatherage, F. E. 1957a. Changes in the water-holding capacity and cationic shifts during heating and freezing and thawing of meat as revealed by a simple centrifugal method for measuring shrinkage. Food Technol. 11: 69.

Wierbicki, E., Cahill, V. R. and Deatherage, F. E. 1957b. Effects of added sodium chloride, potassium chloride, calcium chloride, magnesium chloride, and citric acid on meat shrinkage at 70°C and of added sodium chloride on drip losses after freezing and thawing. Food Technol. 11: 74.

Winegarden, M. W., Lowe, B., Kastelic, J., Kline, E. A., Plagge, A. R. and

Shearer, P. S. 1952. Physical changes of connective tissues of beef during heating. Food Res. 17: 172.

Wismer-Pedersen, J. 1971. Chemistry of animal tissues. Water. In "The Science of Meat and Meat Products," Eds. Price, J. F. and Schweigert, B. S. W. H. Freeman and Co., San Francisco.

Woodburn, M. and Ellington, A. E. 1967. Choice of cooking temperature for stuffed turkeys. II. Microbiological safety of stuffing. J. Home Econ. 59: 186.

Ziprin, Y. A. and Carlin. A. F. 1976. Microwave and conventional cooking in relation to quality and nutritive value of beef and beef-soy loaves. J. Food Sci. 41: 4.

6. Fruits and Vegetables

Many parts of different plants are used as fruits or vegetables. They may be the roots, tubers, bulbs, stems and shoots, leaves, flowers and fruits, or pods and seeds. An understanding of the plant cell and compounds contributing to the texture, color, and flavor is important to a thorough understanding of the influence of various treatments on these quality attributes.

STRUCTURE

The structure of fruits and vegetables can be studied at several levels. Some understanding of the cellular structure as well as the microstructure is basic to the study of the influence of various treatments on the texture, color, and flavor of fruits and vegetables.

Cells

Plant tissues are composed of small units, the cells. The components of a plant cell are represented schematically in Figure 6.1. Cell constituents are divided into two groups, protoplasmic and nonprotoplasmic.

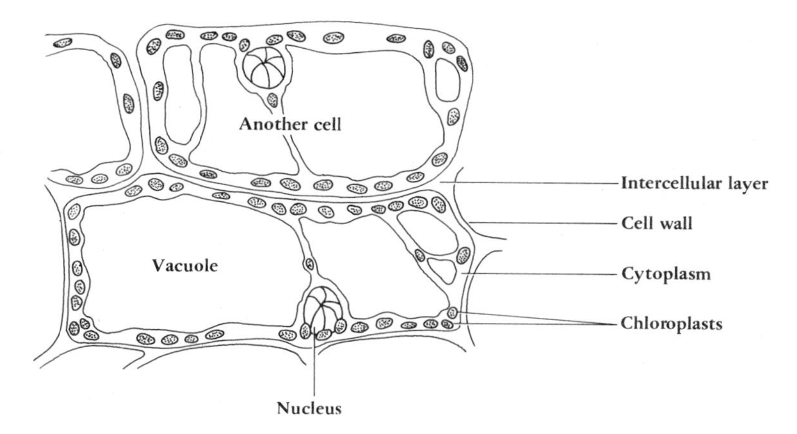

Figure 6.1. Food-synthesizing cells of a green plant.

The *protoplasm* is the living active part of the cell and includes several components. The *nucleus* is the regulator of metabolic activities of the cell. The *cytoplasm* is an undifferentiated part of the protoplasm surrounding the nucleus and forming a rather thin layer inside the cell wall. Cytoplasm, a transparent substance, contains a high proportion of water and organic and inorganic substances in colloidal dispersion or, in some cases, true solution. The plasma membrane, or *plasmalemma,* is a thin membrane on the outer surface of the cytoplasm. In addition, a membrane system, the *endoplasmic reticulum,* forms cavities or sacs in the cytoplasm. The plasmic membranes are selectively permeable (Fahn, 1974) and may serve as a system for segregating and transporting metabolites and provide sites for orderly distribution of enzymes (Esau, 1965).

Within the cytoplasm of certain types of cells are organized bodies called *plastids.* Three types of plastids are recognized: leucoplasts, chloroplasts, and chromoplasts. *Leucoplasts* are plastids that are lacking in pigmentation and are concerned with food storage. Most leucoplasts produce and store starch. These specialized leucoplasts, called *amyloplasts,* occur in potatoes, peas, beans, and other starch-forming tissues. Starch is laid down within the plastid in concentric layers. The pattern of the layers is characteristic of the plant and may be an aid in the microscopic identification of starch from various sources. The *chloroplasts* that occur in green plants contain chlorophyll, the green pigment essential for carbohydrate synthesis. The chlorophyll is located in granules, or *grana,* that give the chloroplasts a granular appearance under a light microscope (Esau, 1965). Chlorophyll is located between layers of lipids and proteins in the chloroplasts with hydrophobic portions oriented toward lipid layers (Clydesdale et al., 1970). *Chromoplasts* contain xanthophylls or carotenes and usually are orange or yellow in color. They occur in such vegetables as carrots and sweet potatoes. The *mitochondria* are metabolically important organelles of the cytoplasm. Enzymes responsible for adenosine triphosphate (ATP) are found in the mitochondria (Fahn, 1974).

The nonprotoplasmic components of the cell are referred to as *ergastic components* and include cavities called *vacuoles,* which contain cell sap. In the mature plant, a vacuole is large in relation to the size of the nucleus and the cytoplasm. *Cell sap* is a watery substance containing many substances such as sugars, salts, organic acids, polysaccharides, phenolic derivatives, flavones, and the red or blue pigments called anthocyanins. These substances are in true solution or are colloidally dispersed. The substances in the cell sap are nutrients to be used by the protoplasm or products of metabolism.

The water in the vacuole of the cell is responsible, in part, for the texture of fruits and vegetables. In crisp raw fruits and vegetables, the pressure exerted on the inside of the cell wall, *turgor,* is equal to the pressure exerted on the outside of the cell wall. Reduction in the relative humidity of the storage atmosphere may result in loss of crispness due to loss of moisture and thus of turgor. Turgor also is lost as water diffuses through cell membranes during heating.

Cell Walls

Cellular components are enclosed by a wall that varies in firmness with the plant tissue and is responsible for the texture of the tissue. The walls of adjoining cells are held together by the *intercellular layer* or *middle lamella.* This layer of cementing substance is composed of pectin in one or more of its forms.

When the cell is immature, the outer, or primary, wall is produced first. In soft tissues, such as those of some fruits, this is the only wall. The primary wall is composed of cellulose, hemicelluloses, and some pectin. In some tissues, a secondary layered wall is laid down inside the primary wall. Cellulose and hemicelluloses are found in the secondary wall. In some cases, particularly vegetables that seem woody, lignin may be present. It stiffens the plant, making it less flexible. An understanding of the properties of the cell wall constituents is essential to the understanding of changes that occur in fruits and vegetables during ripening, storage, processing, and preservation.

Cellulose Cellulose, which is present in large quantity, is responsible for the firmness of the cell wall. Its purpose is structural, since it does not furnish a reserve food supply for the plant. Like starch, which is discussed in Chapter 9, cellulose is a polysaccharide composed of glucose units. It differs from starch in that the glucose units are combined with β-glucosidic linkages rather than α-glucosidic linkages. Estimates of the number of glucose residues in a molecule of cellulose vary from 500 to 10,000 (Fahn, 1974). The source of cellulose and the method used to determine the molecular size influence the results obtained. The linkage of β-D-glucose units in cellulose is:

Cellulose

Cellulose molecules are combined in an orderly crystalline arrangement to form linear bundles called elementary *fibrils,* which have diameters of about 100 Å (Esau, 1965). Hydrogen bonding and Van der Waal's forces link the molecules (Isherwood, 1970). Between 10 and 40 of the elementary fibrils are organized into flattish ribbonlike structures called *microfibrils.*

Hemicelluloses *Hemicelluloses,* which are not very similar chemically to cellulose, also are found in the cell wall. They are insoluble in water but soluble in alkali. They are hydrolyzed more readily than cellulose by both alkali and dilute acid. The ease of hydrolysis of the hemicelluloses by alkali is responsible for the mushiness of vegetables heated in water containing baking soda. The amount of hemicellulose in vegetables is reduced by cooking (Kung, 1944).

Hemicelluloses vary from plant to plant and are not completely defined in all cases. They are polysaccharides, among which are found xylans, galactans, mannans, glucomannans, and arabinogalactans. The hemicelluloses from apples, tomatoes, pears, and citrus fruits contain xylose. Other residues, such as D-glucuronic acid or L-arabinose may be attached to a xylan chain. The cell walls of many fruits contain mannans (Isherwood, 1970).

Lignin Another important constituent of some cell walls is *lignin.* Wood contains large quantities of lignin, as much as 50% of its dry weight. Other plant material, including certain vegetables that are mature and rather firm, also contains some lignin. All three layers of the cell wall may contain lignin. Once deposited in the cellulose framework of the cell wall, it is not reutilized by the plant. It resists the action of chemicals, enzymes, and bacteria. Therefore, as expected, it does not change during cooking (Kung, 1944).

The term *lignin* covers a group of related compounds that are not carbohydrates. Lignin molecules are polymers of phenylpropene derivatives and are found in several forms (Brown, 1961).

Gums Another group of cell wall carbohydrates that may be present is composed of the *gums.* Formation of the gums seems to be stimulated by the presence of microorganisms or by the occurrence of disease or mechanical damage to the cell. Generalizations regarding the composition of these polysaccharides are difficult. They may be composed of a mixture of several sugars or sugar derivatives, including D-galactose, L-arabinose, L-rhamnose, D-mannose, D-xylose, D-glucuronic acid, and others (Isherwood, 1970). Gums may swell to many times their original volume in water. Therefore, gums such as gum arabic, gum karaya, and gum tragacanth are used frequently in formulated foods.

Pectic substances Pectic substances are probably present in all higher plant tissues, although sometimes the quantity is so small that they are detected with difficulty. Apples and the albedo of citrus fruits, which contain an abundance of pectic material, furnish the pectin that is purified and sold commercially. The pectic substances are pectic acid, pectinic acid, pectin, and protopectin.

Pectic acid Pectic acid is the simplest of the pectic substances. The molecule is a polyuronide, composed of galacturonic acid units combined by α-1-4-glycosidic linkages. The formulas for the sugar α-D-galactose and galacturonic acid, the uronic acid derived from it, are:

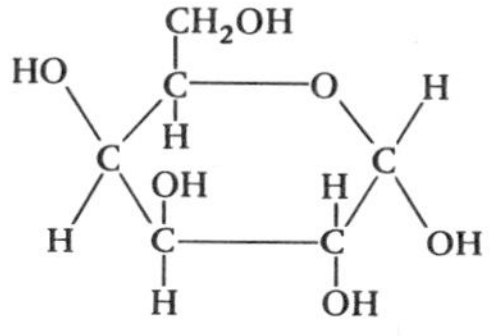

α-D-galactose

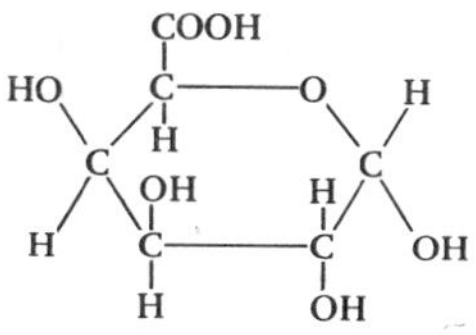

α-D-galacturonic acid

Pectic acid is soluble in water and contains an abundance of carboxyl (—COOH) groups, making it acidic and capable of salt formation. The molecular weights of pectic acid and the other pectic substances vary widely according to the source of the substance and the methods used for isolation and molecular weight determination. Molecular weights of 30,000 to 300,000 have been reported in the literature (Pilnik and Voragen, 1970).

Pectinic acids The pectinic acids are similar to pectic acid except that some of the carboxyl groups are esterified with methyl groups. Like pectic acid, pectinic acids are capable of reacting with metallic ions to form salts. Indeed, many of the pectic substances exist in plants as calcium or magnesium salts. The term *pectin* designates those pectinic acids that are capable of forming jelly with sugar and acid. Pectins are esterified to varying degrees, with the remainder of the carboxyl groups present uncombined or combined to form salts.

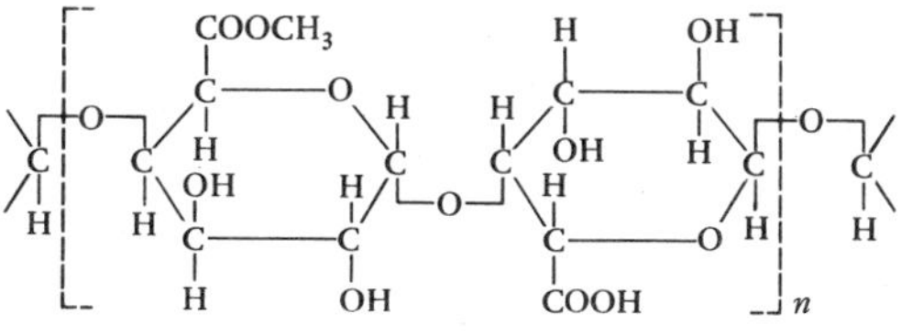

Pectin

Protopectin The nature of protopectin is not fully understood. It is a large molecule from which pectin is formed by restricted hydrolysis. It is present in the cell wall but is probably not chemically associated with cellulose, as some workers have believed. In contrast to the other pectic substances, it is insoluble in water.

Changes during ripening Changes in the textural characteristics of fruits occur during ripening and overripening. These changes are related to changes in the proportions of the pectic substances. For example, during the normal ripening that occurs in the early storage period as apples are stored at 4°C, soluble pectin increases at the expense of protopectin, while hardness decreases (Ulrich et al., 1954). The importance of the proportions of pectic substances to the texture of fruit has been shown in studies on peaches. As tissue firmness decreased with ripening of several varieties of freestone peaches, pectinic acid content increased from initial concentrations of 24–50% to a final concentration of over 70% of the total pectic substances, pectic acid fluctuated at levels of less than 10%, and protopectin decreased significantly (Shewfelt et al., 1971). Ripe melting-fleshed peaches contain lower molecular weight pectins (Chang and Smit, 1973). The enzyme polygalacturonase was shown to be active during the ripening period, when firmness of Elberta peaches decreased rapidly. The decrease in firmness was associated with a decrease in the molecular weight of the pectin resulting from depolymerization of the polygalacturonide chains (Pressey et al., 1971). Hinton and Pressey (1974) more recently reported that cellulase also may be important in ripening because it also increases in activity during ripening.

Changes during cooking and processing Changes in the proportions of pectic substances during the steaming of carrots and parsnips were studied by Simpson and Halliday (1941). The changes resembled those occurring during ripening of fruit in that pectin increased at the expense of protopectin, and total pectic substances decreased, suggesting a degradation of pectin. Hughes and coworkers (1975) reported that potatoes reach a cooked stage when a specific amount of cell wall pectic material has been solubilized. This solubilization decreases adhesion between cells.

Heat processing of fruits and vegetables is an important method of preservation. Studies have been conducted to determine techniques for minimizing the decrease in firmness that occurs with heating. As the pectic substances in cell walls break down, softening of the wall and subsequent cell separation occur. The presence of divalent ions increases firmness of canned fruit (Deshpande et al., 1965), canned tomatoes (Hsu et al., 1965), and cooked carrots (Sterling, 1968). The divalent ions form cross-links between carboxyl groups of pectinic acid molecules, resulting in increased rigidity of the middle lamella and primary cell wall. The presence of monovalent ions prevents cross-link formation and results in decreased firmness (Sterling, 1968). The enzyme pectin methyl esterase (PME) may serve an important role in the firming of tissues with divalent ions. PME in potatoes is active in the range of 50–70°C and catalyzes the removal of methyl groups from pectin molecules, increasing the number of free carboxyl groups. As plant tissue is heated, intracellular ions such as calcium and magnesium may come in contact and react with cell wall components such as the free carboxyl groups to form bridges that strengthen the tissue so that it may resist degradation during heating (Bartolome and Hoff, 1972). A like role of PME in firming of canned tomatoes (Hsu et al., 1965) and in preheating firming of potatoes (Reeve, 1972) has been described. Another important factor in loss of firmness during heating is the pH of the heating medium. As pH increases from 3 to 8, the firmness of carrots decreases because of increased cell separation (Sterling, 1968). Tomato juice consistency was higher if the pH was reduced to below 3 before heating of the tomatoes. The effectiveness of the added acid was attributed to its inhibitory action on pectinolytic enzymes (Wagner et al., 1969). Recently, Davis et al. (1976) reported that the effect of heating varies with the type of tissue, such as phloem and xylem. This may be attributable to differences in the amount and kinds of chemical components of tissues.

Types of Cells and Tissues

The preceding discussion was concerned with cells and plant tissue in general. A plant contains several types of cells combined into tissues with very specific functions. Extensive discussion of all cells and tissues is beyond the scope of this book. However, some understanding of this topic is essential to comprehension of research literature on plant foods. Therefore, cells and tissues of importance to this field of study are briefly described below. For further information, the student is referred to a basic plant anatomy book (Esau, 1965; Fahn, 1974).

Cells form tissues that serve several roles in plants. Mechanical protection is provided by several types of cells constituting the *epidermis.* A cuticle or layer of fatty material sometimes is deposited on the epidermis. The bloom of some fruits such as grapes (Cutter, 1969) or blueberries (Powers et al., 1958) is caused by deposits of wax on the surface of the cuticle. In some plant structures the outer protective layer is the *periderm.* For example, the outer skin of a potato is peridermic tissue (Hsu and Jacobson, 1974). *Parenchyma* cells are components of the tissues of several plant structures including the cortex or tissue located between vascular and dermal tissues, leaves, vascular systems, and secretory or excretory structures. Functions of parenchyma include photosynthesis, storage, and wound healing. *Collenchyma* cells have thicker walls than parenchyma cells and function as support tissue for growing organs as well as in photosynthesis. *Sclerenchyma* tissue is hard tissue containing sclereids, cells with heavily lignified walls. Stone cells are sclereids and are responsible for the gritty texture of some pears (Cutter, 1969). *Xylem* and *phloem* tissues are components of the vascular systems found throughout a plant. Functions of these tissues include conduction of water (xylem), conduction of food (phloem), storage of food, and support. Parenchyma and sclereid cells are located within these tissues. Tracheid and vessel members are water-conducting cells, whereas sieve cells and sieve tube members, are food-conducting cells.

The characteristic texture of a fruit or vegetable depends on the presence and the relative proportions and arrangement of the various types of cells. For example, raw celery is fibrous because of the arrangement of collenchyma, sclerenchyma, and xylem (Reeve, 1970). Additional discussion of structure in relation to texture is presented by Reeve (1970).

COLOR

Much of the appeal of fresh fruits and vegetables is in their varied colors. The attractive color of the raw fruit or vegetable is, however, subject to changes under various conditions associated with use, often resulting in unattractiveness. In this section the nature of the pigments and factors affecting changes in the pigments are considered.

Chlorophyll

The green pigment of plants, chlorophyll, is contained in the chloroplasts, which are present in many parts of a plant. Chlorophyll is present in the leaves because their large surface area is ideal for absorption of the sun's rays and the exchange of gases necessary for photosynthesis. *Photosynthesis* is a process by which CO_2 and water are converted to sugar. In the process, light energy is converted to chemical energy, which makes possible both plant and animal life.

The formula for chlorophyll is shown in Figure 6.2. The structure was confirmed by complete synthesis (Woodward et al., 1960) and is similar to that of heme, shown in Figure 5.1. Both have four pyrrole rings, connected to form a porphyrin nucleus. Resonance of the conju-

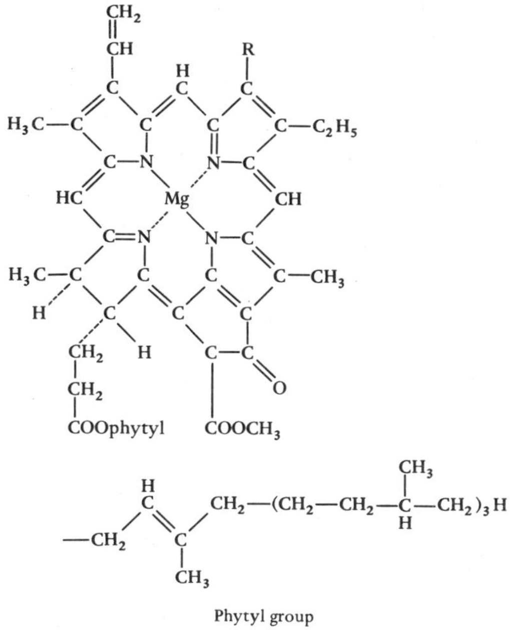

Phytyl group

Figure 6.2. Chlorophyll. In chlorophyll a, R is $-CH_3$; in chlorophyll b, R is $-CHO$.

gated double bonds is responsible for the color of chlorophyll. Chlorophyll contains a nonionic magnesium atom in the center of the porphyrin ring. Two ester groups, one phytyl and one methyl, are part of the molecule. The presence of the phytyl group is responsible for the insolubility of chlorophyll in water. The insolubility of chlorophyll in water explains why it cannot be dissolved in cell sap of the plant or in cooking water. Chlorophyll is soluble in fat and solvents such as ethyl ether, ethanol, acetone, chloroform, carbon disulfide, and benzene.

All higher plants and most lower plants contain two types of chlorophyll, a and b, in the ratio of about three parts of chlorophyll a to one part chlorophyll b. As indicated in Figure 6.2, a formyl group (—CHO) is substituted in chlorophyll b for a methyl group (—CH_3) of chlorophyll a. Chlorophyll a is blue-green in color, whereas chlorophyll b is yellow-green.

Chemical reactions Chlorophyll is a complex molecule and can participate in several reactions. The magnesium of chlorophyll is easily replaced by two hydrogen atoms in the presence of mild acids such as oxalic or acetic to produce a compound called *pheophytin.* Pheophytin a is a grayish-green and pheophytin b is a dull yellowish-green. The mixture results in an olive-green to olive-brown color. The effect of this reaction is seen in green vegetables that have been canned or overcooked. If copper is present when acid is released during heating, the metal replaces magnesium, forming a bright green derivative of chlorophyll. Pickles boiled in a copper kettle have a bright green color for this reason, but cannot be used because the amount of copper they contain may be toxic.

The derivatives formed by the removal of the phytyl group by the enzyme chlorophyllase or by alkali are water soluble, which accounts for the greenish color sometimes seen in water that has been used for cooking green vegetables. Chlorophyllase activity results in formation of phytyl alcohol and chlorophyllide, a green, water-soluble derivative of chlorophyll. This enzyme occurs in many vegetables at varying levels. Spinach is rich in chlorophyllase, but the amount changes both with the season and with the variety (Weast and Mackinney, (1940). Some vegetables such as snap beans may not contain chlorophyllase (Jones et al., 1963). Chlorophyllase is more resistant to heat than many enzymes. It is quite active in water as hot as 66–75°C, but is destroyed by boiling (Clydesdale and Francis, 1968). Chlorophyllase also removes the phytyl group from pheophytin to form pheophorbide, which is similar in color to pheophytin. Pheophorbide may be formed during the brining of cucumbers (White et al., 1963) and blanching of okra, cucumber, and turnip greens at 82°C (Jones et al., 1963). The enzyme is rapidly inactivated in those vegetables at 100°C and in spinach at 84.5°C (Resende et al., 1969). Treatment of chlorophyll with alkali removes both the phytyl and methyl groups by saponification, thus forming chlorophyllin and two alcohols, methanol and phytol. Chlorophyllin is green and water soluble.

Changes during heating The changes that occur in green pigments of vegetables can be related to the properties of chlorophyll. In the raw vegetable, chlorophyll is protected from acid in the cell sap by its location in the chloroplasts. The first color change observed on dropping a green vegetable into boiling water is a brightening of the green color, probably caused by expulsion of air and collapse of the intercellular spaces (Mackinney and Weast, 1940). During cooking, the chloroplasts become shrunken and clumped in the center of a mass of coagulated protoplasm. The chlorophyll remains in the chloroplasts but is no longer protected by the plastid membranes from the acid-containing cell sap (Mackinney and Weast, 1940). Consequently, the dull olive-green pheophytins may be formed.

The extent of the color change depends on the acidity of the cooking medium, the pH of the vegetable, the chlorophyll content, and the time and temperature of cooking. Organic acids are produced in the plant primarily as intermediates of the Kreb's tricarboxylic acid cycle. Large quantities of acids, predominately citric and malic, may accumulate in the vacuoles of the cells of fruits or vegetables. The proportion of each differs from plant to plant. The pH of all common vegetables (AHEA, 1975) is less than 7 because of the acids that are present in the cell sap.

Both volatile and nonvolatile acids are released during the cooking of vegetables. Therefore, measurements of the pH of a distillate collected as vegetables are cooked and that of the cooking water before and after cooking may be made in vegetable cookery studies (Halliday and Noble, 1943). The amount of nonvolatile acid remaining in the water used for cooking cabbage has been found to exceed the amount of volatile acid in the distillate (Masters and Garbutt, 1920). The color of green vegetables is sometimes better if the vegetable is cooked without a cover so that the volatile acids can escape from the pan.

The acids released from the cell vacuole during heating cannot affect the color of vegetables if they are neutralized by the cooking water. The pH of water may be decreased by boiling if carbon dioxide (which has dissolved in the water to form carbonic acid) is released, if bicarbonates are changed to carbonates with the release of carbon dioxide, or if hydrogen sulfide is lost. Water that is heavily chlorinated may become more acidic on boiling due to the formation of hydrochloric acid. Several of these reactions may take place simultaneously. In food experimentation, the source and treatment of water should be controlled carefully to avoid variation in results due to impurities or variation in pH.

The amount of acid that can be neutralized by an alkaline cooking water depends on the water's alkalinity and volume. Large amounts of water may result in a desirable green color because they neutralize or at least dilute plant acids. To study the effects of pH, frozen green beans, peas, Brussels sprouts, broccoli, lima beans, and spinach were cooked in buffers of varying pH (Sweeney and Martin, 1961). As the pH increased from 6.2 to 7.0, the green color of the cooked vegetables improved, but buffers of pH greater than 7 caused marked deterioration in flavor and little further improvement in color. Baking soda sometimes is added to increase the alkalinity of cooking water, but its use usually is not recommended because it is difficult to avoid adding an excess of soda. Sodium bicarbonate not neutralized by the acids in the cooking water adversely affects the flavor and texture of the vegetable. The use of other additives has been studied in attempts to retain the color of cooked green vegetables. The addition of a small quantity of a mixture of calcium carbonate and magnesium carbonate to the cooking liquid resulted in an increase in the pH of green beans held on a steam table after heating. The increased pH reduced conversion of chlorophyll to pheophytin and increased preference scores for the beans (Sweeney, 1970). Chlorophyll retention also was improved by the use of ammonium bicarbonate in small volumes of water in the cooking of fresh and frozen green beans, Brussels sprouts, broccoli, and kale, and in fresh cabbage. Some of the undesirable textural changes associated with the addition of ammonium bicarbonate were prevented by the addition of calcium acetate (Odland and Eheart, 1974). Color retention is not as difficult in frozen as in fresh vegetables. The blanching process facilitates loss of the acids so that the pH of the vegetable is increased.

However, retention of color during blanching is a problem. The use of ammonium bicarbonate improved chlorophyll retention when green vegetables were blanched in water (Eheart and Odland, 1973a), in steam (Odland and Eheart, 1975), or by microwaves (Eheart and Odland, 1973b).

The susceptibility of green vegetables to color change during cooking is affected by their chlorophyll content and their pH, as indicated previously. Frozen vegetables containing more chlorophyll initially retained a higher percentage when cooked, and those of comparatively high pH, spinach and peas, retained more pigment than green beans and Brussels sprouts, which had lower pH levels (Sweeney and Martin, 1961).

Color is best if green vegetables are heated through as rapidly as possible and cooked for only a short time. The conversion from chlorophyll to pheophytin is so rapid that from 50 to 75% of the chlorophyll is likely to be lost during the usual cooking period. No chlorophyll remained in snap beans after 1 hr, whereas 62% remained after 10 min and 28% after 20 min (Mackinney and Weast, 1940). Percentages of chlorophyll retained in broccoli cooked for 5, 10, and 20 min were 82.5, 58.9, and 31.3 respectively (Sweeney and Martin, 1958). Much of this loss is of chlorophyll a. Chlorophyll b has been found to be relatively stable during cooking (Sweeney and Martin, 1961).

The dramatic change of color of green vegetables during the canning process also is a result of conversion of chlorophyll to pheophytin (Gold and Wickel, 1959).

Carotenoids

Most of the yellow and orange colors and some of the red colors of fruits and vegetables are due to the carotenoids that are located in the chromoplasts of the cells. In addition to their occurrence as the sole or predominating pigments in some fruits and vegetables, they always accompany chlorophyll in the ratio of about three or four parts chlorophyll to one part carotenoid. The carotenoids are insoluble in water and soluble in fats and organic solvents. Various members of this large group of related compounds differ in solubility characteristics, offering a convenient method of separating the carotenoids into two groups.

The carotenes contain only hydrogen and carbon and are soluble in petroleum ether. Xanthophylls, oxygen-containing carotenoids, are soluble in alcohol.

Structure Carotenoids usually contain 40 carbon atoms in configurations such as those for β-carotene, α-carotene, and lycopene shown in Figure 6.3. The conjugated double bonds are responsible for the intense color of foods containing carotenoids. Although many different *cis* and *trans* configurations of the structure are possible, most naturally occurring carotenoids appear to have an all *trans*configuration (Zechmeister and Deuel, 1952). As the number of double bonds increases, the hue becomes redder. Lycopene, with two additional double bonds, is redder than β-carotene. A decrease in the number of conjugated double bonds increases yellowness. Consequently, α-carotene is less orange than β-carotene.

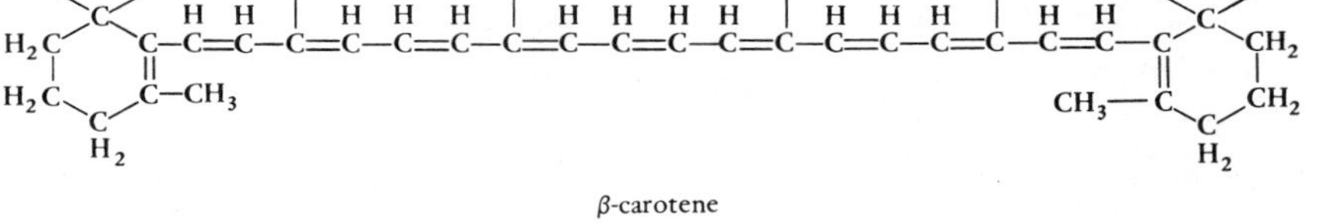

β-carotene

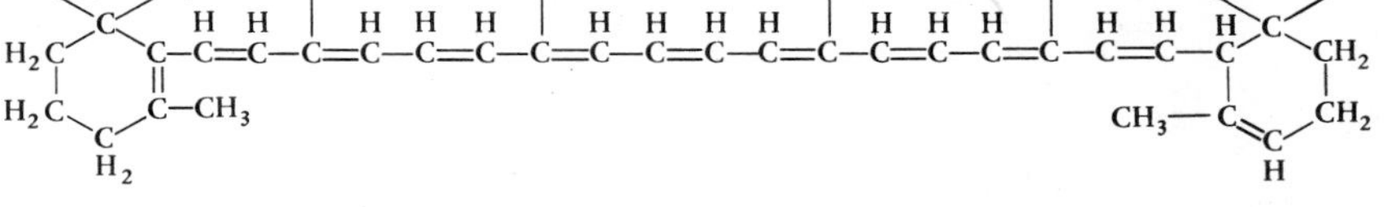

α-carotene

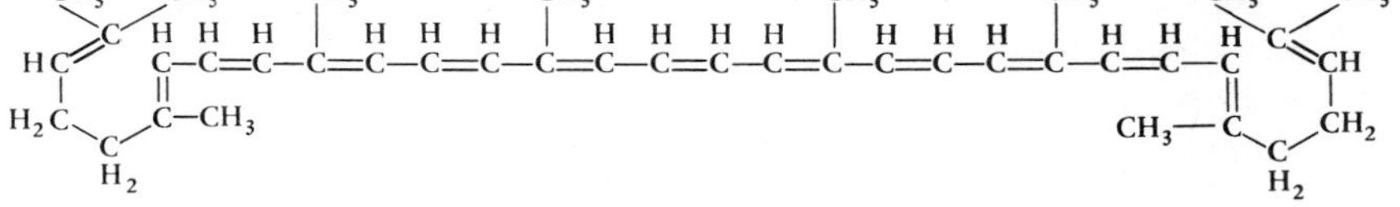

Lycopene

Figure 6.3. Structures of β-carotene, α-carotene, and lycopene.

Lutein and zeaxanthin are xanthophylls that are similar in structure to α- and β-carotene except for the addition of two hydroxyl groups. Cryptoxanthin contains only one hydroxyl group and is an important pigment in yellow corn, mandarin oranges, and paprika.

In addition to their contributions to the appealing colors of fruits and vegetables, most carotenoids are precursors of vitamin A. The symmetrical β-carotene molecule forms two molecules of vitamin A. Carotenoids, such as α- and γ-carotene, in which half of the molecule is exactly like half of the β-carotene molecule, form one molecule of vitamin A. Lycopene has an open ring rather than the closed ring structure that characterizes β-carotene and is therefore not a precursor of vitamin A. Most xanthophylls also are not precursors of vitamin A; the exception is cryptoxanthin, which has only one hydroxyl group.

Changes during preparation, processing, and storage Ordinary cooking methods have little effect on the color or nutritive value of carotenoids. The pigments are little affected by acid, alkali, the volume of water, or the cooking time. Eheart and Gott (1964) reported complete retention of the carotene in peas heated in water in a conventional oven and with or without water in a microwave oven. Although total carotene content does not change as vegetables are heated in water, there is a shift in the visualized color. For example, the orange of carrots may become yellow and the red of tomatoes may become orange-red. Borchgrevink and Charley (1966) attributed this reduction in color intensity in carrots cooked in several ways to an increase in *cis* isomers of β-carotene during cooking. Carrots that appeared to be orange-red had a higher concentration of all *trans*-β-carotene than those carrots that appeared yellow. However, DellaMonica and McDowell (1965) reported that color changes cannot be explained in terms of isomerization. Shifts in the color of carrots are caused also by the solution of carotene in cellular lipids after release from disintegrated chromoplasts (Purcell et al., 1969). Crystals of lycopene are formed in tomatoes during heating so that shift in color is not as pronounced in tomatoes as in carrots and sweet potatoes.

Noble (1975) reported that a conversion of *trans*-lycopene to *cis*-lycopene was not responsible for the loss of redness that occurs as tomato paste is concentrated. This loss was attributed to an actual degradation of lycopene during the extended heating process.

The high degree of unsaturation of the carotenoids makes them susceptible to oxidation, with resulting loss of color, after the food containing them has been dried. Loss of or reduction in color is probably a result of the reaction of peroxides and free radicals, oxidation products of lipids, with the carotenoids (Labuza, 1973). Precooked, dehydrated carrot flakes canned in a nitrogen atmosphere lost little β-carotene, whereas flakes canned in an oxygen atmosphere lost approximately 75% of their β-carotene content (Stephens and McLemore, 1969). Carotene can be protected from oxidation during dehydration by the blanching of vegetables (Dutton et al., 1943) or the sulfuring of fruit.

Flavonoids

Although similar in chemical structure, the members of the flavonoid group are diverse in properties. They include two major groups of related compounds, the anthocyanins and the anthoxanthins. All are phenolic compounds, and possess a basic structure consisting of a $C_6—C_3—C_6$ skeleton.

Two aromatic rings are joined by an aliphatic three-carbon chain. Hydroxyl (—OH), methoxyl ($—OCH_3$), or sugar groups are attached at various points on the skeleton. Most naturally occurring flavonoids are glycosides, meaning that a sugar moiety is present. The combinations of and location of added groups influence the color of the pigment.

The flavonoids are widely distributed in plants. Within the cell, these water-soluble compounds are dissolved in the cell sap in the vacuole. The chemistry of flavonoids is complex. However, a basic knowledge of the chemistry is fundamental to a complete understanding of fruit and vegetable colors.

Anthocyanins Most of the red, purple, and blue colors of fruits and vegetables are caused by anthocyanin pigments. In some fruits, such as some varieties of cherries, apples, and plums, the anthocyanins occur in the cells of the skin but not in those of the flesh. Processing the fruit distributes the anthocyanins throughout the tissues, as demonstrated histologically for canned cherries (Griswold, 1944). Other fruits and vegetables containing anthocyanins include raspberries, blueberries, grapes, strawberries, blackberries, peach skins, red potato skins, radishes, eggplant, and red cabbage. The anthocyanin content of many fruits and vegetables has not been completely identified (Fuleki, 1969). Even so, over 140 naturally occurring anthocyanins have been identified (Francis, 1975).

Structure Anthocyanins contain hydroxyl groups at positions 3, 5, and 7 and are glycosides, in contrast to the corresponding sugar-free anthocyanidins, which are rare in nature (Blank, 1947). The sugar moiety usually is attached to the hydroxyl at position 3. Sugars found in the glycosides may be glucose, rhamnose, galactose, arabinose, fructose, and xylose. The sugar portion is responsible for the solubility (Shirkhande, 1976). Substitutions on ring B result in formation of various anthocyanins. The formula for one common anthocyanin, cyanidin, is shown below:

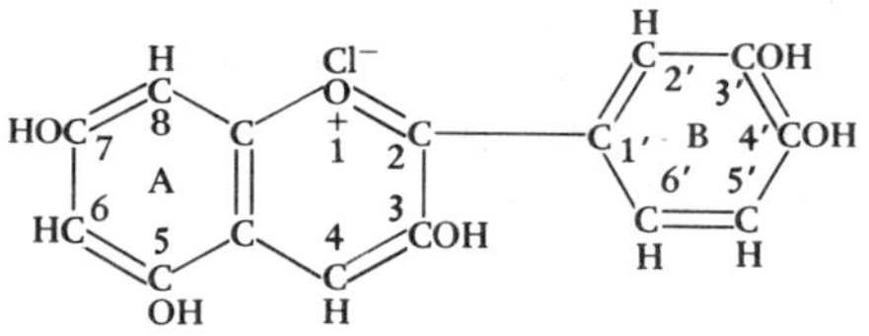

Cyanidin chloride

Derivatives of cyanidin, such as the corresponding glycoside, cyanin, occur in apples, cherries, cranberries, currants, elderberries, purple figs, peaches, plums, raspberries, rhubarb, and purple turnips. Cyanidin has two hydroxyl groups attached to the phenyl ring in positions 3′ and 4′. Pelargonidin, a strawberry pigment, has a hydroxyl group at position 3′ and delphinidin, a pigment found in pomegranate, blueberries, and eggplant, has three hydroxyl groups that are located in the 3′. 4′. and 5′ positions. Other related anthocyanins have methoxyl groups in place of one or more of the hydroxyl groups of the molecule. The color of the anthocyanin varies with variations in molecular structures. As hydroxylation increases, blueness increases. Thus delphinidin is bluer than pelargonidin (Bate-Smith, 1954). As methylation of hydroxyl groups increases, redness increases (Shirkhande, 1976).

Color of the anthocyanins also is influenced by the presence of other phenolic compounds. Other flavonoids form complexes with the glycosides of commonly occurring anthocyanins at pH levels ranging from 2 to 5. This interaction increases the blueness of the anthocyanins (Asen et al., 1972). Carotenoids or chlorophyll may be found in tissue containing anthocyanins.

Changes in preparation and storage Anthocyanins are electron deficient. Therefore, they are very reactive and may undergo detrimental changes during processing and subsequent storage. Color varies with the pH. The molecule assumes the configuration of a cation in acid and is red in color. The color is most intense at very low pH values (Francis, 1975). Lowering the pH of anthocyanin solutions also decreases the lightness as measured by a color difference meter (Van Buren et al., 1974). The molecule is uncharged at a neutral pH and is violet in color. In an alkaline medium, the anionic form exists and results in a blue color. The juice of some fruits and vegetables becomes greenish as alkali is added. This color probably is caused by the presence of flavones or flavonols with the anthocyanins. With the addition of alkali, the flavones or flavonols turn yellow while the anthocyanins turn blue, and a mixture of the two colors appears green. Such a color change can be seen in red cabbage.

Anthocyanins may be degraded by oxidation, hydrolysis, or polymerization (Francis, 1975). Factors influencing degradation include temperature, pH, other cell constituents, enzymes, and the presence of metals. Loss of pigment may be a serious problem in strawberry products such as preserves and grape juice. Furthermore, loss may be very rapid. For example, in strawberry preserves, the half-life of red pigment is about 8 wk at 20°C (Francis, 1975). At higher temperatures, the stability is reduced. Therefore, consideration of storage temperature is important.

Destruction of the pigment increases as the pH increases, as shown in frozen strawberries (Wrolstad et al., 1970) and Concord grape juice (Skalski and Sistrunk, 1973).

Hydrolysis of the 3-glycosidic group may be caused by an enzyme. The anthocyanidin is unstable to oxidative degradation. Phenolases may be indirectly involved through enzymatic oxidation of catechol to a quinone that oxidizes the anthocyanin. The involvement of enzymes has not been described completely (Jurd, 1972). The products of ascorbic acid oxidation, as well as sulfur dioxide, and furfural and hydroxymethylfurfural, products of sugar degradation, exert a detrimental effect on the anthocyanins (Shirkhande, 1976).

Metal salts may have a stabilizing effect on color of products containing anthocyanins. The addition of stannic and stannous salts and of aluminum chloride stabilized the color of strawberry puree (Sistrunk and Cash, 1970). Wrolstad and Erlandson (1973) reported that this stabilization of color was not due to stabilization of the anthocyanins. Formation of an insoluble colored complex with another compound may be responsible.

The adjacent hydroxyl groups attached to the B ring of cyanidin and delphinidin may form complexes with metals that are very stable at a high pH. The complexes formed may be green, slate, or blue in color. Aluminum chloride complexes with cyanidin-3-glucoside at a pH of 3.0–3.5 to shift color from red to blue-violet (Asen et al., 1969).

Enamel-lined cans are used for foods containing anthocyanins. If an unlined can or a can with an imperfection in the enamel lining is used, the acid of the fruit reacts with the metal of the can to form a salt. The anthocyanin then combines with the metal ion, thus releasing the acid to continue its attack on the can (Culpepper and Caldwell, 1927). After the tin has been removed in this way, underlying iron enters into a similar reaction. A pinhole is formed that may make microbial invasion of the contents possible.

Because of their acidity, fruits containing anthocyanins usually do not suffer undesirable color changes in cooking. Red cabbage is the anthocyanin-containing vegetable most often cooked. When steamed or boiled in tap water, red cabbage turns a bluish color that many people consider unpleasant. The color can be changed to an attractive reddish color by the addition during cooking of acid in some form, such as vinegar or apples.

Anthoxanthins *Anthoxanthins* is a term given to a group of compounds including flavones, flavonols, and flavonones. Skeletal structures for the three are shown in Figure 6.4. The sugar moiety of the glycosidic form usually is attached on the A ring at position 7, as shown in the structure diagram.

The pigments of this group are almost colorless or pale yellow. Like the anthocyanins, they are water-soluble and occur in the vacuoles of plant cells. They may occur alone in light-colored vegetables such as potatoes and yellow-skinned onions, or with other pigments such as anthocyanins. They are so widely distributed that it is exceptional to find a plant in which anthoxanthins are not present. They frequently occur in complex mixtures. The formula for the flavonoid quercetin, which is probably the most widely distributed pigment of this group, is shown below:

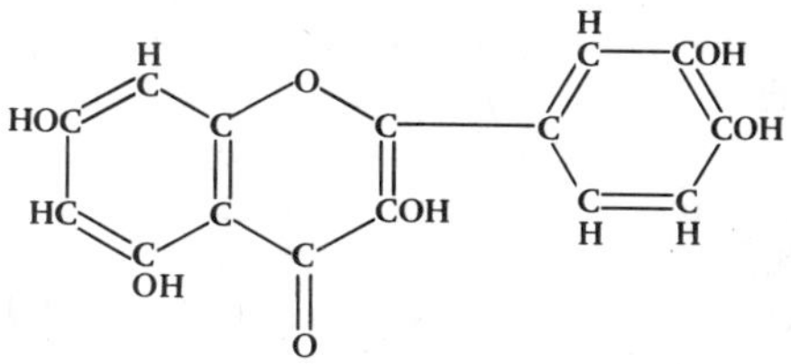

Quercetin

Comparison of this formula with that of cyanidin shows that the hydroxyl groups are similarly located but that quercetin is more highly oxidized than the related anthocyanidin. With its derivatives, quercetin occurs in asparagus, yellow-skinned onions, grapes, apples, the rind of citrus fruits, and tea. The anthoxanthins turn yellow in the presence of alkali. They also have the ability to chelate metals, which also may result in discoloration. Iron salts cause a brownish discoloration. Traces of iron salts in alkaline tap water may react with anthoxanthins and other closely related phenolic compounds to produce some of the yellow-to-brown discoloration often seen in cooked white vegetables. Yellow-skinned onions cooked in alkaline water are especially likely to show such a change. Similar results are seen if an aluminum pan is used. The color of white vegetables often can be improved by adding an acidic compound such as cream of tartar to the cooking water, but at the expense of some firming of the tissues.

Blackening of potatoes after cooking is attributed to formation of a dark-colored complex between iron and chlorogenic acid in potatoes with low organic acid content. Organic acids such as citric acid chelate metals so that they are not available to react with chlorogenic acid (Heisler et al., 1964). Cauliflower may become discolored because flavonol glycosides complex with ferrous or stannous ions. Thus lacquered cans are used for cauliflower (Chandler, 1964). Asparagus contains a flavonol, rutin, which precipitates as yellow crystals from the liquid of asparagus canned in glass containers. This precipitate is not

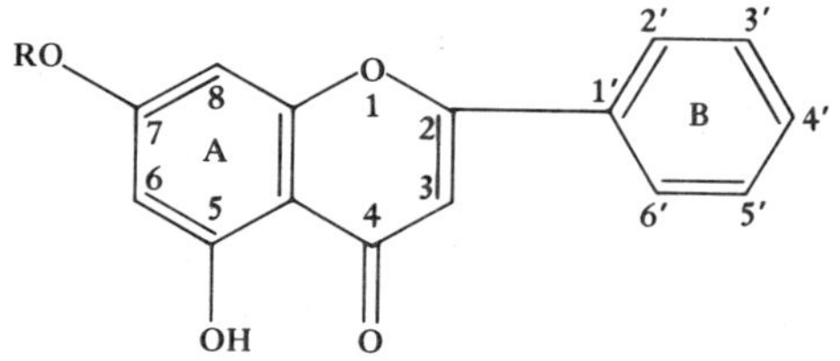

RO, OH, O, 1, 2, 3, 4, 5, 6, 7, 8, A, B, 1′, 2′, 3′, 4′, 5′, 6′

Flavone

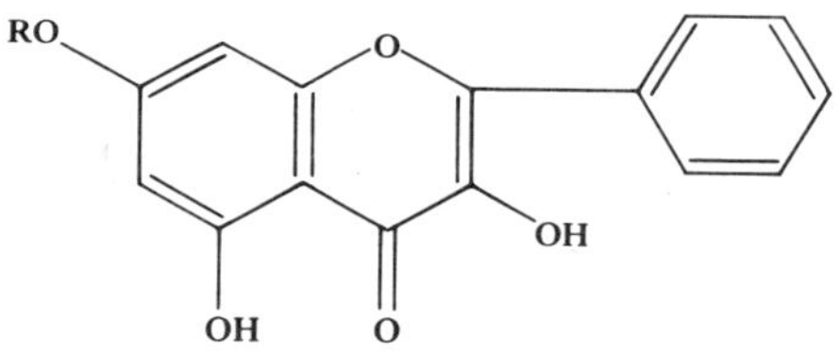

RO, OH, O, OH

Flavonol

RO, OH, O, H, H, H

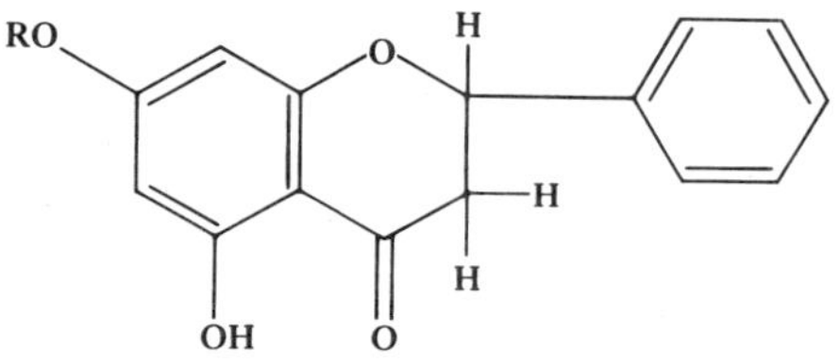

Flavonone

Figure 6.4. Structures of flavone, flavonol, and flavonone.

found in asparagus canned in tin containers because rutin forms a light yellow complex with stannous ions. The complex is more soluble than rutin so precipitation does not occur (Dame et al., 1959). Sweet potatoes form yellow complexes with tin and dark greenish complexes with iron. Use of tin cans rather than enamel-lined cans with sweet potatoes provides tin to compete with the iron for complex formation and prevents darkening (Scott et al., 1974; Twigg et al., 1974).

Leucoanthocyanins Another group of flavonoid compounds important to the color of some canned foods is composed of *leucoanthocy-*

anins, colorless phenolic compounds sometimes referred to as *proanthocyanins.* These compounds are converted to anthocyanin when subjected to boiling hydrochloric acid. Some products may become pink or red as a result of this conversion. The pink color that sometimes develops in canned pears was identified as a tin-cyanidin complex. The compound forming the complex with tin was formed by the oxidation of leucoanthocyanins (Chandler and Clegg, 1970). Leucoanthocyanins also may be responsible for the pinking of sauerkraut (Gorin and Jans, 1971) and applesauce (Singleton, 1972).

Betalains

The *betalains* are responsible for the color of the red beet (Mabry and Dreiding, 1968). This group includes the betacyanine or red pigments and the betaxanthin or yellow pigments. The major red pigment is betanin. Its structure is shown below:

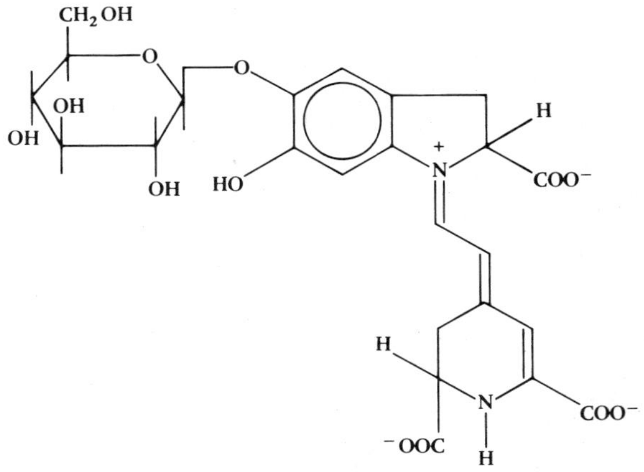

Betanin

The color of betanin depends on pH. The red color is associated with a pH of 4 to 7. At a pH level below 4, the color shifts to violet. Above pH 10, the color shifts to yellow (von Elbe et al., 1974a; von Elbe,

1975). With the banning of Red No. 2, a previously widely used red food coloring, the betalains have been the subject of much study as potential coloring agents (von Elbe et al., 1974b; Pasch et al., 1975; von Elbe, 1975).

Enzymatic Browning

Phenolic compounds other than those previously discussed may be important in the color of fruits and vegetables, not because of their contribution to color but because of their role in discoloration. The term *tannins* is used frequently for a group of polyphenolic compounds that serve as substrates for enzymatic browning and also contribute astringency to foods (Bate-Smith, 1954). These compounds serve as the substrates for the enzyme ortho-diphenol:oxygen oxido-reductase. The enzyme is more commonly known as polyphenol oxidase or polyphenolase. Substrates for this enzyme are diphenolic compounds, including catechin as shown below:

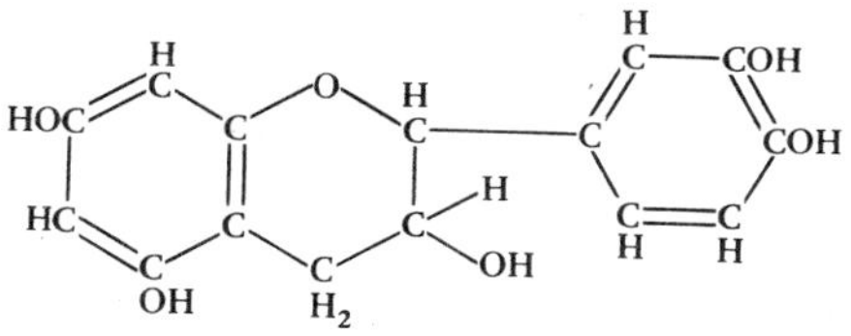

Catechin

Others include leucoanthocyanins, chlorogenic acid, caffeic acid, dicatechol, tyrosine, dihydroxyphenylalanine, and many others (Mathew and Parpia, 1971).

Reaction Several reactions are involved in formation of the yellow and darker-colored compounds associated with enzymatic browning, which occurs rapidly when the fruit or vegetable is cut or otherwise injured. In the first reaction in the series, an *o*-dihydroxyphenolic compound is oxidized in the presence of polyphenoloxidase and atmospheric oxygen to form an *o*-quinone and water.

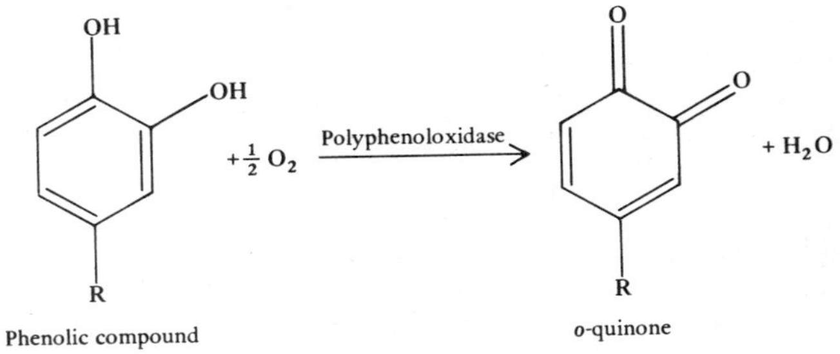

$$\text{Phenolic compound} + \tfrac{1}{2}O_2 \xrightarrow{\text{Polyphenoloxidase}} o\text{-quinone} + H_2O$$

Reaction for first step in browning

The o-quinones are polymerized to form the complex, colored compounds. The polymerization reactions do not require oxygen or the presence of enzymes (Berk, 1976).

Control Three components, the substrate, the enzyme, and oxygen, are necessary for the initial reaction of browning. Control of the reaction must involve elimination or limitation of one of the components. Elimination of the substrate is a problem for geneticists. The Sunbeam peach, which contains little phenolic compound, does not brown when sliced (Kertesz, 1933). Obviously, attention to control of the other two components is more practical; however, several methods may be used to prevent enzymatic browning. Polyphenol-oxidase is denatured and therefore inactivated by a heat treatment such as blanching. Rapid blanching is important because slow blanching may result in activation rather than inactivation of the enzyme (Tate et al., 1964). However, blanching will give a cooked flavor and soft texture that may not be desirable in fruits that are to be frozen or potatoes that will be used to make potato chips.

If heating is not desirable, other conditions unfavorable to activity of the enzyme can be promoted. Sulfur dioxide prevents browning through inhibition of the enzyme system, as do sodium chloride and organic acids. If susceptible foods are soaked in dilute bisulfite or sodium chloride solutions after peeling and slicing, enzyme activity may be inhibited (Berk, 1976). Acids such as citric, malic, or tartaric may be effective because phenyloxidase activity is greatest at a pH of 7 and decreases as the pH is lowered below 4.

Exclusion of oxygen is another approach to prevention of browning. Holding vegetables in water or a salt solution after slicing and holding fruits in sirup will help to prevent browning by limiting oxygen.

Oxidation to quinones, if the previous measures are not successful, may be reversed, thus preventing polymerization to dark-colored compounds. This possibly is another facet of the protective role of sulfite. Ascorbic acid also acts as a reducing agent in prevention of browning. Its effectiveness persists until the supply is exhausted. The use of these two chemicals on apples prior to refrigeration and freezing and on peaches prior to canning has been effective (Ponting and Jackson, 1972; Ponting et al., 1972; Luh and Phithakpol, 1972). Further discussion of these techniques is found in Chapter 7.

FLAVOR

The characteristic flavor of a fruit or vegetable is probably due to a mixture of many compounds, some of which are present in very small amounts. The volatile flavoring components are important; among them are organic acids, aldehydes, alcohols, and esters.

Phenolic Compounds

Phenolic compounds such as catechin and leucoanthocyanins are responsible for the astringent or "puckery" taste of some foods. Whereas a slightly astringent taste is desirable in tea, coffee, cocoa, and certain fruits, an undesirable degree of astringency can be observed in certain underripe fruits such as persimmons and bananas. Time and temperature of heating during preparation of juice extracts from red currant and red raspberry fruits influence the tannin content of the juice. As tannin content increases, bitterness and astringency scores also increase (Watson, 1973). Another flavonoid compound of significance is *naringin,* the 7-rhamnoglucoside of naringenin, a flavone. Naringin is responsible for the bitterness of grapefruit (Horowitz and Gentili, 1961).

Sugars

Sugars are responsible in part for the superior flavor of freshly harvested vegetables. The loss of sugar during storage was more than twice as fast in sweet corn and peas as in asparagus. It was delayed by refrigeration (Platenius, 1939). Sugars, including glucose and fructose, increase in potatoes stored at refrigerator temperatures (4°C) (Eskin et al., 1971). The metabolic processes of the plant are responsible for changes in concentration of sugar both during ripening and after harvest. Variation in the sugar present, both quantitative and qualitative, contributes to the variation in taste of different fruits (Whiting, 1970). During storage, sugar content may increase or decrease depending on the fruit or vegetable.

Acids

All fruits and vegetables are acidic, a factor of undoubted importance to their taste. Plants vary in the kind of acid that they contain and in the amount, as reflected by pH. Acidity varies with the maturity of the plant, frequently decreasing as fruit ripens. Plants contain various organic acids, some of which are important in the intermediary metabolism of both plants and animals. Individual fruits and vegetables usually contain several acids, with one or more predominating.

For example, malic is most abundant in apples, pears, peaches, lettuce, cauliflower, green beans, and broccoli, whereas citric is most abundant in citrus fruits, tomatoes, leafy vegetables, strawberries, and cranberries. Many other acids, including acetic, butyric, lactic, pyruvic, fumaric, oxalic, and benzoic, are found in some fruits and vegetables at low levels. Organic acids in fruits are listed by Ulrich (1970) and Aurand and Woods (1973).

Sulfur Compounds

Sulfur compounds are known to be important contributors to the flavor of two groups of vegetables—the *Allium* genus of the onion family and the *Brassica* genus of the Cruciferae family. The typical odor of garlic is due to allicin, which is produced by the action of allinase, an enzyme of garlic, on S-allylcysteine sulfoxide (alliin), as shown on page 189 (Stoll and Seebeck, 1951).

$$2\,CH_2{=}CH{-}CH_2{-}\overset{\overset{\displaystyle O}{\|}}{S}{-}CH_2{-}\underset{\underset{\displaystyle NH_2}{|}}{CH}{-}COOH + H_2O \xrightarrow{\text{Alliinase}}$$

Alliin (S-allylcysteine sulfoxide)

$$O{=}S(CH_2{-}CH{=}CH_2){-}S{-}CH_2{-}CH{=}CH_2 + 2\,CH_3{-}C({=}O){-}COOH + 2\,NH_3$$

Allicin Pyruvic acid Ammonia

Reaction of alliin to form allicin, pyruvic acid, and ammonia

The odor of diallyl thiosulfinate is typical of garlic but is not unpleasant. It is an intermediate in the formation of the more volatile, unpleasant-smelling diallyl disulfide ($CH_2 = CH{-}CH_2{-}S{-}S{-}CH_2{-}CH = CH_2$), garlic oil.

Schwimmer and Weston (1961) described the production of volatile sulfur compounds from S-methyl-L-cysteine sulfoxide and S-propyl-L-cysteine sulfoxide by the action of an allinase type enzyme. Hydrogen sulfide and n-propyl mercaptan ($CH_3{-}CH_2{-}CH_2{-}SH$) are produced from raw onions as they are crushed (Niegisch and Stahl, 1956). Onions become increasingly mild as they are boiled because of solution, vaporization, and further degradation of the sulfur compounds. The lachrymatory factor of onions has been identified by Spåre and Virtanen (1963). Propenylsulphenic acid is produced from (+)-S-(prop-1-enyl)-L-cysteine sulfoxide.

The characteristic flavor of raw cabbage is due to allyl isothiocyanate formed by the hydrolysis of sinigrin, a glucoside, and several related compounds (Bailey et al., 1961). Sinigrin is split by the enzyme myrosinase as shown in the equation on page 190.

When dehydrated cabbage is reconstituted with water for slaw, it does not have the flavor of fresh raw cabbage even though it contains sinigrin. If myrosinase is added during rehydration, the flavor of fresh cabbage is restored, indicating that the precursor of the flavor component is not changed by processing, but the enzyme myrosinase is inactivated (Bailey et al., 1961). A study by Schwimmer (1963) involv-

$$CH_2{=}CH{-}CH_2{-}N{=}C\begin{matrix} \diagup OSO_3K \\ \diagdown S{-}C_6H_{11}O_5 \end{matrix} + H_2O \xrightarrow{\text{Myrosinase}}$$

Sinigrin

$$CH_2{=}CH{-}CH_2{-}N{=}C{=}S + C_6H_{12}O_6 + KHSO_4$$

Allyl isothiocyanate ("mustard oil") Glucose Potassium acid sulfate

Reaction of sinigrin and water

ing the use of enzymes from various sources suggested that the products produced from a common precursor differ to give the characteristic flavors. The use of an onion enzyme preparation with dehydrated cabbage gives a flavor similar to that of onions. Dimethyl disulfide, the main constituent responsible for the odor of cooked cabbage, is formed from the sulfur-containing amino acid, S-methyl-L-cysteine sulfoxide. The presence of this precursor has been shown in cabbage, turnip, cauliflower, kale, and white mustard but not in watercress or radish (Synge and Wood, 1956). The amino acid is the precursor of dimethyl disulfide, a volatile sulfur compound (Dateo et al., 1957). Hydrogen sulfide also is formed when cabbage is cooked. These two volatile sulfur compounds were obtained from fresh, dehydrated, and red cabbage as well as from sauerkraut, cauliflower, broccoli (Dateo et al., 1957), and rutabaga (Hing and Weckel, 1964). When members of the cabbage family are overcooked, decomposition of the sulfur compounds continues, as indicated by the finding in one of the earliest studies on sulfur-containing vegetables that the amount of hydrogen sulfide evolved from cabbage and cauliflower increases with the cooking period (Simpson and Halliday, 1928). Maruyamu (1970) suggested that with prolonged cooking, gradual loss of pleasant volatile components occurs, allowing less pleasant sulfur components to predominate. The change in taste that occurs on cooking cabbage is to be contrasted to the decreased intensity of taste that occurs on cooking onions. It furnishes additional evidence of differences in the sulfur compounds involved. Lea (1963) suggested that flavors in fresh, uncooked foods are developed because of enzymatic reactions, whereas nonenzymatic reactions are responsible for flavors in cooked foods.

METHODS OF COOKING VEGETABLES

Of the many methods of vegetable cookery in common use, perhaps the most common is *boiling* in an amount of water that varies from that clinging to the vegetable after washing to more than enough to cover. If the amount of water is less than enough to cover the vegetable, the saucepan is covered during cooking to retain the steam that cooks the vegetable, but if enough water is used to cover the vegetable after it has wilted in the boiling water, it is possible to cook either with or without a cover. In "waterless" cooking, only the water clinging to the vegetable after washing or a minimum amount necessary to prevent scorching is used. A heavy pan with a tight fitting lid usually is recommended for this method. *Panning* is a method in which the finely cut vegetable is cooked with a small amount of fat in a covered pan. *Steaming* over boiling water and *baking* also are used for cooking vegetables. The *pressure saucepan* makes it possible to cook vegetables rapidly in a small amount of water. Microwave heating also permits rapid cooking in a small amount of water.

There are many factors to consider in selection of a cooking method, and the method best for one factor may not be preferred for another. Factors to be considered include texture, color, flavor, and nutritive value.

Studies of the influence of cooking method on the palatability factors—texture, color, and flavor—in general indicate the advisability of cooking vegetables in a small amount of water, which amounts to about half the weight of the vegetable, or 125–250 ml for four servings. Charles and Van Duyne (1954) reported that vegetables cooked in this proportion of water scored higher in appearance, color, and flavor than those cooked in the minimum amount of water required to prevent scorching. Vegetables cooked in a minimum of water may be off-colored and unattractive (Fisher and Dodds, 1952). In a comparison of "waterless" cooking, boiling, and pressure cooking of broccoli, cabbage, cauliflower, turnips, and rutabagas, Gordon and Noble (1964) reported that flavor and color were best and ascorbic acid retention poorest with the boiling-water method, whereas the pressure saucepan retained more ascorbic acid in all vegetables except turnips but with less desirable color and flavor.

When vegetables were cooked in an open pan with enough water to cover the vegetable at all times, the color of green vegetables was greener and the flavor of vegetables of the cabbage family milder than when they were cooked in steam as in a pressure saucepan, steamer, or tightly covered pan (Gordon and Noble, 1959, 1960). Whether a milder or stronger flavor is preferred is, of course, a personal matter.

Fresh broccoli boiled in a moderate amount of water (300 g for 454 g of vegetable) was at least as satisfactory as that cooked in larger amounts of water or by methods using less water, such as steaming or pressure cooking (Sweeney et al., 1959; Gilpin et al., 1959). Results from another laboratory indicate that broccoli, Brussels sprouts, Lima beans, peas, and soybeans were more palatable when cooked in a pressure saucepan, but asparagus, cauliflower, and green beans were preferred when cooked in a tightly covered pan (Van Duyne et al., 1951).

Frozen broccoli can be cooked successfully by boiling, steaming, or pressure cooking. Boiling in a small amount of water, equivalent to about 125 ml of water for about 285 g of broccoli, produced a better product than cooking in twice as much water (Sweeney et al., 1960).

Microwave heating of vegetables also has been studied in relation to the quality of cooked vegetables. In an early study, Chapman and coworkers (1960) reported that electronically cooked fresh broccoli is slightly better than broccoli cooked by conventional boiling. The advantage of one method over the other for frozen broccoli was not established. Ascorbic acid retention in both fresh and frozen broccoli was greater with electronic cookery. Bowman and coworkers (1975) found that microwave heating did not significantly affect the ascorbic acid content of 13 of 16 vegetables when compared with boiling.

COOKING OF DRIED LEGUMES

The method used to cook the many varieties of dried legumes used for food must ensure softening of the seed coat and the cotyledons. In an early study on cooking dried beans, Snyder (1936) reported that during soaking, water enters the hilum. Preliminary soaking shortens the cooking period. Dawson and coworkers (1952) reported that the soaking time could be reduced from overnight to 1 hr by boiling for 2 min prior to soaking. Time required to cook dried beans increases with storage at high temperatures (Burr et al., 1968).

Some varieties of beans contain a higher than normal amount of potassium phytate, a water-soluble salt of phytic acid (Crean and Haisman, 1963). Calcium from the middle lamella complexes with phytic acid, making the calcium unavailable for the formation of insoluble salts with pectic substances. This reaction, therefore, aids in tenderization during cooking. The proportion of calcium to phytic acid is important to quality of peas (Rosenbaum et al., 1966). There must be enough phytic acid to complex with calcium in the pea as well as any that might be available in the cooking water.

EXPERIMENTS

I. Plant Pigments

The effects of different hydrogen ion concentrations on the colors of several plant pigments are studied by experiments with plant extracts and cooked vegetables.

A. *Preparation of plant extracts*

1. Red fruit juice. Dilute grape juice 1:1 with distilled water.
2. Red cabbage extract. Blend 100 g of finely cut red cabbage with 100 ml of distilled water for 1–2 min. Filter the extract through qualitative filter paper.
3. Spinach extract. Prepare dried spinach by spreading 400–500 g of washed leaves on a baking sheet and drying in an oven at 66°–93°C (150°–200°F). Spinach will stain plastic counter tops; therefore, this experiment should be done on a chemically resistant surface. Weigh 10 g of dried spinach and grind to a powder with a mortar and pestle. Add 40 ml of acetone mixed with 10 ml of water, grind, and filter through qualitative filter paper.
4. Onion extract. Cook chopped yellow-skinned onions in distilled water in a beaker. Filter the extract through qualitative filter paper.

B. Procedures

1. Effect of hydrogen ion concentration on color of various pigments.
   a. Anthocyanins. Put 5 ml of buffer solutions of pH 3, 5, 7, and 9 into each of four test tubes of the same diameter. To the pH 3 buffer, add with a medicine dropper enough pigment extract (grape juice or red cabbage extract) to produce an observable but not dark color. Add the same number of drops to the other buffer solutions. Observe the color, immediately and after 1 hr.
   b. Chlorophyll. Repeat the procedure outlined for anthocyanins using the spinach extract.
   c. Anthoxanthins. Put 5 ml of onion extract into each of three test tubes. To one add 2 ml of 10% NaOH and to another add 2 ml of 10% HCl. The third sample should serve as a control. If no color is evident in 15 min, heat tubes in a beaker of water and record color change.
2. Effect of added acid or alkali on color of cooked vegetables.
   a. Boiled green beans. Wash about 200 g of green beans, break into 3-cm pieces, and divide into three equal portions. Put 100 ml of tap water into each of three 1-liter pans. Add 0.5 g of baking soda to one of the pans. Add 15 ml of vinegar to another pan. Bring the water to a boil and drop one sample of beans into each pan. Cook covered for 10 min. Drain liquid into a beaker and put beans on a white plate. Measure the pH of the liquid.
   b. Boiled red cabbage. Cut about 200 g of red cabbage into strips about 5 mm wide and divide into three equal portions. Follow procedures outlined in 2a except cook covered for 25 min.

C. Results

1. Experiment B1 (a, b, and c). Describe the color of each extract.
   - Compare the colors seen in red fruit juice and red cabbage

at various hydrogen ion concentrations. Compare the pH range over which color changes were noted for these two foods. Apply the results with the spinach extract to the cooking of green vegetables. Explain how pH influences the color of anthoxanthins

2. Experiment B2 (a and b). Score color, flavor, and texture of each vegetable using an appropriate scale. (See the Appendix.)

   Compare the effects of soda and vinegar on the color of cooked vegetables with the results in Experiment B1 (a and b). What effects do alkali and acid have on the texture of cooked vegetables?

II. Vegetable Cookery

In this experiment, the characteristics of vegetables cooked by several methods are compared. The effects of these methods on vegetables containing various pigments, and on vegetables of the cabbage and onion families, with their special flavor problems, are studied.

A. *Procedures*

1. Preparing the vegetable. Prepare each vegetable as directed below. Be careful to mix the prepared vegetable well so that the sample taken for each method is representative of the whole. Save a sample of each prepared, raw vegetable for comparison with the cooked vegetable.
   a. Green beans. Wash, cut off each end, break into 3-cm pieces, mix, and weigh 180 g samples.
   b. Cabbage. Remove any undesirable outer leaves. Wash and cut in half, shred coarsely. If more than one head of cabbage is used, mix well. Weigh 230 g samples.
   c. Carrots. Wash, scrape off the peel and slice crosswise into 4-cm pieces, mix, and weigh 180 g samples.
   d. Onions. Peel, wash, and partially quarter from the bud end. Weigh samples of approximately 230 g. Weights will not be exact because of the size of the onions, but will be closer if small uniform onions have been used than if the vegetables are large and irregular in size.

2. Cooking the vegetable. Cook each vegetable by the six methods described in Table 6.1. Because all vegetables should be freshly cooked and uniformly hot when tasted, plan the work so that for each vegetable all cooking methods will be finished at the same time.

   Use 1-liter pans for all methods except the pressure saucepan. For each method, put tap water in the pan, cover, and bring to a boil. Add the vegetable immediately to avoid boiling the water away, and cover unless the vegetable is to be cooked uncovered. As soon as the water has returned to the boiling point, start timing and reduce heat to the minimum required to keep the water boiling. Special care must be taken in some of the methods not to boil the water away and thus burn the vegetable. If necessary to prevent burning, add a small, measured amount of boiling water to the vegetable during cooking.

   Since cooking times vary with several factors, such as the maturity of the vegetable, the size of the pieces into which it is cut, and the conditions under which it has been held since it was harvested, the cooking times given in Table 6.1 are only approximations. All vegetables should be uniformly cooked until just tender except the vegetables in Method 5, which are cooked for a longer time. Record the exact cooking times for all methods. Add salt to cooked, drained vegetables. Stir thoroughly.

B. *Results*

Arrange all the cooked and the corresponding raw vegetables on a large surface. Score the flavor, color, and texture of each vegetable using an appropriate scale. (See the Appendix.)

C. *Questions*

What effect does the volume of cooking water have on the green color of vegetables? What effect does covering the pan have on the green color of vegetables? How does this effect compare with that of the volume of cooking water? What effect does the volume of cooking water have on the flavor of cabbage? Of onions? Do variations in cooking methods have

Table 6.1 Methods for cooking vegetables

Cooking method	Green beans	Cabbage	Carrots	Onions
Weight of vegetable for each method, g	180	230	180	230
Amount of salt for each method, g	1.5	1.5	1.5	1.5
1. Minimum water, covered pan[a]				
Volume water, ml	70	70	70	70
Time, min	25–35	10–15	15–20	30–40
2. Half weight of water, covered pan				
Volume water, ml	90	115	90	115
Time, min	20–25	8–10	15–18	25–35
3. Water to cover, uncovered pan				
Volume water, ml	250	400	250	375
Time, min	20–25	8–10	15–18	25–35
4. Water to cover, covered pan				
Volume water, ml	250	400	250	375
Time, min	20–25	8–10	15–18	25–35
5. Water to cover, covered pan				
Volume water, ml	250	400	250	375
Time, min	60	60	60	60
6. Pressure saucepan				
Volume water, ml	70	70	70	70
Time after pressure is up, min[b]	2.5	1.5	1.5	4

[a]Choose a heavy pan for this method if possible.
[b]At the end of the cooking period, put the saucepan into cold water to reduce the pressure quickly.

as much effect on carrots as on the other vegetables tested? How do vegetables cooked in a pressure saucepan compare with those cooked by other methods? How does a long cooking period affect the color, texture, taste, and odor of vegetables?

SUGGESTED EXERCISES

1. Compare the texture and color of white or cream-colored vegetables cooked in water with added soda, with added acid, and without addition.
2. Add a few drops of red fruit juice to tap water with added soda, soap, or acid, or without addition. Explain the colors observed.
3. Cook a green vegetable other than spinach in water with and without soda. Compare the color of the samples and the cooking water.
4. Compare the color of the following samples of peas: fresh raw, fresh dipped in a large amount of rapidly boiling water for 1 min, fresh cooked by several methods, frozen cooked by several methods, commercially canned with traces of alkali added, commercially canned by conventional methods. Estimate the proportions of chlorophyll and pheophytin in these samples, assuming that the cooked frozen peas contain 50% and the standard canned peas contain 100% of their green pigment in the form of pheophytin. The same exercise can be done with green beans, except that this vegetable is not canned with traces of alkali, and the cooked frozen beans contain 60–85% of their chlorophyll as pheophytin.
5. Cook vegetables in covered and uncovered pans, in small and large amounts of water. Compare the vegetables with respect to color and determine the pH and titratable acidity of the water drained from the vegetables.
6. Compare the speed of heat penetration and quality of potatoes baked with and without aluminum foil. Speed of heat penetration can be studied by means of a meat thermometer or a thermocouple and temperature recorder.
7. Study the effect of various metals on pigments by adding small amounts of compounds such as ferric chloride, aluminum chloride,

and stannous or stannic chloride to 5-ml samples of the vegetable or fruit extracts.

8. Compare the palatability of the same vegetable fresh, stored, frozen, canned, and dehydrated if possible.
9. Compare the quality of dried beans soaked and cooked by different methods. Include distilled and tap water with and without added (0.08% solution) sodium bicarbonate (Dawson et al., 1952).
10. Compare the palatability of several vegetables cooked by conventional methods and cooked in a microwave oven.

REFERENCES

AHEA. 1975. "Handbook of Food Preparation." American Home Economics Assoc., Washington, DC.

Asen, S., Norris, K. H. and Stewart, R. N. 1969. Absorption spectra and color of aluminum-cyanidin-3-glucoside complexes as influenced by pH. Phytochemistry 8: 653.

Asen, S., Stewart, R. N. and Norris, K. H. 1972. Co-pigmentation of anthocyanins in plant tissues and its effect on color. Phytochemistry 11: 1139.

Aurand, L. W. and Woods, A. E. 1973. "Food Chemistry." Avi Publ. Co., Westport, CT.

Bailey, S. D., Bazinet, M. L., Driscoll, J. L. and McCarthy, A. I. 1961. The volatile sulfur components of cabbage. J. Food Sci. 26: 163.

Bartolome, L. G. and Hoff, J. E. 1972. Firming of potatoes: biochemical effects of preheating. J. Agric. Chem. 20: 266.

Bate-Smith, E. C. 1954. Flavonoid compounds in foods. Adv. Food Res. 5: 261.

Berk, Z. 1976. "Braverman's Introduction to the Biochemistry of Foods." Elsevier Scientific Publ. Co., New York.

Blank, F. 1947. The anthocyanin pigments of plants. Botan. Rev. 13: 241.

Borchgrevink, N. C. and Charley, H. 1966. Color of cooked carrots related to carotene content. J. Amer. Dietet. Assoc. 49: 116.

Bowman, F., Berg, E. P., Chuang, A. L., Gunther, M. W., Trump, D. C. and Lorenz, K. 1975. Vegetables cooked by microwaves vs. conventional methods. Retention of reduced ascorbic acid and chlorophyll. Microwave Energy Appl. Newsletter 8(3): 3.

Brown, S. A. 1961. Chemistry of lignification. Science 134: 305.

Burr, H. K., Kon, S. and Morris, H. J. 1968. Cooking rates of dry beans as influenced by moisture content and temperature and time of storage. Food Technol. 22: 336.

Chapman, V. J. Putz, J. O., Gilpin, G. L., Sweeney, J. P. and Eisen, J. N. 1960. Electronic Cooking of fresh and frozen broccoli, J. Home Econ. 52: 161.
Chandler, B. V. 1964. Discoloration in processed cauliflower. Food Preserv. Quart. 24: 11.
Chandler, B. V. and Clegg, K. M. 1970. Pink discoloration in canned pears. I. —Role of tin in pigment formation. J. Sci. Food Agric. 21: 315.
Chang, Y. S. and Smit, C. J. B. 1973. Characteristics of pectins isolated from soft and firm fleshed peach varieties. J. Food Sci. 38: 646.
Charles, V. R. and Van Duyne, F. O. 1954. Palatability and retention of ascorbic acid of vegetables cooked in a tightly covered saucepan and in a "waterless" cooker. J. Home Econ. 46: 659.
Clydesdale, F. M., Fleishman, D. L. and Francis, F. J. 1970. Maintenance of color in processed green vegetables. Food Prod. Dev. 4(5): 127.
Clydesdale, F. M. and Francis, F. J. 1968. Chlorophyll changes in thermally processed spinach as influenced by enzyme conversion and pH adjustment. Food Technol. 22: 793.
Crean, D. E. C. and Haisman, D. R. 1963. The interaction between phytic acid and divalent cations during the cooking of dried peas. J. Sci. Food Agric. 14: 824.
Culpepper, C. W. and Caldwell, J. S. 1927. Behavior of anthocyan pigments in canning. J. Agric. Res. 35: 107.
Cutter, E. G. 1969. "Plant Anatomy: Experiment and Interpretation. Part I. Cells and Tissues." Addison-Wesley Publ. Co., Reading, MA.
Dame, C. Jr., Chichester, C. O. and Marsh, G. L. 1959. Studies of processed all-green asparagus. IV. Studies on the influence of tin on the solubility of rutin and on the concentration of rutin present in the brines of asparagus processed in glass and tin containers. Food Res. 24: 28.
Dateo, G. P., Clapp, R. C., MacKay, D. A. M., Hewitt, E. J. and Hasselstrom, T. 1957. Identification of the volatile sulfur components of cooked cabbage and the nature of the precursors in the fresh vegetable. Food Res. 22: 440.
Davis, E. A., Gordon, J. and Hutchinson, T. E. 1976. Scanning electron microscope studies on carrots: effects of cooking on the xylem and phloem. Home Econ. Res. J. 4: 214.
Dawson, E. H., Lamb, J. C., Toepfer, E. W. and Warren, H. W. 1952. Development of rapid methods of soaking and cooking dry beans. Tech. Bull. 1051, USDA, Washington, DC.
DellaMonica, E. S. and McDowell, P. E. 1965. Comparison of beta-carotene content of dried carrots prepared by three dehydration processes. Food Technol. 19: 1597.
Deshpande, S. N., Klinker, W. J., Draudt, H. N. and Desrosier, N. W. 1965.

Role of pectic constituents and polyvalent ions in firmness of canned tomatoes. J. Food Sci. 30: 594.
Dutton, H. J., Bailey, G. F. and Kohake, E. 1943. Dehydrated spinach. Changes in color and pigments during processing and storage. Ind. Eng. Chem. 35: 1173.
Eheart, M. S. and Gott, C. 1964. Conventional and microwave cooking of vegetables. J. Amer. Dietet. Assoc. 44: 116.
Eheart, M. S. and Odland, D. 1973a. Quality of frozen green vegetables blanched in four concentrations of ammonium bicarbonate. J. Food Sci. 38: 954.
Eheart, M. S. and Odland, D. 1973b. Use of ammonium compounds for chlorophyll retention in frozen green vegetables. J. Food Sci. 38: 202.
Esau, K. 1965. "Plant Anatomy," 2nd ed. John Wiley and Sons, Inc., New York.
Eskin, N. A. M., Henderson, H. M. and Townsend, R. J. 1971. "Biochemistry of Foods." Academic Press, New York.
Fahn, A. 1974. "Plant Anatomy," 2nd ed. Pergamon Press, New York.
Fisher, K. H. and Dodds, M. L. 1952. Ascorbic acid and ash in vegetables cooked in stainless steel utensils. J. Amer. Dietet. Assoc. 28: 726.
Francis, F. J. 1975. Anthocyanins as food colors. Food Technol. 29(5): 52.
Fuleki, T. 1969. Anthocyanins of strawberry, rhubarb, radish and onion. J. Food Sci. 34: 365.
Gilpin, G. L., Sweeney, J. P., Chapman, V. J. and Eisen, J. N. 1959. Effect of cooking methods on broccoli. II. Palatability. J. Amer. Dietet. Assoc. 35: 359.
Gold, H. J. and Weckel, K. G. 1959. Degradation of chlorophyll to pheophytin during sterilization of canned green peas by heat. Food Technol. 13: 281.
Gordon, J. and Noble, I. 1959. Effect of cooking method on vegetables. J. Amer. Dietet. Assoc. 35: 578.
Gordon, J. and Noble, I. 1960. Application of the paired comparison method to the study of flavor differences in cooked vegetables. Food Res. 25: 257.
Gordon, J. and Noble, I. 1964. "Waterless" vs. boiling water cooking of vegetables. J. Amer. Dietet. Assoc. 44: 378.
Gorin, N. and Jans, J. A. 1971. Discoloration of sauerkraut probably caused by a leucoanthocyanidin. J. Food Sci. 36: 943.
Griswold, R. M. 1944. Factors influencing the quality of home-canned Montmorency cherries. Tech. Bull. 194, Michigan State College Agric. Exp. Stn., East Lansing.
Halliday, E. G. and Noble, I. 1943. "Food Chemistry and Cookery." Chicago Univ. Press, Chicago.
Heisler, E. G., Siciliano, J., Woodward, C. F. and Porter, W. L. 1964. After

cooking discoloration of potatoes. Role of the organic acids. J. Food Sci. 29: 555.

Hing, F. S. and Weckel, K. J. 1964. Some volatile components of cooked rutabaga. J. Food Sci. 29: 149.

Hinton, D. M. and Pressey, R. 1974. Cellulase activity in peaches during ripening. J. Food Sci. 39: 783.

Horowitz, R. M. and Gentili, B. 1961. Phenolic glycosides of grapefruit: a relation between bitterness and structure. Arch. Biochem. Biophys. 92:191.

Hsu, C. P., Deshpande, S. N. and Desrosier, N. W. 1965. Role of pectin methylesterase in firmness of canned tomatoes. J. Food Sci. 30: 583.

Hsu, D. L. and Jacobson, M. 1974. Macrostructure and nomenclature of plant and animal food sources. Home Econ. Res. J. 3: 25.

Hughes, J. C., Faulks, R. M. and Grant, A. 1975. Texture of cooked potatoes. Relationship between compressive strength, pectic substances and cell size of Redskin tubers of different maturity. Potato Res. 18: 495.

Isherwood, F. A. 1970. Hexosans, pentosans and gums. In "The Biochemistry of Fruits and Their Products," Ed. Hulme, A. C. Academic Press, New York.

Jones, I. D., White, R. C. and Gibbs, E. 1963. Influence of blanching and brining treatments on the formation of chlorophyllides, pheophytins, and pheophorbides in green plant tissue. J. Food Sci. 28: 437.

Jurd, L. 1972. Some advances in the chemistry of anthocyanin-type plant pigments. In "The Chemistry of Plant Pigments," Ed. Chichester, C. O. Academic Press, New York.

Kertesz, Z. I. 1933. The oxidase system of a non-browning yellow peach. Tech. Bull. 219, New York Agric. Exp. Stn., Geneva.

Kung, L. C. 1944. Complex carbohydrates of some Chinese foods. J. Nutr. 28: 407.

Labuza, T. P. 1973. Effects of processing, storage, and handling on nutrient retention in foods. Effects of dehydration and storage. Food Technol. 27(1): 20.

Lea, C. H. 1963. Some aspects of recent flavour research. Chem. and Ind. p. 1406.

Luh, B. S. and Phitakpol, B. 1972. Characteristics of polyphenoloxidases related to browning in cling peaches. J. Food Sci. 37: 264.

Mabry, T. J. and Dreiding, A. S. 1968. The betalains. Adv. Phytochem. 1: 145.

Mackinney, G. and Weast, C. A. 1940. Color changes in green vegetables. Ind. Eng. Chem. 32: 392.

Maruyamu, F. T. 1970. Identification of dimethyl trisulfide as a major aroma component of cooked Brassicaceous vegetables. J. Food Sci. 35: 540.

Masters, H. and Garbutt, P. 1920. An investigation of the methods employed for cooking vegetables, with special reference to the losses incurred. Part II. Green vegetables. Biochem. J. 14: 75.

Mathew, A. G. and Parpia, H. A. B. 1971. Food browning as a polyphenol reaction. Adv. Food Res. 19: 75.

Niegisch, W. D. and Stahl, W. H. 1956. The onion: gaseous emanation products. Food Res. 21: 657.

Noble, A. C. 1975. Investigation of the color changes in heat concentrated tomato pulp. J. Agric. Food Chem. 23: 48.

Odland, D. and Eheart, M. S. 1974. Ascorbic acid retention and organoleptic quality of green vegetables cooked by several techniques using ammonium bicarbonate. Home Econ. Res. J. 2: 241.

Odland, D. and Eheart, M. S. 1975. Ascorbic acid, mineral and quality retention in frozen broccoli blanched in water, steam and ammonia steam. J. Food Sci. 40: 1004.

Pasch, J. H., von Elbe, J. H. and Sell, R. J. 1975. Betalaines as colorants in dairy products. J. Milk Food Technol. 38: 25.

Pilnik, W. and Voragen, A. G. J. 1970. Pectic substances and other uronides. In "The Biochemistry of Fruits and Their Products," Ed. Hulme, A. C. Academic Press, New York.

Platenius, H. 1939. Effect of temperature on the rate of deterioration of fresh vegetables. J. Agric. Res. 59: 41.

Ponting, J. D. and Jackson, R. 1972. Pre-freezing processing of Golden Delicious apple slices. J. Food Sci. 37: 812.

Ponting, J. D., Jackson, R. and Watters, G. 1972. Refrigerated apple slices: preservative effects of ascorbic acid, calcium and sulfites. J. Food Sci. 37: 434.

Powers, M. J., Reeve, R. M., Leinbach, L. R., Talburt, W. F. and Brekke, J. E. 1958. Clumping in canned blueberries. Food Technol. 12: 99.

Pressey, R., Hinton, D. M. and Avants, J. K. 1971. Development of polygalacturonase activity and solubilization of pectin in peaches during ripening. J. Food Sci. 36: 1070.

Purcell, A. E., Walter, W. M. Jr. and Thompkins, W. T. 1969. Relationship of vegetable color to physical state of the carotenes. J. Agric. Food Chem. 17: 41.

Reeve, R. M. 1970. Relationships of histological structure to texture of fresh and processed fruits and vegetables. J. Texture Studies 1: 247.

Reeve, R. M. 1972. Pectin and starch in preheating firming and final texture of potato products. J. Agric. Food Chem. 20: 1282.

Resende, R., Francis, F. J. and Stumbo, C. R. 1969. Thermal destruction and regeneration of enzymes in green beans and spinach puree. Food Technol. 23: 63.

Rosenbaum, T. M., Henneberry, G. O. and Baker, B. E. 1966. Constitution of

leguminous seeds. VI.—The cookability of field peas (Pisum sativum L). J. Sci. Food Agric. 17: 237.
Schwimmer, S. 1963. Alteration of the flavor of processed vegetables by enzyme preparations. J. Food Sci. 28: 460.
Schwimmer, S. and Weston, W. J. 1961. Enzymatic development of pyruvic acid in onion as a measure of pungency. J. Agric. Food Chem. 9: 301.
Scott, L. E., Twigg, B. A. and Bouwkamp, J. C. 1974. Color of processed sweet potatoes: effects of can type. J. Food Sci. 39: 563.
Shewfelt, A. L., Paynter, V. A. and Jen, J. J. 1971. Textural changes and molecular characteristics of pectic constituents in ripening peaches. J. Food Sci. 36: 573.
Shirkande, A. J. 1976. Anthocyanins in foods. CRC Critical Reviews in Food Sci. Nutr. 7: 193.
Simpson, J. and Halliday, E. G. 1928. The behavior of sulphur compounds in cooking vegetables. J. Home Econ. 20: 121.
Simpson, J. and Halliday, E. G. 1941. Chemical and histological studies of the disintegration of cell-membrane materials in vegetables during cooking. Food Res. 6: 189.
Singleton, V. L. 1972. Common plant phenols other than anthocyanins, contributions to coloration and discoloration. In "The Chemistry of Plant Pigments," Ed. Chichester, C. O. Academic Press, New York.
Sistrunk, W. A. and Cash, J. N. 1970. The effect of certain chemicals on the color and polysaccharides of strawberry puree. Food Technol. 24: 473.
Skalski, C. and Sistrunk, W. A. 1973. Factors influencing color degradation in Concord grape juice. J. Food Sci. 38: 1060.
Snyder, E. B. 1936. Some factors affecting the cooking quality of the pea and Great Northern types of dry beans. Res. Bull. 85, Nebraska Agric. Exp. Stn., Lincoln.
Spåre, C. G. and Virtanen, A. I. 1963. On the lachrymatory factor in onion (Allium cepa) vapours and its precursor. Acta. Chem. Scand. 17: 641.
Stephens, T. S. and McLemore, T. A. 1969. Preparation and storage of dehydrated carrot flakes. Food Technol. 23: 1600.
Sterling, C. 1968. Effect of solutes and pH on the structure and firmness of cooked carrot. J. Food Technol. 3: 367.
Stoll, A. and Seebeck, E. 1951. Chemical investigations on alliin, the specific principle of garlic. Adv. Enzymol. 11: 377.
Sweeney, J. P. 1970. Improved chlorophyll retention in green beans held on a steam table. Food Technol. 24: 490.
Sweeney, J. P. and Martin, M. 1958. Determination of chlorophyll and pheophytin in broccoli heated by various procedures. Food Res. 23: 635.
Sweeney, J. P. and Martin, M. E. 1961. Stability of chlorophyll in vegetables as affected by pH. Food Technol. 15: 263.

Sweeney, J. P., Gilpin, G. L., Staley, M. G. and Martin, M. E. 1959. Effect of cooking methods on broccoli. I. Ascorbic acid and carotene. J. Amer. Dietet. Assoc. 35: 354.

Sweeney, J. P., Gilpin, G. L., Martin, M. E. and Dawson, E. H. 1960. Palatability and nutritive value of frozen broccoli. J. Amer. Dietet. Assoc. 36: 122.

Synge, R. L. M. and Wood, J. C. 1956. (+)-(S-methyl-L-cysteine S-oxide) in cabbage. Biochem. J. 64: 252.

Tate, J. N., Luh, B. S. and York, G. K. 1964. Polyphenoloxidase in Bartlett pears. J. Food Sci. 29: 829.

Twigg, B. A., Scott, L. E. and Bouwkamp, J. C. 1974. Color of processed sweet potatoes: effect of additives. J. Food Sci. 39: 565.

Ulrich, R. 1970. Organic acids. In "The Biochemistry of Fruits and Their Products," Ed. Hulme, A. C. Academic Press, New York.

Ulrich, R., Paulin, A. and Mimault, J. 1954. Recherches sur la refrigeration des pommes. Revue generale du Froid 31: 675.

Van Buren, J. P., Hrazdina, G. and Robinson, W. B. 1974. Color of anthocyanin solutions expressed in lightness and chromaticity terms. Effect of pH and type of anthocyanin. J. Food Sci. 39: 325.

Van Duyne, F. O., Owen, R. F., Wolfe, J. C. and Charles, V. R. 1951. Effect of cooking vegetables in tightly covered and pressure saucepans: retention of reduced ascorbic acid and palatability. J. Amer. Dietet. Assoc. 27: 1059.

von Elbe, J. H. 1975. Stability of betalaines as food colors. Food Technol. 29(5): 42.

von Elbe, J. H., Maing, I.-Y. and Amundson, C. H. 1974a. Color stability of betanin. J. Food Sci. 39: 334.

von Elbe, J. H., Klement, J. T., Amundson, C. H., Cassens, R. G. and Lindsay, R. C. 1974b. Evaluation of betalain pigments as sausage colorants. J. Food Sci. 39: 128.

Wagner, J. R., Miers, J. C., Sanshuck, D. W. and Becker, R. 1969. Consistency of tomato products. 5. Differentiation of extractive and enzyme inhibitory aspects of the acidified hot break process. Food Technol. 23: 247.

Watson, E. K. 1973. Tannins in fruit extracts as affected by heat treatments. Home Econ. Res. J. 2: 112.

Weast, C. A. and Mackinney, G. 1940. Chlorophyllase. J. Biol. Chem. 133: 551.

White, R. C., Jones, I. D. and Gibbs, E. 1963. Determination of chlorophylls, chlorophyllides, pheophytins, and pheophorbides in plant material. J. Food Sci. 28: 431.

Whiting, G. C. 1970. Sugars. In "The Biochemistry of Fruits and Their Products," Ed. Hulme, A. C. Academic Press, New York.

Woodward, R. B., Ayer, W. A., Beaton, J. M., Bickelhaupt, F., Bonnett, R., Buchschacher, P., Closs, G. L., Duntler, H., Hannah, J., Hauck, F. P., Ito, S.,

Langemann, A., Le Goff, E., Leimgruber, W., Lwowski, W., Sauer, J., Valenta, Z. and Volz, H. 1960. The toal synthesis of chlorophyll. J. Amer. Chem Soc. 82: 3800.

Wrolstad, R. E. and Erlandson, J. A. 1973. Effect of metal ions on the color of strawberry puree. J. Food Sci. 38: 460.

Wrolstad, R. E., Putnam, T. P. and Varseveld, G. W. 1970. Color quality of frozen strawberries: effect of anthocyanin, pH, total acidity and ascorbic acid variability. J. Food Sci. 35: 448.

Zechmeister, L., Deuel, H. J. Jr., Inhoffen, H. H., Leemann, J., Greenberg, S. M. and Ganguly, J. 1952. Stereochemical configuration and provitamin A activity. Arch. Biochem. Biophys. 36: 80.

7. Food Preservation

The term *food preservation* most frequently makes the listener think of canning or freezing. However, it is a more inclusive term than that. The purpose of food preservation is to prevent or at least delay enzymatic and/or microbial changes in food so that the food is available at some future time. That time may be later today, tomorrow, next week, or several months from now. A discussion of food preservation should involve enzymes and microorganisms and the basic principles of preventing their activity in food. The principles are employed in several basic techniques of food preservation, including use of high temperatures, use of low temperatures, removal of water, and the addition of inhibitory substances.

Enzymes, which were discussed in Chapter 2, are readily destroyed by heat, as in canning or in blanching vegetables before freezing or dehydration. If enzymes are not inactivated before freezing, they are likely to cause deterioration because freezing slows but does not stop their action.

MICROORGANISMS AND FOOD-BORNE ILLNESS

Prevention of the growth of microorganisms or their destruction is basic to most methods of food preservation. Yeasts and molds are killed easily by heat, but the ease of destruction of bacteria varies with the type of bacteria and with their condition. Actively growing bacteria are destroyed more easily than are dormant bacteria in the spore form. For all methods of preservation it is important that the raw food contain as few microorganisms as possible.

Microorganisms that cause food-borne illnesses are of interest because methods used to preserve foods should ensure the destruction of these undesirable microorganisms as well as of microorganisms that cause spoilage. Microorganisms of the genera *Staphylococcus* and *Salmonella* and the *Clostridium* species *C. perfringens* and *C. botulinum* are of particular interest. The first three are most frequently involved in outbreaks of food-borne illness but the fourth, *C. botulinum,* is the most dangerous.

Outbreaks of staphylococcal intoxication caused by certain strains of *S. aureus* are caused more often by inadequate refrigeration of vulnerable foods than by improper use of other methods of food preservation. Staphylococcus organisms usually are involved when several people become ill 1–6 hr after eating a meal together. When vulnerable foods such as cooked meats, cream-filled pastries, potato salad, macaroni salad, and salad dressing that have been contaminated with these microorganisms, often by handling, are allowed to stand without proper refrigeration, the organisms produce an extracellular substance, or enterotoxin. When the toxin is ingested it causes nausea, vomiting, and diarrhea. Recovery is rapid and complete. The main reservoir of the organism is the nose. Secondary sources are infected wounds. Thus, care must be taken not to cough or sneeze into food and to wash the hands when starting to prepare food and after using a handkerchief or touching the hair or face. To prevent the growth of and toxin production by *S. aureus* organisms that may have found their way into food, food should be cooled promptly after cooking. Toxin production occurs in the temperature range of 16–46°C. Therefore, hot food should be kept hotter than 46°C and cold food colder than 16°C. It is possible to destroy the microorganisms in food. Woodburn and coworkers (1962) reported that two million cells per gram were destroyed when chicken products in plastic pouches were boiled for 10 min or heated by microwaves for 2 min. However, it is not possible to destroy the toxin by heating; therefore, prevention depends on destruction of the microorganisms or inhibition of their growth.

The food-borne illness caused by certain members of the genus *Salmonella* is a food infection. The illness is caused by ingestion of large numbers of the organism rather than by a preformed toxin. Symptoms may include nausea, vomiting, diarrhea, headache, and chills and usually develop 12–24 hr after ingestion of the contaminated food. The main source of the organism is the intestinal tract of animals such as birds, reptiles, farm animals, and humans. Thus, personal sanitation is extremely important in the control of the disease. Eggs, poultry, meat, and mixtures containing these foods are the most common vehicles. Salmonella grows best between 0 and 46°C. Therefore, important aspects of control are to keep foods above or below this range and to properly cook vulnerable foods to assure destruction of the organisms.

Clostridium perfringens is the causative agent in another food intoxication. It occurs most frequently in mass feeding situations involving meat dishes such as pot pies, stews, sauces, and gravies that are cooked one day, then cooled, and reheated prior to serving. Slow cooling rates for large quantities of foods of these types are responsible for their involvement (Jay, 1978). The organisms grow best at temperatures between 37 and 45°C. Survival during cooking of roasts at low temperatures is possible (Sundberg and Carlin, 1976), so adequate cooling during holding is imperative. Control depends primarily on rapid cooling and reheating of vulnerable foods. Microwave reheating of precooked chicken may not be adequate to destroy all cells of *C. perfringens* (Craven and Lillard, 1974). Bryan (1975) predicted that unless carefully controlled, outbreaks of this food-borne illness will increase as fast-food operations and banquet services featuring roast beef increase. The symptoms usually appear 10–20 hr after food containing the organisms is ingested and include abdominal pain, diarrhea, and nausea without vomiting. Large numbers of the organism must be ingested. When they reach the intestine, a toxin is produced (Volk and Wheeler, 1973).

Improper methods of food preservation may result in a less common but more serious form of food intoxication called *botulism.* The organism, *Clostridium botulinum,* is the most heat resistant of the organisms responsible for food-borne illnesses. It is a spore-forming anaerobic bacterium that produces an extremely poisonous toxin. Although the toxin is destroyed by heat, the spores are more heat resistant than most bacteria. No harm results from consuming the spores of the organism, but if the toxin is ingested it damages the central nervous system, resulting in double vision and difficulty in swallowing and in speaking, possibly followed by death due to respiratory failure (Frazier, 1967). There may or may not be digestive disturbances. The mortality rate from this illness in the United States was over 65% at one time, but recent advances in detection have helped to lower it to 22% in 1974 and 11% in 1975 (CDC, 1976). *C. botulinum* occurs in the soil of many areas and contaminates food grown in these areas. This organism is of great importance in the study of canning methods because if it is not destroyed by the canning process, it produces its deadly toxin under the anaerobic conditions of the can. Further information on food-borne illnesses may be found in food microbiology textbooks (Frazier, 1967; Jay, 1978).

METHODS OF PRESERVATION

Methods of preserving food for future use are many. Some are practical for a home situation; others are not. Both types will be discussed, with more emphasis on those methods that can be used in a home situation. An understanding of the methods available is important for professionals because interest in home production and preservation of food has increased in recent years. In the discussion that follows, scientific principles are emphasized and reasons for some of the procedures commonly used are discussed. Specific directions for preserving various foods by several techniques are available in bulletins published by the United States Department of Agriculture (USDA). Directions for drying are given in state experiment station and extension publications (Miller et al., 1975).

Canning

Canning involves the use of one of the two most commonly employed agents of food preservation, high temperature. In canning, microorganisms are destroyed by heat, probably because of denaturation of their protein. The reentry of microorganisms is prevented by a hermetically sealed container. Complete sterility is not always attained in canning. For success in commercial canning, the heat process must be sufficient to destroy pathogenic microorganisms. Microorganisms likely to cause spoilage under ordinary conditions of storage also are destroyed. Dormant nonpathogenic microorganisms may remain but will not reproduce because environmental conditions are not favorable (Lund, 1975). The term "commercially sterile" or "effectively sterile" is more accurately applied to canned foods than the term "sterile."

Recommendations for home canning (USDA, 1975a; Hamilton et al., 1976) as well as procedures established for commercial canning vary with the particular food being canned because the thermal conditions needed to produce "commercial sterility" vary. Lund (1975) reports that five factors affect the conditions needed. These factors are the nature of the food including its pH, the heat resistance of the microorganisms, the heat-transfer characteristics of the food and the container, the number of microorganisms present, and the projected conditions of storage for the processed food.

Different methods are recommended for home canning of acid and low-acid foods because the pH of food affects the temperature required to destroy any spores of *C. botulinum* that might be present in the food. A pH of 4.5 is the dividing line between low-acid and acid foods. The division is based on the fact that *C. botulinum* will not grow and produce toxin below a pH of slightly above 4.5 (Stumbo, 1973). Early workers (Esty and Meyer, 1922) studied the pH-temperature relationship by suspending spores of *C. botulinum* in buffers of varying pH, heating at 100°C, and determining the length of time necessary to destroy the spores. Only a short time is required at a low pH or at a pH higher than that normally associated with food, but a much longer time is necessary around neutrality. The pH values of a number of foods are shown in Table 7.1. Fruits and tomatoes are acid foods, whereas vegetables and foods of animal origin are low-acid foods.

Because of the comparative ease of destruction of *C. botulinum* in acid foods, they can be processed at the temperature of boiling water, as in the water-bath method. Low-acid foods, however, require a long time for processing at 100°C, and there is no certainty that all the spores will be destroyed at this temperature even after a long period because of varying conditions affecting heat penetration. As the temperature of heating a buffer of pH 7 containing *C. botulinum* spores increased from

Table 7.1 Average pH values for some common foods

pH	Food	pH	Food
0		.	
.		5.0	Pumpkins, carrots
2.9	Vinegar	5.1	Cucumbers
3.0	Gooseberries	5.2	Turnips, cabbage, squash
.		5.3	Parsnips, beets, snap beans
3.2	Rhubarb, dill pickles	5.4	Sweet potatoes
3.3	Apricots, blackberries	5.5	Spinach, aged meat
3.4	Strawberries	5.6	Asparagus, cauliflower
3.5	Peaches	5.8	Mushrooms; meat, ripened
3.6	Raspberries, sauerkraut	.	
3.7	Blueberries	6.0	Tuna
3.8	Sweet cherries	6.1	Potatoes
3.9	Pears	6.2	Peas
.		6.3	Corn
4.2	Tomatoes	.	
4.4	LOWEST ACIDITY FOR PROCESSING AT 100°C (212°F)	7.0	Meat, before aging
4.6	Figs, pimentos		

SOURCE: Pritchard, 1974.

100 to 120°C, the time required to destroy the spores decreased from 300 min to 4 min. In order to increase the temperature to above 100°C, pressure must be increased beyond that of the atmosphere. Therefore, to assure destruction of *C. botulinum* with heating for a reasonable length of time, the pressure canner is recommended in USDA publications for the home canning of low-acid foods (USDA, 1972a; USDA, 1975a).

Low-acid foods More research has been done on canning low-acid than acid foods because of the health aspects of the problem. The early research was conducted by commercial canners and the methods thus developed were adapted for household use. These procedures are not ideal because the equipment and methods used for canning at home differ from those used commercially. One of the main differences is the use of glass jars, which heat and cool more slowly than the tin cans used commercially. Additional processing takes place as glass jars are removed from the pressure cooker and cooled at room temperature; tin cans, on the other hand, are cooled rapidly in water. The research on the home canning of low-acid foods that serves as a basis for today's recommendations was done by the USDA (Toepfer et al., 1946) and several experiment stations.

For studies to determine the effectiveness of a processing procedure, food is inoculated with spores of *C. botulinum* or putrefactive anaerobe (PA) 3679. PA 3679 is significantly more heat resistant than *C. botulinum,* nontoxic, and easy to assay. Therefore, it is used to determine whether or not a process is adequate (Lund, 1975). The destruction of the organism is taken as a measure of the adequacy of the process because more heat is required for its destruction than for most, if not all, of the other organisms that may be present.

In canning low-acid vegetables and meat, care must be taken to observe every precaution given in the directions and to have the pressure canner in good working order (USDA, 1976a). Either the hot pack or the raw pack can be used for most vegetables (USDA, 1975a) and for meat and poultry (USDA, 1972a). For some vegetables that are more difficult to pack closely in the jar, the hot pack method is recommended. Cold meats can be packed into the jars and the open jars heated before sealing and processing (USDA, 1972a). This procedure achieves the same results as packing hot meat into the jars but is easier. The prepared jars are placed in a pressure canner and processed for the recommended time.

An argument that sometimes is used against the recommendation that all low-acid foods be processed in a pressure canner is that *C. botulinum* is rarely if ever reported in the soil of some areas. However, considering the possible danger when contaminated food has not been adequately processed, the wisdom of recommending a canning process inadequate to destroy any pathogens that may be present is debatable. It is possible that vegetables shipped from another area may be used for canning. This reasoning is the basis for the recommendation that the pressure canner always be used for low-acid foods.

Canned foods that show evidence of spoilage or of pressure within the container should be rejected. Unfortunately, however, development of the toxin of *C. botulinum* in canned foods cannot always be detected by changes in appearance and odor. It is therefore important not to taste suspected food for signs of spoilage. A minute amount of the toxin can be fatal. If the food or the canning process is under suspicion, the food can be made safe by boiling it actively for at least 15 min. The toxin, which is much more sensitive to heat than the spores, is inactivated by this procedure.

Acid foods Acid foods, or those having a pH of 4.5 or less (see Table 7.1), are processed more easily than low-acid foods because acid aids in the destruction of certain microorganisms by heat, as explained previously, and because the microorganisms that usually cause spoilage of such foods are destroyed easily by heat. Acid foods can be processed safely if their internal temperature, as determined by a thermometer inserted into the contents of the can during processing, reaches 82.2°C (Ball, 1938).

Adequacy of processing method is determined as for low-acid food processes by using those microorganisms responsible for spoilage. *Bacillus coagulans,* the microorganism responsible for flat sour in tomato products, is used as a test organism (Lund, 1975). *Flat sour* is a condition characterized by a slight change in pH and the development of off odors and flavors. Commercially, the development of flat sour can be of economic significance.

Fruits and tomatoes can be packed into jars or cans either raw or hot (USDA, 1975a). The hot pack has the advantage that more food can be packed into a jar, but it was reported that the palatability (Gilpin et al., 1951; Dawson et al., 1953) of certain fruits and vegetables is better, and their vitamin content (Dawson et al., 1953; Thiessen, 1949) at least as good, when they are packed raw as when they are packed hot. The raw pack, sometimes called the cold pack, is especially good for fruits with a delicate texture, such as berries. Fruits packed without added sugar are processed for the same length of time as those packed with sugar or sirup. Because sugar helps retain the color and flavor of fruits, it usually is omitted only when the intake of sugar must be restricted.

After fruit is in the jar and partially sealed, it usually is processed in a container of boiling water deep enough to come 2.5–5 cm above the tops of the jars. Harris and Davis (1976) studied the effect of various water levels in a boiling water bath canner on the rate of heating of tomato juice in quart jars. They concluded that the tomato juice heated as fast with 6.4 cm of water in the canner as with higher levels up to 2.5 cm above the jar. The amount of energy required to preheat water was reduced by approximately 75% with the use of smaller amounts of water.

Sometimes steam is used for processing instead of water. Steam under pressure is effective, but fruit is likely to be overcooked if processed under pressure.

Some canning directions have suggested that jars of food be processed in the oven, but this procedure is not recommended because the temperature of the liquid inside a jar cannot exceed about 100°C unless the jar is sealed. If the seal is tight and the temperature of the oven is above the boiling point of the liquid, pressure will build up in the jar and will break either the seal or the jar. Serious accidents have been caused by explosions of glass jars in oven canning. Another argument against oven canning is that the slower rate of heat transfer from the air of the oven makes the processing times longer than those in water.

The open kettle method also has been used for canning acid foods. In this method, the food is cooked in a pan, packed boiling hot into sterile jars, and sealed completely without further processing. This method is recommended only for jams and jellies because of the difficulty of handling the jars and tops without contamination by microorganisms, and because organisms may drop in as the jars are filled. It is safer to process the jars in boiling water, if only for a short time, after they are filled.

The texture of such foods as tomatoes and apple slices is firmer if a calcium salt is added in canning. The firming action is attributed to the insolubilization of the pectic acid of the fruit by formation of calcium pectate (Kertesz, 1942). Whole tomatoes usable for salads can be prepared by adding 5 ml of a calcium chloride solution containing 64 g of anhydrous calcium chloride per 275 ml of water to whole tomatoes in a quart jar before filling the jars with tomato juice and processing (Kertesz, 1942). The recommendation of Gould and Gray (1974) is for 1/6 tsp of calcium chloride and 1/3 tsp of sodium chloride per quart jar. When calcium chloride is used commercially, cans are labeled "traces of calcium salts added."

A controversy regarding the pH of modern tomato varieties has developed. Some investigators (Fields et al., 1977; Gould and Gray, 1974; Leonard et al., 1960; Pray and Powers, 1966; Schoenemann et al., 1974) have indicated that some tomatoes may have pH levels of greater than 4.5 whereas USDA (1976a) has reported that this is not the case. Some states (Gould and Gray, 1974) have recommended that citric acid be added to lower pH. The Food and Drug Administration (FDA) permits the use of organic acids to lower pH of tomatoes for commercial processing (FDA, 1966). The use of lemon juice, vinegar, or acetic acid in home situations may not be effective in lowering the pH significantly (USDA, 1976a). One state (Fields et al., 1977), on the basis of finding that several jars of tomatoes canned by consumers had pH levels of 4.5 or higher, has recommended that tomatoes be processed at 5 lb pressure for 10 min. Although the USDA has not, on the basis of the previously discussed facts, changed its recommendation for home canning of tomatoes (USDA, 1975a), a situation of potential significance has been recognized (USDA, 1976a). The acidity of tomatoes can be lowered with products produced by molds that grow on the surface of improperly canned tomatoes. The pH may be raised to a point where *C. botulinum* spores, if present, can germinate, multiply, and produce toxin. Therefore, any moldy tomato product should be discarded in its entirety.

Heat penetration rates The canning process must, of course, heat food in all parts of the can to the temperature necessary to destroy microorganisms that cause spoilage or food-borne illness. Studies of heat penetration are made by inserting a thermocouple or thermometer into the can and following the change in temperature as a function of time. The temperature of the slowest heating point should be monitored. This point may or may not be in the center of the can depending on the nature of the food, which in turn determines whether the food will be heated by conduction or convection (Lund, 1975).

It has been found that heat penetrates a food packed in brine or sirup—for example, peas or peaches—more rapidly than a viscous food like pumpkin. Foods in a watery medium are heated by *convection currents* that rapidly equalize the temperatures on the outside and the center of the jar. Viscous foods are heated rather slowly by *conduction,* in which transfer of heat takes place between adjacent molecules rather than by circulation of liquid. For similar reasons, heat penetrates better if brine or sirup is added than if the food is packed too closely into the container. The processing times shown in canning directions depend on packing the food in a certain way, since the times would be longer for food packed more closely than specified in the directions and longer for food packed cold rather than hot. Because heat penetrates the center of large containers more slowly than small containers, and glass more slowly than metal, separate processing times are given for jars and for cans of various sizes. Lists of processing times for home canning involving the use of a small number of jar sizes are short (USDA, 1975a) as compared to processing data for commercial operations, where many different can sizes may be used as described by Stumbo et al. (1975). That heat is conducted more slowly by air than by water or steam at the same temperature has been discussed in relation to oven canning. Increasing the temperature of steam by applying pressure will accelerate the rate of heat penetration.

Freezing

Freezing involves the use of the other most commonly employed agent of food preservation, low tempeature. In this section, the principles of freezing will be discussed with emphasis on the influence of freezing on the palatability of various foods. For further information on freezing, the reader is referred to USDA publications on freezing (USDA 1975b, 1976b, c, d), to locally available state publications, and to two volumes on commercial freezing, one on fresh foods (Tressler et al., 1968b) and the other on cooked and prepared foods (Tressler et al., 1968a).

The success of freezing as a method of food preservation depends on the fact that low temperatures destroy some microorganisms and prevent the growth of others. Thus frozen food often contains fewer bacteria than the corresponding fresh product, but the food is not rendered sterile (Borgstrom, 1955). Freezing is lethal to some microorganisms because as water freezes it is no longer available to cells, resulting in dehydration, solute concentration, pH changes, and perhaps denaturation of cellular proteins (Jay, 1978). Whether the bacterial count of the thawed food is low also depends on the conditions of thawing. Since some microorganisms may survive the freezing process, thawing and holding at elevated temperatures may provide opportunity for growth. Although *C. botulinum* endures freezing well, it does not form toxin because it is unable to grow at freezing temperatures or under aerobic conditions. The action of enzymes is slowed but not stopped by freezing temperatures. They present a problem in the freezing of vegetables, which will be discussed later.

Packaging Before freezing, food must be packaged in such a way that it cannot dry out during storage. The first requirement of a packaging material is that it be moisture–vapor-proof, meaning that it will prevent the passage of water or water vapor from the food to the dry air of the freezer. An additional requirement of wrapping material for meat is that it be grease-proof. The cost of a freezer package must be considered in relation to the number of times that it can be used. Some packaging materials, such as meat wrappings, are used only once, whereas others such as plastic boxes with tight-fitting covers can be used many times. Because freezer space is limited, packages designed to make the best possible use of space are advantageous. Brick-shaped packages utilize space much more efficiently than do cylindrical packages, and the ability of the package to stack well is of great importance. Although the glass jars used for canning are moisture–vapor-proof, they have several disadvantages as freezer containers. Some of these disadvantages have been overcome in glass jars designed for freezing.

Rigid containers made of plastic, metal, glass, or wax-treated cardboard are often used for fruits and precooked foods. It is convenient to freeze and reheat precooked foods in the same container, which might be a metal freezer dish, a baking dish, or a casserole. If such containers are not available, the baking pan or casserole can be lined with aluminum foil before cooking and the food removed in the foil after freezing. For meats, a single heavy piece of moisture–vapor-proof paper, such as laminated paper in which at least two materials have been layered, or heavy-duty aluminum foil can be used. Waxed paper and the type of aluminum foil intended for kitchen use are unsatisfactory for freezing because they may have very small holes that will allow moisture and vapors to pass and microorganisms to enter.

Baldwin et al. (1972) reported that wax paper allowed more weight loss from hamburger patties than aluminum foil (0.001 gauge), freezer paper, and polyvinyl-chloride-1 film. Of the wraps studied, aluminum foil provided the greatest protection for palatability at the higher temperatures associated with the freezer compartment of a one-door refrigerator.

Wrapping must be done so that moisture is not lost from folds in the paper. The drugstore wrap, in which interlocking folds are made at the center and ends of the package and then sealed with pressure sensitive tape, usually is recommended. In all methods of packaging, care must be taken to exclude as much air as possible, but to leave space for the expansion of the food in freezing.

Rate of freezing A distinction can be made between rapid and "sharp," or slow, freezing. *Rapid freezing* is attained by direct immersion in a refrigerating medium, by indirect contact with a refrigerant, as in pressing a package of food between refrigerated metal plates, or by freezing in a blast of cold air. These methods usually are available only in commercial plants. *Sharp freezing,* sometimes called slow freezing, is accomplished by placing foods in a room or cabinet having a temperature varying from −15 to −29°C. Usually the air is still, although a fan may be used. Freezing may require from 3 to 72 hr under such conditions (Fennema, 1975; Fennema et al., 1973). It is clear that home freezing is more likely to be sharp than rapid.

The size of ice crystals formed during the freezing process varies with rate of freezing. Slow freezing promotes the formation of few large crystals, whereas fast freezing favors the formation of many small crystals. Rapid freezing minimizes cellular changes (Fennema, 1966). Brown (1967) reported that large crystals may rupture cell walls of plant tissues, resulting in inferior texture. When green beans are frozen, there is a critical rate for quality retention, but that rate may not be as rapid as that of liquid nitrogen freezing. Liquid nitrogen freezing of carrots resulted in cracking (Wolford and Nelson, 1971), supporting the idea that there is a critical rate for freezing.

Although rapid freezing has advantages, the widespread use of home freezers suggests that sharp-frozen foods are satisfactory. Indeed, early research indicated no significant differences between the palatability and nutritive value of most foods when frozen rapidly or slowly (Lee et al., 1946, 1949, 1950, 1954). More recently, Baker and coworkers (1976) reported that chickens frozen in a home type freezer did not differ from chickens frozen in an air-blast freezer with respect to tenderness, juiciness, flavor, overall acceptability, thiobarbituric acid (TBA) values, and shear values. Drip losses were greater for chickens frozen at the slower rate. There is no doubt, however, of the importance of freezing the food as soon as possible and as quickly as possible after its preparation for the freezer in order to avoid undesirable changes in palatability, nutritive value, and bacterial count. It also is important to make the best use of the freezing equipment by placing packages against freezing plates or coils, allowing some space between packages, and avoiding stacking while the food is being frozen. No more food should be placed in the freezer than will freeze within 24 hr.

Storage A storage temperature of −18°C usually is recommended for frozen foods. Because food is not completely inert at this temperature, deteriorative changes during storage may be greater than changes occurring during blanching, freezing, and thawing (Fennema, 1975). Variation in storage temperature influences the degree of deterioration. For example, it may take two to five times as long for a given amount of deterioration to occur if the storage temperature is lowered 3°C. Certain products, such as precooked frozen foods and orange juice, deteriorate more rapidly than others, such as peas and raspberries (DeEds, 1959). Some temperature fluctuations are inevitable in a freezer. The temperature rises when the door of the freezer is opened or when unfrozen foods are introduced unless a separate freezing compartment is available. Fluctuations also are caused by variations in room temperature and by the operation of the freezer itself because the temperature must rise to a certain point before the compressor operates and must drop a few degrees before the compressor stops. This process is known as *cycling.* Accumulated frost within a package of frozen food represents moisture sublimed from the food during temperature rises and deposited as frost during temperature drops. The effects of temperature fluctuations are not great if they occur below −18°C and if the packages are well wrapped. Cycling tends to have greater consequences in the frozen food compartment of a refrigerator than in a freezer.

Specific changes that occur during freezer storage are of a chemical or physical nature since microorganisms are not viable at such low temperatures. Migration of water and recrystallization occur during storage. The process of recrystallization may minimize differences between rapidly and slowly frozen foods. After several months of storage at −18°C, crystals in rapidly frozen fruit were essentially the same size as crystals in slowly frozen fruit (Lee et al., 1949). Recrystallization is particularly damaging to frozen desserts, breads, and foods thickened with starch (Fennema, 1966). Another physical change that may occur is freezer burn. Freezer burn occurs on the surface of improperly packaged foods and results from the sublimitation of ice during temperature recycling. Kaess and Weidemann (1969) reported that freezer burn in beef liver and muscle is reduced by a reduction in freezing rate and dipping the samples in a sodium chloride or glycerol solution prior to freezing. Slower freezing promotes formation of a layer of compact cells on the surface that may delay the development of freezer burn (Fennema et al., 1973). Use of suitable packaging material is a must in a home situation.

The period during which a food can be stored before quality is lost varies with the food and how it was processed and packaged as well as with storage temperatures. Table 7.2 shows approximate maximum storage times at −18°C for home-frozen and commercially frozen foods. Table 7.2 indicates that storage times vary with the type of meat and are shorter if the meat is ground or cooked than if it is frozen as larger pieces and prior to cooking. Storage times for variety meats, smoked meats, and fish also are comparatively short, as are those for most precooked and baked foods.

Thawing Thawing may be a greater cause of damage than freezing (Fennema, 1975). The time required for thawing is considerably longer than that for freezing. The food rapidly warms to the temperature of change of state and remains there for an extended period (Fennema, 1966). However, since the time required for thawing is short in relation to the normal length of frozen storage, changes during thawing may not be as great as those occurring during storage. Fennema (1975) based this conclusion on the observation that food can be frozen and immediately thawed with little loss of quality.

In some cases it may be advantageous to cook food from the frozen state to avoid the potential damage of slow thawing. Frozen vegetables usually are cooked without prethawing, and cooking times are shorter than for fresh vegetables because the tissue of the vegetable has been softened both by blanching and by freezing. Cooking meat from the frozen state is possible and is mentioned in Chapter 5. Fruits must be thawed under milder conditions to prevent undesirable textural changes. It is important to remember that not all microorganisms in a food are destroyed by freezing and are, therefore, ready to multiply

Table 7.2 Storage life recommendations to maintain high quality frozen food stored at −18°C or lower

Food	Months
Fresh meats	
Beef and lamb roasts and steaks	8–12
Veal and pork roasts	4–8
Chops, cutlets	3–6
Variety meats	3–4
Ground beef, veal, or lamb and stew meats	3–4
Ground pork	1–3
Sausage	1–2
Cured, smoked, and ready-to-serve meats	
Ham—whole, half, or sliced	1–2
Bacon, corned beef, frankfurters, and weiners	Less than 1
Ready-to-eat luncheon meats	Freezing not recommended
Cooked meat and meat dishes	2–3
Fresh poultry	
Chicken and turkey	12
Duck and goose	6
Giblets	3
Cooked poultry	
Cooked poultry dishes and cooked poultry slices or pieces covered with gravy or broth	6
Fried chicken	4
Sandwiches of poultry meat and cooked slices or pieces not covered with gravy or broth	1
Fresh fish	6–9
Commercially frozen fish	
Shrimp and fillets of lean type fish	3–4
Clams, shucked and cooked fish	3
Fillets of fatty type fish and crab meat	2–3
Oysters, shucked	1
Fruits and vegetables, most	8–12
Home-frozen citrus fruits and juices	4–6
Milk products	
Cheddar cheese (450 g or less, no more than 2.5 cm thick)	6 or less

Table 7.2 *continued*

Food	Months
Butter or margarine	2
Frozen milk desserts, commercial	1
Prepared foods	
Cookies	6
Cakes, prebaked	4–9
Combination main dishes and fruit pies	3–6
Breads, prebaked, and cake batters	3
Yeast bread dough and pie shells	1–2

SOURCE: Allen and Crandall, 1974.

when the food is thawed. Thawing in a refrigerator keeps the bacterial count at a lower level than thawing at room temperature (Borgstrom, 1955).

Refreezing thawed foods Refreezing thawed foods should be avoided if possible because it results in reduced quality and increased bacterial count. However, there are times when food thaws because of uncontrollable situations. The USDA has published guidelines for determining whether or not to refreeze food (USDA, 1972b). Each package should be examined individually. In general, refreezing food usually is safe if there still are ice crystals in the food and no portion of the food has become warm. Even if thawing is complete, fruit can be safely refrozen. Vegetables, fish, and cooked foods should be discarded if any part of the food is warmer than 4°C because it is difficult to tell by odor if these foods have spoiled.

Meat and poultry that show no signs of spoilage as indicated by poor odor or color can be refrozen. Increased cooking losses will occur if chicken or beef is repeatedly frozen and thawed (Baker et al., 1976; Locker and Daines, 1973). However, Baker and coworkers (1976) concluded that chicken can be repeatedly refrozen if refrozen as soon as it has thawed. Sensory panel scores for tenderness, juiciness, and flavor were not affected by thawing and refreezing five times. Moisture content and shear values were not affected. Chickens were kept in freezer storage for 4 to 5 da between thawings.

The necessity for refreezing thawed food sometimes can be avoided by the addition of dry ice when the normal operation of the freezer has been interrupted. Refrozen food should be used first because of possible bacterial growth and enzyme activity during thawing.

Vegetables Most vegetables can be frozen successfully. However, there are exceptions. Salad vegetables that are to be eaten raw are not frozen successfully because they lose their crisp texture, and hence their appeal, in the process of blanching and freezing. These vegetables are lettuce, endive, cress, parsley, celery, cucumbers, onions, radishes, green peppers, cabbage, and tomatoes. Some of these vegetables, such as green peppers, onions, and tomatoes can be frozen for cooking. Methods for commercial freezing of tomatoes to retain their characteristic texture have been studied. Cryogenic freezing, or freezing by exposure to extremely cold freezants such as liquid nitrogen (Fennema et al., 1973), have been less than successful. Hoeft et al. (1973) concluded that because tomatoes frozen with cryogenic techniques are firmer than fresh tomatoes, new varieties suitable for the process must be developed.

Although blanching vegetables before freezing reduces the number of microorganisms, removes some air from the tissues, makes them more compact, and enhances their color, its most important function is to inactivate enzymes that would otherwise cause deterioration in palatability, color, and ascorbic acid content during storage. Enzymes that are isolated from the substrate in intact tissue may come in contact with the substrate as cellular structures are disrupted in the freezing process (Brown, 1976). One of the principal reasons for off flavors is rancidity of the fatty material in unblanched frozen vegetables. Rancidity development will be discussed in Chapter 8. In vegetables frozen without blanching, marked off flavors develop rapidly by the end of the freezing period in some vegetables and after storage for 3 wk in others (Noble and Winter, 1952). For example, flavor profiles for green beans change with freezing to include earthy, sour, and bitter notes (Moriarty, 1966). In addition to the development of rancidity and other off flavors, loss of chlorophyll and carotene from green vegetables may occur during storage (Lee and Wagenknecht, 1951).

Either steam or boiling water can be used for blanching. Steam retains the vitamin content of the vegetables better than boiling water, but may be more difficult to produce with household equipment. Ideal blanching times for vegetables are just long enough to inactivate the enzymes, but not long enough to cook the vegetables, because most vegetables retain quality better in frozen storage when blanched than when cooked (Paul et al., 1952). It is essential that vegetables be cooled promptly after blanching. This usually is done at home by immersing the vegetables in ice-cold water (USDA, 1976b).

The use of various additives in blanching and cooling vegetables for home freezing was evaluated by Hudson and coworkers (1974a, b). The addition of salt (1.2%) to blanching water improved retention of chlorophyll in spinach and Brussels sprouts (Hudson et al., 1974a). On the basis of sensory evaluation scores it was recommended that 1.2% NaCl be added to the blanching water for runner beans and Brussels sprouts, 1.2% sucrose for peas and Brussel sprouts, and 2% NaCl for spinach (Hudson et al., 1974b). Microwave heating in combination with water and steam blanching resulted in better flavor, chlorophyll, and ascorbic acid retention than conventional use of water or steam (Dietrich et al., 1970). However, microwave blanching was less desirable for broccoli than conventional blanching (Eheart, 1970). Other studies regarding blanching were discussed in Chapter 6. Research on blanching techniques indicated that there may be ways to improve the quality of frozen vegetables.

Fruits Enzymes cause frozen fruits to turn brown during frozen storage. The enzyme responsible for this color change, polyphenol oxidase, was discussed in Chapter 6. As with vegetables, the enzymes of fruits can be inactivated by blanching, but this usually is not done because it gives the fruit a cooked flavor and soft texture. When such changes are not important, as with apple slices for pie, fruit is sometimes blanched. The enzymes also can be inactivated by sulfur dioxide, a treatment that is used before dehydration but not usually before freezing.

Several methods for prevention of enzymatic browning are possible, including the use of sulfur dioxide or ascorbic acid alone or in combination with citric or malic acid. As alternatives, sugar or sirup added to fruit before freezing helps to prevent browning by slowing the action of enzymes and by protecting the fruit from air. Sirup usually is preferred to sugar for fruit that is to be eaten without cooking because it retains the texture of the fruit better and excludes air more effectively than sugar. Sugar draws liquid from the fruit by osmosis, forming a sirup and causing the tissues to shrink. Although fruit is less attractive when packed with sugar than with sirup, it may be preferred for cooking or making jam because it contains no added water and occupies less space in the freezer than the sirup pack. Fruit is best if cut directly into sirup or sugar to prevent oxidation, and if excess air is excluded from the package. Hudson et al. (1975a, b) explored methods for improving the quality of home-frozen strawberries. Calcium lactate, ascorbic acid, citric acid, and tartaric acid were added alone or in selected combinations to a 60% (w/w) sugar sirup used with strawberries held in storage for up to 3 mo. Citric acid gave the best color and overall rating whereas calcium lactate and ascorbic acid improved firmness. Therefore, it was postulated that a combination of calcium ascorbate and citric acid might be best (Hudson et al., 1975a). In a study of the effects of sirup temperatures, soaking time, freezing medium, and other factors on quality of strawberries, Hudson and coworkers (1975b) concluded that berries were best if frozen prior to the addition of a cold 60% (w/w) sugar solution and thawed rapidly. Ponting and Jackson (1972) evaluated prefreezing techniques for Golden Delicious apples. Apples were of highest quality when soaked in a 20–30% sugar solution containing 0.2–0.4% $CaCl_2$ and 0.2–1.0% ascorbic acid or 0.02% SO_2.

Frozen fruits are best if thawed in the unopened package because they are protected from the air. The texture of the fruit, which is otherwise rather soft after thawing, is improved by leaving a few ice crystals in the tissue.

Meat, poultry, and fish Unlike fruits and vegetables, meat and poultry require no special preparation for freezing other than wrapping. During frozen storage, slowly occurring changes are similar to those that take place more rapidly at ordinary refrigerator temperatures.

Early workers (Hiner and Hankins, 1951) reported that tenderness increases with freezing and storage as a result of enzyme activity and the formation of ice crystals. However, Smith et al. (1969) reported that rib steaks that had been frozen were neither more nor less tender than steaks that had not been frozen. Law et al. (1967) reported similar results in a study of boneless loin steaks. Tenderness of pork loin chops may increase (Kemp et al., 1976) or decrease with freezing (Bannister et al., 1971). Tenderness of lamb may decrease or remain unchanged with freezing (Smith et al., 1968). Lind et al. (1971) reported that slowly frozen lamb chops were less tender than extremely rapidly frozen chops. Reports in the literature regarding the influence of freezer storage on tenderness conflict. This is not surprising considering the complexity of this food material and the number of factors that influence tenderness. In many cases, the small differences in tenderness attributed to freezing and freezer storage may not be of practical significance.

The influence of freezing on flavor of meat and poultry also has been studied. Scores for flavor and other quality attributes may decrease in 6 wk for ground beef (Baldwin et al., 1972) and 1–4 wk for pork chops (Bannister et al., 1971). Carbonyl compounds accumulate as autoxidation of the lipids of the muscle proceeds (Awad et al., 1968). If the muscle tissue is stored at refrigerator temperatures, autoxidation proceeds rapidly. Freezer storage reduces the rate of the reaction but does not stop it. Freezer stability of muscle tissue depends on the degree of unsaturation of the fat. Thus fish will develop oxidative rancidity more rapidly than avian or mammalian muscles (Lundberg, 1962). To protect fish from exposure to oxygen, an essential reactant for autoxidation, it usually is frozen in brine or coated with ice. Other factors influencing the rate of autoxidation include the length of storage before and after freezing, the storage temperature, the presence of oxygen, and the presence of pro-oxidants and antioxidants. These factors were reviewed by Fennema et al. (1975). Of particular importance to this discussion are the roles of heme pigments and salt in the development of rancidity. Heme pigments of meat accelerate rancidity. The intimate mixture of muscle pigments and fat that occurs in ground meats may explain why they become rancid more readily during storage than cuts frozen whole. This effect of the heme pigments also may partially explain why the dark meat of turkey, which owes its color to myoglobin, becomes rancid faster than the white meat. The effect of salt in accelerating the rancidity of ground meat has not been explained. Salt is best omitted from ground meats prepared for freezing, but other spices commonly used in making sausage may delay the onset of rancidity (Watts, 1954). The presence of soy protein in beef patties inhibited rancidity development during freezer storage for 12 mo (Kotula et al., 1976).

A reddish fluid called drip exudes from the meat during thawing.

Rapidly frozen meat loses less drip than slowly frozen meat if the surface-to-volume ratio is large (Hiner et al., 1945; Khan, 1966). In general, as the length of the storage period increases, the amount of drip increases (Awad et al., 1968; Wierbicki et al., 1957).

Darkening of the tissue surrounding the bone in frozen poultry is due to liberation of the hemoglobin from the bone marrow during freezing and thawing (Brant and Stewart, 1950). It results from freezing immature broilers (Spencer et al., 1961) and can be reduced by roasting pieces of chicken for 30 min or more before freezing.

Eggs Eggs are removed from the shell before freezing because expansion usually breaks the shell. If egg yolks are frozen without any addition, waxy lumps of egg solids remain after thawing and prevent smooth blending with other ingredients. Causes for this were discussed in Chapter 3. The addition of corn sirup, sugar, or salt to whole eggs or egg yolks before freezing prevents this problem. No addition is made to the whites. Properly prepared home-frozen eggs are satisfactory for use in plain cakes and custards (Jordan et al., 1952a) and sponge cakes (Jordan et al., 1952b).

Precooked and prepared foods Tbe convenience of frozen precooked foods is appealing to many. Whether these are produced commercially or in the home, certain problems and techniques associated with various products must be considered in order to ensure good quality. As indicated in Table 7.2, the storage times for most prepared and precooked foods are short. Detailed information on freezing precooked foods is presented by Tressler et al. (1968a).

Breads Yeast loses gassing power when frozen; therefore, yeast breads are best frozen after they are baked, then thawed and heated to serving temperature in the oven (Meyer, 1955). Although storage times are much shorter for the unbaked than for the baked product, it also is possible to freeze dough, then thaw and shape it. This procedure is more successful than freezing the shaped dough (Meyer et al., 1956). Leftover muffins, biscuits, and popovers can be frozen, but the batters or doughs for these products do not freeze successfully. Sandwiches made with fillings that freeze well can be prepared in quantity and frozen; fillings like meat, cheese, and peanut butter are suitable, but jelly should be omitted because it soaks into bread, cooked egg white because it toughens, vegetables because they wilt, and mayonnaise because it separates. Maximum storage times for the individual ingredients should be considered.

Pies Fruit pies can be frozen either before or after baking. Freezing before baking is preferred because of the tendency for crusts of prebaked pies to become soggy (Meyer, 1955). If fruits that are subject to enzymatic browning are used, blanching prior to freezing or the addition of ascorbic acid may prevent color changes (Tressler et al., 1968a). Custard and cream pies tend to curdle and meringues to toughen when frozen and stored. Tinklin and Perry (1966) reported that chocolate pies prepared and frozen at home were more acceptable than commercially prepared pies or pies made from a mix. The only pie to show signs of separation of liquid was the one made from a mix. Lemon (Snow and Briant, 1960) or pumpkin filling can be frozen and put into a shell after thawing. Commercially, stabilizers and "freeze-resistant" starches are used in cream pies to prevent the separation that occurs after freezing (Palmer, 1968).

Whether the pie has been frozen before or after baking, it is best put hard frozen into a heated oven to avoid a soggy bottom crust. Pie shells can be frozen either baked or unbaked, but it may be more convenient to roll pastry into sheets, stack them separated with paper on a piece of cardboard, wrap, and freeze. The pastry sheets are thawed for about 10 min and shaped for the pan (Briant and Snow, 1957).

Cakes Shortened cakes can be frozen successfully either before or after baking. Cake batter is better if frozen in cartons and thawed before baking than if frozen in cake pans and baked without thawing. It is best to use pure vanilla extract and hydrogenated shortening rather than synthetic vanilla and butter or lard in frozen cakes and batters. Spice cakes are better frozen prebaked than as batter. Sponge cakes can be frozen prebaked, but not as batter because of their high egg-yolk content. Angel cake batter can be frozen successfully if it is put in a baking pan, frozen immediately after mixing, and not allowed to thaw before baking (Meyer, 1955). Baked angel food cakes also freeze well. All types of cakes deteriorated during a month of frozen storage but were still superior to day-old unfrozen cakes (Pence and Heid, 1960).

Combination dishes Many combination main dishes can be frozen successfully (USDA, 1976c). The preparation of such dishes at home allows the cook to prepare favorite recipes in quantity when time allows and freeze them in one-meal quantities. Usually these foods are frozen after cooking, preferably slightly undercooked because of the softening effect of freezing and reheating. A few foods such as ham loaf mixture are, however, better frozen raw and cooked for serving than frozen precooked (West et al., 1959). As with fruit pies, dishes with pastry crusts are preferred frozen unbaked. Gravies and white sauces tend to separate on thawing, but stirring on reheating will often recombine the ingredients. This tendency can be reduced by substituting waxy rice or waxy cornstarch for wheat flour or regular cornstarch, but these special products are not on the retail market at present. The separation of gravies also can be reduced by removing much of the fat before preparing the gravy. Hard-cooked egg whites are omitted from combination dishes to be frozen because they become rubbery and granular as a result of mechanical damage by ice crystals (Davis et al., 1952).

Cooked meat Short-time storage (1–4 da) of cooked chicken or turkey in a freezer may result in little change in flavor whereas storage for the same period of time in a refrigerator results in the development of rancidity (Jacobson and Koehler, 1970). However, as the period of frozen storage lengthens, cooked meat acquires first a stale and then a rancid flavor. This tendency can be decreased by excluding as much oxygen as possible (Hanson, 1960). This can be done to some extent by using a tight-fitting, vapor-proof package, but the most effective means available to consumers is to exclude air from the meat by covering it with sauce, gravy, or broth so that a solid pack is formed. For this reason, leftover roasts usually are frozen without slicing, or if sliced are covered with gravy to avoid drying out and the development of stale flavors in the freezer (Meyer, 1955). Broiled or fried meats lose crispness in freezing and develop off flavors more rapidly than meats packed in gravy. Foods coated with a batter and then frozen tend to lose that coating when reheated. Partially cooking the food prior to freezing minimizes this problem (Hanson and Fletcher, 1963).

Precautions Precooked foods are potential health hazards because many of them are good media for the growth of bacteria (Borgstrom, 1955). There are many opportunities for contamination of such foods during preparation, especially if they must be handled. Prompt chilling, preferably in ice or cold water, and freezing are important. Leftovers are best frozen immediately. Fortunately, at post-thaw refrigerator temperatures, bacteria that cause off flavors and odors grow more readily in such foods as meat pies than do bacteria of public health significance. Thus the food is likely to become unacceptable before it becomes a health hazard (Burr and Elliott, 1960). It is important not to leave such foods at room temperature after thawing and to heat them thoroughly before they are eaten.

Dehydration

One of the oldest methods of food preservation is *dehydration,* or removal of water by evaporation or sublimation (Karel, 1975a). Removal of water inhibits growth of microorganisms and retards the activity of enzymes. Vegetables are blanched before dehydration in order to inactivate enzymes, and light-colored fruits often are treated with sulfur dioxide to prevent browning.

Dehydrated foods can be prepared by drying, a natural process in which water is removed from food during the ripening of grain and the sun-drying of foods, or by dehydration in which heat is applied artificially. Commercial methods of drying include spray-drying, drum-drying, evaporation, and freeze-drying. These methods are described in detail by Karel (1975a, b).

Fruits and vegetables can be dried in the home using an oven, a vegetable dehydrator, or the sun. Details are given by Miller and coworkers (1975). As with any preservation method, it is important to remember that drying does not improve the quality, so only fresh, top-quality foods should be used.

Freeze-drying has several advantages over other drying methods. In freeze-drying, raw or cooked food is frozen and exposed to a vacuum so that ice from the food is removed as vapor without becoming a liquid. So little heat is applied that the food does not reach a high temperature. The spongelike products retain their original size and shape and rehydrate more readily and usually are superior in flavor, texture, and nutritive value to dried foods. Rehydration rate can be increased by soaking cooked carrots in a sodium chloride solution prior to processing (Curry et al., 1976). Changes that do occur in quality have been attributed to the freezing temperature and prior blanching (Rahman et al., 1971). Freeze-dried foods are much lighter in weight than equivalent amounts of fresh or frozen food and can be shipped without refrigeration.

Freeze-drying does have the disadvantage of being expensive. The need for especially protective packaging adds to the cost of freeze-dried foods. Being porous and even drier than conventionally dried foods, freeze-dried foods take up water rapidly. In addition, the low water content of freeze-dried foods makes them particularly susceptible to deteriorative chemical changes involving oxygen. Evacuation of the package prior to sealing or replacement with nitrogen provides protection from moisture and oxygen.

Use of Additives

Another method of reducing the amount of water available to microorganisms is to add large amounts of salt or sugar, because the water tied up by these solutes cannot be used for microbial growth. This is done in salting fish and in making jellies and jams, which will be discussed below. The amounts of salt added in canning vegetables and of sugar in canning fruits do not act as preservatives because the amounts used are far below those required to inhibit microbial growth.

Many foods can be conveniently preserved by a wide variety of chemicals either added to the food or developed in it. Examples of added preservatives are sugar and salt, the nitrites used in smoked meats, sodium benzoate, sodium propionate, sulfur dioxide, sorbic acid, and spices. Preservatives can be developed in food by the controlled growth of microorganisms to produce fermentation that may be of the alcoholic, acetic acid, or lactic acid type. Examples of foods and beverages produced or preserved by this method are alcoholic beverages, sour milk, and sauerkraut.

Jellies and Jams

Jellies and jams are preserved by their sugar content, which is high enough to prevent the growth of microorganisms, except for mold growth on the surface. Access of molds to homemade jellies and jams is prevented by a paraffin seal or canning jar lid. The commercial products usually are protected by vacuum caps and may be pasteurized after the containers are filled.

Jelly is an example of a class of substances known to the colloid chemist as gels. *Gels* are semirigid and elastic substances formed by colloidal solutions, or *sols.* In the case of fruit jelly, gelation of pectin is brought about by the addition of sugar in the presence of acid. Pectin also can be gelled by alcohol or glycerine. Sugar, alcohol, and glycerine precipitate pectin by acting as dehydrating agents. The rigidity of gels indicates that they have structure. It seems probable that hydrogen bonding is responsible, at least in part, for the rigidity of fruit jellies.

Jellies are clear substances since they are made of fruit juice or a water extract of fruit. Jams, however, contain all or most of the insoluble solids of the fruit because whole, crushed, macerated, or pureed fruit is used in their manufacture. Technically, jams and preserves are identical, except the term *preserves* is used for products containing the whole fruit.

Pectin, acid, and sugar Three essential ingredients of most jellies are pectin, sugar, and acid. The relationships among the ingredients are important to the quality of the product.

Pectin Purified pectin is made from apple pomace and the white inner skin, or *albedo,* of citrus fruit. The latter is more abundant. Apple pectin makes a more elastic jelly than citrus pectin, whereas jelly made from citrus pectin is broken more easily. Pectins with high molecular weight and a relatively high proportion of methyl ester groups have the best jelly-forming ability. The pectins of fruits other than apple and citrus have not been studied extensively, but they are known to vary in quality. For example, strawberries are said to have weaker pectin than many other fruits. The quality of the pectin is indicated by its ability to carry sugar when made into jelly and is expressed commercially by grade. Jelly grade is the proportion of sugar that one part of pectin is capable of turning, under prescribed conditions, into a jelly with suitable characteristics. If, for example, 1 lb of pectin will carry 150 lb of sugar to make a standard jelly, it is a 150-grade pectin. This grade of pectin is in wide use commercially. The procedure for grading pectin was developed by a committee of the Institute of Food Technologists (IFT, 1959). The chemistry of pectin was discussed in Chapter 6.

Sugar The amount of sugar used in a jelly depends on the amount and quality of the pectin used. When the sugar content of a mixture is increased or the pectin content decreased, a weaker jelly will result. If too little sugar is added, the effect is the same as for an excess of pectin, and the jelly will be tough. In a normal finished jelly, the concentration of sugar is about that of a saturated sugar solution.

Acid The third essential ingredient for jelly is acid. It was recognized early by Tarr (1923) that the amount of acid necessary depends not on total but on active acidity or pH. Apparently the acid acts by neutralizing the charge on the carboxyl groups of pectin, thus increasing the tendency of the molecules to associate, and hence to form a gel (Kertesz, 1951). Jelly formation usually is possible only below pH 3.5. As the pH is decreased below 3.5, the firmness of the jelly increases until an optimum pH, usually in the range of 2.6 to 3.4, is reached (Kertesz, 1951). At pH values below the optimum, syneresis or weeping occurs (Tarr, 1923). The exact value of the optimum pH depends on the amounts of pectin and sugar and the presence of buffer salts. Acids such as vinegar, lemon juice, lime juice, citric acid, lactic acid, malic acid, and tartaric acid often are added in making jellies and jams. The time of adding the acid is another variable. Acid that is present during boiling hydrolyzes some of the sugar to invert sugar, which helps prevent crystallization of the sucrose as the jelly is stored. The presence of acid in the boiling jelly mixture may have the unfortunate effect of degrading the pectin by hydrolysis. For this reason, commercial manufacturers often add acid after jelly has cooked. Small lots of jelly cook so quickly that pectin hydrolysis seldom presents a problem.

Water In addition to the three essential ingredients in jelly making discussed above, a fourth, water, is taken for granted. In a natural fruit jelly, there are traces of other components, such as salts, proteins, and starches. Such components are not essential as shown by a formula (Cox and Higby, 1944) for test jelly: 450 ml of distilled water, 5.2 g of 150-grade pectin, and 775 g of sucrose. These ingredients are cooked until the batch weighs 1200 g and poured into four glasses, each of which contains 2.0 ml of tartaric acid solution (488 g of acid per liter of solution). This product is sometimes called a "synthetic" jelly because only purified ingredients are used in making it.

Jelly without added pectin If jelly is to be made without the addition of pectin, a fruit or combination of fruits rich in pectin must be selected. The use of some underripe fruit is advised since pectic substances and acidity decrease as fruit ripens. Directions for extracting juice are given in a USDA (1975c) publication on jelly making. There are several ways to judge the pectin content of juice in order to estimate the jellying power of the juice. In the alcohol test, 95% ethyl alcohol is added slowly to an equal volume of juice. If the mixture forms a stiff jelly, equal volumes of sugar and juice can be used for jelly. If the mass is soft and jellylike, only 150 g of sugar should be used per 240 ml of juice, and if small separate lumps of pectin precipitate, only 100 g of sugar should be used.

Another method of estimating the pectin content of fruit juice is to study its viscosity. Viscosity is a fairly accurate means of determining the quantity of pectin in juice because the large pectin molecules increase the viscosity of the fluid. Baker (1934) developed a simple viscometer called the *Jelmeter,* which is a modified pipet, for indirectly estimating the pectin content of fruit juice. Its use is based on the principle that sugar to be added to juice for jelly is proportional to the pectin content of the juice, which in turn determines the viscosity of the juice. Juices that flow through the narrow opening most slowly contain the largest amounts of pectin and hence require the largest amounts of sugar. The Jelmeter is calibrated directly in cups of sugar to be used per cup of juice. This method is a useful guide to making jelly on a small scale. It is not, however, an accurate reflection of pectin content because pH, salts, sugars, proteins, and starches also influence viscosity (Baker, 1934). Possibly the most accurate but also the most time consuming method of determining the jellying power of a fruit juice is to prepare a series of jellies from the juice, using a different proportion of sugar for each small batch. After the jellies have stood for 24 hr, it is easy to decide which jelly has the best consistency, and hence what proportions should be used for the remainder of the juice.

Methods of determining when jelly has cooked long enough can be divided into two groups: measurement of gelation and measurement of solids content. Gelation can be measured by chilling a small amount of boiling jelly in a refrigerator or by the *sheeting-off test,* in which the mixture is allowed to drip from a large, cool spoon. If the sirup separates into two streams of drops that sheet off together, the jelly is done. The success of these crude tests depends on the experience of the operator.

The other tests are measures of soluble-solids content. Measurements of solids are useful because if all the ingredients and other conditions are right, the mixture will form a jelly when it reaches a certain solids content. Juices poor in pectin or acid may not form jelly until the solids content is higher than normal, and juice rich in these constituents may gel at an unusually low solids content. The measures of soluble solids used in making jelly are boiling point, refractive index, and kettle yield. *Boiling point* is the only one of these methods available for home cooking. It is a useful guide, especially if applied to a standardized mixture that is cooked at a certain rate so that the extent of hydrolysis is controlled. Of course, the hydrolysis of sucrose caused by the fruit acid will increase the boiling point of the mixture without having a corresponding effect on soluble solids concentration. Boiling point is an accurate guide if a series of test jellies cooked to different boiling points is prepared and judged, and the best temperature is chosen. A check also can be made with the sheeting-off test. The boiling point ordinarily used for jelly is 104–105°C, or, more accurately, 4–5°C above that of water in order to correct for variation in atmospheric pressure and for inaccuracies in the thermometer. A boiling point 4.8°C above that of water corresponds to the 65% soluble-solids content required for jellies that are to be sold. The *refractometer* is used widely commercially, since the refractive index is the best measure of the soluble-solids content of jellies and jams. Convenient hand refractometers reading directly in soluble solids are available. Another method of measuring total solids content, sometimes called the *kettle yield* method, is to cook until the batch has a certain weight. Insoluble as well as soluble solids are, of course, measured if they are present, as in jam.

Jelly with added pectin If pectin is to be added in making jelly or jam, fruit is not selected and extracted for maximum yield of pectin. Fully ripe fruit is used and the extraction period is shorter than when no pectin is added; indeed, juice may be pressed from some fruits without cooking. Possibly to compensate for the lower acid content of fully ripe fruits, acid is added to most, if not all, pectins sold in retail packages. Directions for the use of pectin call for very short boiling of the combined ingredients—often 1 min. These directions make it unnecessary to decide when the mixture has been adequately cooked.

Gilpin and coworkers (1957) reported that small cooking lots of jelly, made with 2 cups of fruit juice, were nearest optimum in firmness, while medium- and large-sized lots, made with 4 and 6 cups of fruit juice, were slightly better in flavor. Either fast or slow heating was satisfactory except that fast heating gave a slightly more desirable product when jelly was made in large lots. The heating time that gave the best results varied with the size of the cooking lot. For small amounts of jelly, boiling for 1 min was most satisfactory, while medium and large lots generally were improved when boiled for 2 or 3 min.

Fruit spreads of good color and fresh taste can be made without cooking and stored for short periods in a refrigerator or for longer periods in a freezer. In such formulas, an uncooked puree made from fresh or frozen fruits is combined with added pectin. The proportions used and the lack of evaporation during preparation result in a lower proportion of fruit than in true jellies and jams.

Low-ester pectins As explained in Chapter 6, some of the carboxyl groups of pectin are esterified with methyl groups, while others are uncombined or combined with metals to form salts. Pectin in which all the carboxyl groups are methylated contains 16.3% methoxyl, a degree of methylation that is theoretically possible but has not been attained. Pectins containing 8% methoxyl have the highest grade and those with more or less methoxyl are of lower quality (Baker, 1948). However, fruit jellies can be made from pectins of widely differing methoxyl content if the molecular weight of the pectin is in the normal range.

Pectin can be demethylated by enzymes, acid, or alkali. If the process is continued until the pectin contains only 2.5–4.5% methoxyl, a low-ester pectin of desirable properties is produced (Kertesz, 1951). The advantage of this pectin is that it will gel with little or no sugar if calcium or a similar divalent ion is present. Such ions cause ionic bonding by reaction of a calcium ion with two carboxyl groups, as in the formula shown in Figure 7.1. Gel strength depends on pectin and calcium concentration (Lopez and Li, 1968).

Low-ester pectins, also called low-methoxyl pectins, make it possible to gel salads and desserts that contain nearly any desired amount of sugar, including none. They can be used commercially to make dessert mixes (Sheperd et al., 1951; Black and Smit, 1972a, b), canned salads, and other products. Added calcium is unnecessary for milk desserts but must be added in making salads and desserts from fruit or vegetable juices. Although some of the prepared products have appeared on the market, low-ester pectin is not available at present in retail packages.

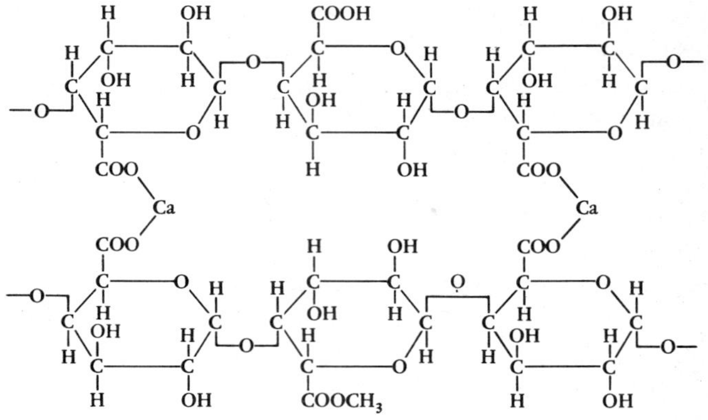

Figure 7.1. Low-ester pectin precipitated by calcium.

SUGGESTED EXERCISES

1. Determine the effect of variation in amount of sugar on the quality of jelly. Follow instructions in "How to Make Jellies, Jams, and Preserves at Home" (USDA, 1975c) for preparation of juice for apple jelly without pectin. Add acid to the juice as specified.

   Prepare four lots of jelly using 240 ml of juice in each and the following levels of sugar: (a) 100 g, (b) 150 g, (c) 200 g, and (d) 250 g. Prepare according to the basic directions, heating the jelly to 104°C.

   Record the volume of each jelly. Measure the percentage of sag after 24 hr. Evaluate the jellies using a score card designed by the class.

   What is the best level of sugar for the fruit juice used? What is the relationship between the level of sugar and the volume of the jelly? Why? What effect does too little sugar have on the quality of the jelly? Too much sugar?
2. Using the level of sugar selected as best in suggested Exercise 1, prepare three jellies, cooking them to 1, 4, and 7°C above the boiling point of water. Evaluate as in 1. Note any differences in the sheeting-off test at the three temperatures.

   What effects do different cooking temperatures have on the quality and volume of the jelly? How did the sheeting-off test compare with temperature as a means of determining whether the jelly had cooked long enough?

3. Study the effects of various factors on the speed of heat penetration during canning by means of a thermometer inserted into the center of the food. A maximum registering thermometer or thermocouple is necessary if a pressure cooker is used. The thermometer is inserted through a cork placed in a hole bored in the center of the jar lid or the top of the can.

   If a metal screwband and flat metal lid are used, a large rubber stopper with a hole in the center to hold the thermometer can be inserted in the screwband in place of the flat metal lid.

   Compare the heat penetration rate of apple sauce and sliced apples in syrup. Compare the rate of heat penetration for whole kernel corn and cream-style corn.
4. Wrap samples of food, such as ground meat patties, in various ways or various materials, freeze them, and weigh them periodically for several weeks. Observe the amount of freezer burn and calculate the loss of weight as percentage of the raw sample weight.
5. Freeze a vegetable without blanching and after the recommended blanching treatment. After storage for at least a month, compare the palatability of the two samples. Samples can be prepared by one class and judged by the following class in order to extend the storage period.
6. Compare the quality of apples or peaches frozen with and without ascorbic acid, stored as long as possible, and thawed under various conditions.
7. Study the effect of rate of freezing on the amount of drip obtained from meat. Small, uniform, weighed pieces of meat can be frozen at different rates by placing them in various parts of the freezer, as well as by freezing in freezers and the frozen-food compartment of a refrigerator and by wrapping in much and little paper. After freezing and storage, unwrap, weigh, and place in large funnels inserted in weighed glass bottles. Cover the funnels with plastic wrap or watch glasses. Allow the samples to thaw in a refrigerator and weigh the bottles containing the drip. Determine the weight of the drip by subtraction. Calculate the amount of drip as a percentage of the fresh sample weight.
8. Compare drying of carrots and apples in a conventional oven and a microwave oven.

9. Using fruit from the same lot, prepare uncooked jam and cooked jam, made with and without pectin. Store as long as possible and compare the quality of the three products.
10. Obtain some low-methoxyl pectin (Sunkist Growers, Ontario, California) and prepare products using the directions supplied by the manufacturer.

REFERENCES

Allen, M. and Crandall, M. L. 1974. The cold facts about freezing. In "Shopper's Guide—1974 Yearbook of Agriculture." USDA, Washington, DC.

Awad, A., Powrie, W. D. and Fennema, O. 1968. Chemical deterioration of frozen bovine muscle at −4°C. J. Food Sci. 33: 227.

Baker, G. L. 1934. A new method for determining the jellying power of fruit juice extractions. Food Ind. 6: 305.

Baker, G. L. 1948. High-polymer pectins and their deesterification. Adv. Food Res. 1: 395.

Baker, R. C., Darfler, J. M., Mulnix, E. J. and Nath, K. R. 1976. Palatability and other characteristics of repeatedly refrozen chicken broilers. J. Food Sci. 41: 443.

Baldwin, R. E., Borchelt, D. and Cloninger, M. 1972. Palatability of ground beef home frozen and stored in selected wraps. Home Econ. Res. J. 1: 119.

Ball, C. O. 1938. Advancement in sterilization methods for canned foods. Food Res. 3: 13.

Bannister, M. A., Harrison, D. L., Dayton, A. D., Kropf, D. H. and Tuma, H. J. 1971. Effects of a cryogenic and three home freezing methods on selected characteristics of pork loin chops. J. Food Sci. 36: 951.

Black, S. A. and Smit, C. J. B. 1972a. The grading of low-ester pectin for use in dessert gels. J. Food Sci. 37: 726.

Black, S. A. and Smit, C. J. B. 1972b. The effect of demethylation procedures on the quality of low-ester pectins used in dessert gels. J. Food Sci. 37: 730.

Borgstrom, G. 1955. Microbiological problems of frozen food products. Adv. Food Res. 6: 163.

Brant, A. W. and Stewart, G. F. 1950. Bone darkening in frozen poultry. Food Technol. 4: 168.

Briant, A. M. and Snow, P. R. 1957. Freezer storage of pie shells. J. Amer. Dietet. Assoc. 33: 796.

Brown, M. S. 1967. Texture of frozen vegetables: effect of freezing rate on green beans. J. Sci. Food Agric. 18: 77.

Brown, M. S. 1976. Effects of freezing on fruit and vegetable structure. Food Technol. 30(5): 106.

Bryan, F. L. 1975. Status of food-borne disease in the United States. J. Environmental Health 38: 74.

Burr, H. K. and Elliott, R. P. 1960. Quality and safety in frozen foods. J. Amer. Med. Assoc. 174: 1178.

Cox, R. E. and Higby, R. H. 1944. A better way to determine the jelling power of pectins. Food Ind. 16: 441.

Craven, S. E. and Lillard, H. S. 1974. Effect of microwave heating of precooked chicken on Clostridium perfringens. J. Food Sci. 39: 211.

CDC. 1976. Botulism in 1975—United States. Mortality and Morbidity Weekly Rpt. 25: 76.

Curry, J. C., Burns: E. E. and Heidelbaugh, N. D. 1976. Effect of sodium chloride on rehydration of freeze-dried carrots. J. Food Sci. 41: 176.

Davis, J. G., Hanson, H. L. and Lineweaver, H. 1952. Characterization of the effect of freezing on cooked egg white. Food Res. 17: 393.

Dawson, E. H., Gilpin, G. L., Warren, H. W. and Toepfer, E. W. 1953. Canning snap beans; precooked and raw pack. J. Home Econ. 45: 165.

DeEds, F. 1959. Time-temperature tolerance in frozen foods. J. Amer. Dietet. Assoc. 35: 128.

Dietrich, W. C., Huxsoll, C. C. and Guadagni, D. G. 1970. Comparison of microwave, conventional and combination blanching of Brussels sprouts for frozen storage. Food Technol. 24: 613.

Eheart, M. S. 1970. Effect of storage and other variables on composition of frozen broccoli. Food Technol. 24: 1009.

Esty, J. R. and Meyer, K. F. 1922. The heat resistance of the spores of B. botulinus and allied anaerobes. XI. J. Infectious Disease 31: 650.

FDA. 1966. Canned tomatoes; standards of identity and quality. Fed. Register 31: 10676.

Fennema, O. 1966. An over-all view of low temperature food preservation. Cryobiology 3: 197.

Fennema, O. R., Powrie, W. D. and Marth, E. H. 1973. "Low-temperature Preservation of Foods and Living Matter." Marcel Dekker, New York.

Fennema, O. 1975. Freezing preservation. In "Principles of Food Science. Part II. Physical Principles of Food Preservation," Eds. Karel, M., Fennema, O. R. and Lund, D. B. Marcel Dekker, New York.

Fields, M. L., Zamora, A. F. and Bradsher, M. 1977. Microbiological analysis of home-canned tomatoes and green beans. J. Food Sci. 42: 931.

Frazier, W. C. 1967. "Food Microbiology," 2nd ed. McGraw-Hill Book Company, Inc., New York.

Gilpin, G. L., Hammerle, O. A. and Harkin, A. M. 1951. Palatability of home-canned tomatoes. J. Home Econ. 43: 282.

Gilpin, G. L., Lamb, J. C. and Staley, M. G. 1957. Effects of cooking procedures on quality of fruit jelly. J. Home Econ. 49: 435.
Gould, W. A. and Gray, E. 1974. Canning tomatoes in the home. Publ. L-170, Cooperative Ext. Ser., Ohio State Univ., Columbus.
Hamilton, L. W., Kuhn, G. D., Rugh, K. A. and Editors of Farm Journal. 1976. "Home Canning. The Last Word." Country Side Press, Farm Journal, Inc., Philadelphia.
Hanson, H. L. 1960. Prepared frozen foods. J. Amer. Dietet. Assoc. 36: 581.
Hanson, H. L. and Fletcher, L. R. 1963. Adhesion of coatings on frozen chicken. Food Technol. 17: 793.
Harris, H. and Davis, L. M. 1976. Use of low water level in boiling water bath canning. Circular 226, Auburn Univ. Agric. Exp. Stn., Auburn, AL.
Hiner, R. L., Madsen, L. L. and Hankins, O. G. 1945. Histological characteristics, tenderness, and drip losses of beef in relation to temperature of freezing. Food Res. 10: 312.
Hiner, R. L. and Hankins, O. G. 1951. Effects of freezing on tenderness of beef from different muscles and from animals of different ages. Food Technol. 5: 374.
Hoeft, R., Bates, R. P. and Ahmed, E. M. 1973. Cryogenic freezing of tomato slices. J. Food Sci. 38: 362.
Hudson, M. A., Sharples, V. J., Pickford, E. and Leach, M. 1974a. Quality of home frozen vegetables. I. Effects of blanching and/or cooling in various solutions on organoleptic assessments and Vitamin C content. J. Food Technol. 9: 95.
Hudson, M. A., Sharples, V. J. and Gregory, M. E. 1974b. Quality of home frozen vegetables. II. Effects of blanching and/or cooling in various solutions on conversion of chlorophyll. J. Food Technol. 9: 105.
Hudson, M. A., Holgate, M. E., Gregory, M. E. and Pickford, E. 1975a. Home frozen strawberries. II. Influence of additives in syrup on sensory assessments and texture measurements. J. Food Technol. 10: 689.
Hudson, M. A., Leach, M., Sharples, V. J. and Pickford, E. 1975b. Home frozen strawberries. I. Influences of freezing medium, fanning, syrup temperature, soaking time, storage time and temperature, and rates of freezing and thawing on sensory assessments. J. Food Technol. 10: 681.
IFT. 1959. Pectin standardization. Final report of the IFT Committee. Food Technol. 13: 496.
Jacobson, M. and Koehler, H. H. 1970. Development of rancidity during short-time storage of cooked poultry meat. J. Agric. Food Chem. 18: 1069.
Jay, J. M. 1978. "Modern Food Microbiology," 2nd ed. D. Van Nostrand Co., New York.
Jordan, R., Luginbill, B. N., Dawson, L. E. and Echterling, C. J. 1952a. The effect of selected pretreatments upon the culinary qualities of eggs frozen

and stored in a home-type freezer. I. Plain cakes and baked custards. Food Res. 17: 1.
Jordan, R., Dawson, L. E. and Echterling, C. J. 1952b. The effect of selected pretreatments upon the culinary qualities of eggs frozen and stored in a home-type freezer. II. Sponge cakes. Food Res. 17: 93.
Kaess, G. and Weidemann, J. F. 1969. Freezer burn of animal tissue. 7. Temperature influence on development of freezer burn in liver and muscle tissue. J. Food Sci. 34: 394.
Karel, M. 1975a. Dehydration of foods. In "Principles of Food Science. Part II. Physical Principles of Food Preservation," Eds. Karel, M., Fennema, O. R. and Lund, D. B. Marcel Dekker, New York.
Karel, M. 1975b. Freeze dehydration of foods. In "Principles of Food Science. Part II. Physical Principles of Food Preservation," Eds. Karel, M., Fennema, O. R. and Lund, D. B. Marcel Dekker, New York.
Kemp. J. D., Montgomery, R. E. and Fox, J. D. 1976. Chemical, palatability and cooking characteristics of normal and low quality pork loins as affected by freezer storage. J. Food Sci. 41: 1.
Kertesz, Z. I. 1942. The use of calcium chloride in the home canning of whole tomatoes. Circ. 195, New York State Agric. Exp. Stn., Geneva.
Kertesz, Z. I. 1951. "The Pectic Substances." Interscience Publishers, Inc., New York.
Khan, A. W. 1966. Biochemical changes in poultry muscle during freezing and storage. Cryobiology 3: 224.
Kotula, A. W., Twigg, G. G. and Young, E. P. 1976. Evaluation of beef patties containing soy protein, during 12-month frozen storage. J. Food Sci. 41: 1142.
Law, H. M., Yang, S. P., Mullins, A. M. and Fielder, M. M. 1967. Effect of storage and cooking on qualities of loin and top-round steaks. J. Food Sci. 32: 637.
Lee, F. A., Gortner, W. A. and Whitcombe, J. 1946. Effect of freezing rate on vegetables. Ind. Eng. Chem. 38: 341.
Lee, F. A., Gortner, W. A. and Whitcombe, J. 1949. Effect of freezing rate on fruits. Food Technol. 3: 164.
Lee, F. A., Brooks, R. F., Pearson, A. M., Miller, J. I. and Volz, F. 1950. Effect of freezing rate on meat. Appearance, palatability, and vitamin content of beef. Food Res. 15: 8.
Lee, F. A. and Wagenknecht, A. C. 1951. On the development of off-flavor during the storage of frozen raw peas. Food Res. 16: 239.
Lee, F. A., Brooks, R. F.,Pearson, A. M., Miller, J. I. and Wanderstock, J. J. 1954. Effect of rate of freezing on pork quality. Appearance, palatability, and vitamin content. J. Amer. Dietet. Assoc. 30: 351.
Leonard, S., Luh, B. S. and Pangborn, R. M. 1960. Effect of sodium chloride,

citric acid, and sucrose on pH and palatability of canned tomatoes. Food Technol. 14: 433.

Lind, M. L., Harrison, D. L. and Kropf, D. H. 1971. Freezing and thawing rates of lamb chops: effects on palatability and related characteristics. J. Food Sci. 36: 629.

Locker, R. H. and Daines, G. J. 1973. The effect of repeated freeze-thaw cycles on tenderness and cooking loss in beef. J. Sci. Food Agric. 24: 1273.

Lopez, A. and Li, L.-H. 1968. Low-methoxyl pectin apple gels. Food Technol. 22: 1023.

Lund, D. 1975. Heat processing. In "Principles of Food Science. Part II. Physical Principles of Food Preservation," Eds. Karel, M., Fennema, O. R. and Lund, D. B. Marcel Dekker, New York.

Lundberg, W. O. 1962. Mechanisms. In "Symposium on Foods: Lipids and Their Oxidation," Eds. Schultz, H. W., Day, E. A. and Sinnhuber, R. O. Avi Publ. Co., Westport, CT.

Meyer, B. 1955. Home freezing of prepared and precooked foods. J. Home Econ. 47: 603.

Meyer, B., Moore, R. and Buckley, R. 1956. Gas production and yeast roll quality after freezer storage of fermented and unfermented doughs. Food Technol. 10: 165.

Miller, M. W., Winter, F. H. and York, G. K. 1975. Drying foods at home. Leaflet 2785, Div. of Agric. Sci., Univ. California, Berkeley.

Moriarty, J. H. 1966. Flavor changes in frozen foods. Cryobiology 3: 230.

Noble, I. and Winter, J. D. 1952. Is blanching necessary when vegetables are to be kept in frozen storage a month or less? J. Home Econ. 44: 33.

Palmer, H. H. 1968. Sauces and gravies; thickened desserts and fillings; whipped toppings; salad dressings and soufflés. In "The Freezing Preservation of Foods. Vol. 4. Freezing of Precooked and Prepared Foods," Eds. Tressler, D. K., Van Arsdel, W. B. and Copley, M. J. Avi Publ. Co., Westport, CT.

Paul, P. C., Cole, B. I. and Friend, J. C. 1952. Precooked frozen vegetables. J. Home Econ. 44: 199.

Pence, J. W. and Heid, M. 1960. Effect of temperature on stability of frozen cakes. Food Technol. 14: 80.

Ponting, J. D. and Jackson, R. 1972. Pre-freezing processing of Golden Delicious apple slices. J. Food Sci. 37: 812.

Pray, L. W. and Powers, J. J. 1966. Acidification of canned tomatoes. Food Technol. 20: 87.

Pritchard, I. 1974. Can's and can'ts for canners. In "Shopper's Guide—1974 Yearbook of Agriculture." USDA, Washington, DC.

Rahman, A. R., Henning, W. L. and Westcott, D. E. 1971. Histological and physical changes in carrots as affected by blanching, cooking, freezing, freeze-drying and compression. J. Food Sci. 36: 500.

Schoenemann, D. R., Lopez, A. and Cooler, F. W. 1974. pH and acidic stability during storage of acidified and nonacidified canned tomatoes. J. Food Sci. 39: 257.

Shepherd, A. D., McCready, R. M. and Owens, H. S. 1951. New, quick, cold-water dessert mix. Food Eng. 23(pt. 2): 44.

Smith, G. C., Spaeth, C. W., Carpenter, Z. L., King, G. T., and Hoke, K. E. 1968. The effects of freezing, frozen storage conditions and degree of doneness on lamb palatability characteristics. J. Food Sci. 33: 19.

Smith, G. C., Carpenter, Z. L. and King, G. T. 1969. Considerations for beef tenderness evaluations. J. Food Sci. 34: 612.

Snow, P. R. and Briant, A. M. 1960. Frozen fillings for quick lemon meringue pies. J. Home Econ. 52: 350.

Spencer, J. V., Sauter, E. A. and Stadelman, W. J. 1961. Effect of freezing, thawing and storing broilers on spoilage, flavor and bone darkening. Poultry Sci. 40: 918.

Stumbo, C. R. 1973. "Thermobacteriology in Food Processing," 2nd ed. Academic Press, New York.

Stumbo, C. R.,Purohit, K. S. and Ramakrishnan, T. V. 1975. Thermal processing lethality guide for low-acid foods in metal containers. J. Food Sci. 40: 1316.

Sundberg, A. D. and Carlin, A. F. 1976. Survival of Clostridium perfringens in rump roasts cooked in an oven at 107° or 177°C or in an electric crockery pot. J. Food Sci. 41: 451.

Tarr, L. W. 1923. Fruit jellies. I. The rôle of acids. Bull. 134, Delaware Univ. Agric. Exp. Stn., Newark.

Thiessen, E. J. 1949. Conserving vitamin C by varying canning procedures in snap beans, tomatoes, peaches, and pears. Food Res. 14: 481.

Tinklin, G. L. and Perry, A. K. 1966. A comparison of quality, acceptability, and cost of frozen chocolate cream pies. J. Home Econ. 58: 808.

Toepfer, E. W., Reynolds, H., Gilpin, G. L. and Taube, K. 1946. Home canning processes for low-acid foods. Developed on the basis of heat penetration and inoculated packs. Tech. Bull. 930, USDA, Washington, DC.

Tressler, D. K., Van Arsdel, W. B. and Copley, M. J. 1968a. "The Freezing Preservation of Foods. Vol. 4. Freezing of Precooked and Prepared Foods." Avi Publ. Co., Westport, CT.

Tressler, D. K., Van Arsdel, W. B. and Copley, M. J. 1968b. "The Freezing Preservation of Foods. Vol. 3. Commercial Food Freezing Operations—Fresh Foods." Avi Publ. Co., Westport, CT.

USDA. 1972a. Home canning of meat and poultry. Home and Garden Bull. 106, USDA, Washington, DC.

USDA. 1972b. What to do when your home freezer stops. Leaflet 321. USDA, Washington, DC.
USDA. 1975a. Home canning of fruits and vegetables. Home and Garden Bull. 8, USDA, Washington, DC.
USDA. 1975b. Home freezing of poultry. Home and Garden Bull. 70, USDA, Washington, DC.
USDA. 1975c. How to make jellies, jams, and preserves at home. Home and Garden Bull. 56, USDA, Washington, DC.
USDA. 1976a. Modern tomato varieties as safe to can as older varieties. News USDA 1397–76, USDA, Washington, DC.
USDA. 1976b. Home freezing of fruits and vegetables. Home and Garden Bull. 10, USDA, Washington, DC.
USDA. 1976c. Freezing combination main dishes. Home and Garden Bull. 40, USDA, Washington, DC.
USDA. 1976d. Freezing meat and fish in the home. Home and Garden Bull. 93, USDA, Washington, DC.
Volk, W. A. and Wheeler, M. F. 1973. "Basic Microbiology," 3rd ed. J. B. Lippincott Co., Philadelphia.
Watts, B. M. 1954. Oxidative rancidity and discoloration in meat. Adv. Food Res. 5: 1.
West, L. C., Titus, M. C. and Van Duyne, F. O. 1959. Effect of freezer storage and variations in preparation on bacterial count, palatability and thiamine content of ham loaf, Italian rice, and chicken. Food Technol. 13: 323.
Wierbicki, E., Kunkle, L. E. and Deatherage, F. E. 1957. Changes in the water-holding capacity and cationic shifts during the heating and freezing and thawing of meat as revealed by a simple centrifugal method for measuring shrinkage. Food Technol. 11: 69.
Wolford, E. R. and Nelson, J. W. 1971. Comparison of texture of carrots frozen by airblast, food freezant-12 and nitrogen vapor. J. Food Sci. 36: 959.
Woodburn, M., Bennion, M. and Vail, G. E. 1962. Destruction of salmonellae and staphylococci in precooked poultry products by heat treatment before freezing. Food Technol. 16(6): 98.

8. Fats and Their Lipid Constituents

Fats are mixtures of lipids. *Lipids* are chemical compounds present naturally in many foods. Lipid mixtures in the form of shortenings, frying fats, and salad oils are used commonly in food preparation and are referred to here as fats. The major functions of a fat in food preparation include the following: (1) to tenderize, (2) to contribute to batter or dough aeration, (3) to serve as a heating medium, (4) to serve as a phase in an emulsion, (5) to contribute to flavor, and (6) to enhance smoothness, body, or other textural properties.

The subject of lipids and fats is characterized by so much interrelatedness that choosing a sequence for discussing the various aspects is not easy. Chemical structure of lipids determines both the physical properties and the chemical reactions that may occur. The relationship between the chemical properties of lipids and the physical properties of lipids and fats underlies the processing methods used for fats. The relationship between chemical and physical properties of lipids also underlies the functional properties of the fats used in food preparation. The chemical structure of lipids will be discussed first because it is basic to the study of all other aspects of lipids and the fats that contain them.

CHEMICAL STRUCTURE OF LIPIDS

Some lipids contain only carbon, hydrogen, and oxygen; others also contain phosphorus and nitrogen; some contain sulfur. Fatty acids are important structural components of most lipids and thus will constitute the starting point for discussion of lipid structure.

Fatty Acids

Most naturally occurring fatty acids have straight carbon chains, even numbers of carbons, and only hydrogen atoms attached to the carbons other than the carboxyl carbon. Fatty acids having branched chains, odd numbers of carbons, or a hydroxyl group in place of a hydrogen occur in trace amounts in some foods. Fatty acids differ considerably as to chain length and degree of saturation. Most fatty acids in foods have chain lengths of 4–24 carbon atoms and most have 0–4 double bonds. The empirical formulas $C_nH_{2n+1}COOH$, $C_nH_{2n-1}COOH$, $C_nH_{2n-3}COOH$, $C_nH_{2n-5}COOH$, and $C_nH_{2n-7}COOH$ represent saturated, monounsaturated, diunsaturated, triunsaturated, and tetraunsaturated fatty acids, respectively. Monounsaturated, diunsaturated, triunsaturated, and tetraunsaturated fatty acids also are referred to as monoenes, dienes, trienes, and tetraenes.

Some common fatty acids with both their common and systematic names are given in Table 8.1. Palmitic and oleic acids are particularly abundant and widely distributed in both plant and animal lipids. Stearic acid is found in both plant and animal lipids but is more abundant in animals than in plants. Butyric acid, although present in rather low concentration, is unique to butterfat. The 6:0–14:0 fatty acids are present in both plant and animal lipids in variable, but relatively small, amounts. The polyunsaturated fatty acids, those with at least two double bonds, are most abundant in seed oils and fish liver oils.

Glycerides

Glycerides are esters of the 3-carbon alcohol glycerol and fatty acids. Glycerol, sometimes called glycerine, has three hydroxyl groups per molecule and thus is described as a polyhydric alcohol.

Table 8.1 Common fatty acids

C:=[a]	Common name	Systematic name	Formula
4 : 0	Butyric	Butanoic	$CH_3(CH_2)_2COOH$
6 : 0	Caproic	Hexanoic	$CH_3(CH_2)_4COOH$
8 : 0	Caprylic	Octanoic	$CH_3(CH_2)_6COOH$
10 : 0	Capric	Decanoic	$CH_3(CH_2)_8COOH$
12 : 0	Lauric	Dodecanoic	$CH_3(CH_2)_{10}COOH$
14 : 0	Myristic	Tetradecanoic	$CH_3(CH_2)_{12}COOH$
16 : 0	Palmitic	Hexadecanoic	$CH_3(CH_2)_{14}COOH$
16 : 1[b]	Palmitoleic	9-Hexadecenoic	$CH_3(CH_2)_5CH{:}CH(CH_2)_7COOH$
18 : 0	Stearic	Octadecanoic	$CH_3(CH_2)_{16}COOH$
18 : 1[b]	Oleic	9-Octadecenoic	$CH_3(CH_2)_7CH{:}CH(CH_2)_7COOH$
18 : 2[b]	Linoleic	9,12-Octadecadienoic	$CH_3(CH_2)_4CH{:}CHCH_2CH{:}CH(CH_2)_7COOH$
18 : 3[b]	Linolenic	9,12,15-Octadecatrienoic	$CH_3CH_2CH{:}CHCH_2CH{:}CHCH_2CH{:}CH(CH_2)_7COOH$
20 : 0	Arachidic	Eicosanoic	$CH_3(CH_2)_{18}COOH$
20 : 4[b]	Arachidonic	5,8,11,14-Eicosatetraenoic	$CH_3(CH_2)_4CH{:}CHCH_2CH{:}CHCH_2CH{:}CHCH_2CH{:}CH(CH_2)_3COOH$
22 : 0	Behenic	Docosanoic	$CH_3(CH_2)_{20}COOH$
24 : 0	Lignoceric	Tetracosanoic	$CH_3(CH_2)_{22}COOH$

[a]Number of carbon atoms:number of double bonds.
[b]*cis*-configuration.

$$\begin{array}{c} H \\ HCOH \\ | \\ HCOH \\ | \\ HCOH \\ H \end{array}$$

Glycerol

A glyceride is a monoglyceride, diglyceride, or triglyceride, depending on whether one, two, or three of the alcohol groups of glycerol are esterified with fatty acid(s). The terms monoacylglycerol, diacylglycerol, and triacylglycerol also are used. Triglycerides are much more abundant than monoglycerides or diglycerides. A triglyceride may be a simple triglyceride, containing three molecules of a single fatty acid, as tristearin,

$$\begin{array}{l} H \\ HC—O—\overset{O}{\overset{\|}{C}}(CH_2)_{16}CH_3 \\ | \\ HC—O—\overset{O}{\overset{\|}{C}}(CH_2)_{16}CH_3 \\ | \\ HC—O—\overset{O}{\overset{\|}{C}}(CH_2)_{16}CH_3 \\ H \end{array}$$

Tristearin

or a mixed triglyceride, containing more than one fatty acid. Most naturally occurring triglycerides are mixed and may or may not be symmetrical. In symmetrical mixed triglycerides, the terminal fatty acids (α-positions) are the same and are different from the middle fatty acid (β-position). Although a triglyceride molecule frequently is represented diagrammatically with all fatty acid chains extending in the same direction,

"Comb-type" representation

the β-chain probably extends more frequently in the opposite direction from the α-chains. The existence of more than one system for designat-

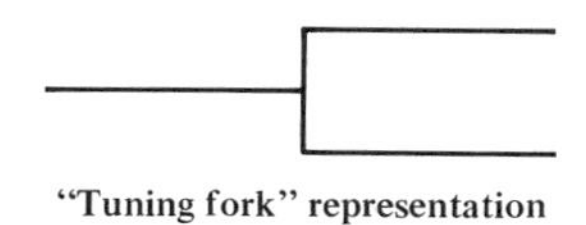

"Tuning fork" representation

ing the three positions should be noted; for example, Gurr and James (1971) prefer use of the numerals 1, 2, and 3.

Triglycerides frequently are referred to as neutral lipids because their molecules do not have free acidic or basic groups. Neutral lipids are by far the most abundant class of lipids. Even the major sources of phospholipid contain more neutral lipid than phospholipid. Natural fats consist largely of mixtures of mixed triglycerides, with traces of monoglycerides, diglycerides, and free fatty acids. Other lipid classes, such as phospholipids and sterols, are present in some fats.

Phospholipids

The major phospholipids are lecithin and the cephalins. Their molecules contain, in addition to glycerol, two fatty acid molecules, a phosphate group and a nitrogenous base, and may be represented diagrammatically as follows:

Fatty acid ——— Fatty acid

Phosphate——Nitrogenous base

Diagrammatic representation of a phospholipid

As in a triglyceride, the fatty acids are in ester linkage with the glycerol. The phosphate is in ester linkage both with the glycerol and with the nitrogenous base. The nitrogenous base is choline in lecithin,

$$HO-CH_2-CH_2-\underset{\substack{|\\HO}}{N}(CH_3)_3$$

Choline

which preferably is called phosphatidyl choline. Either of two nitrogenous bases occurs in a cephalin: ethanolamine or serine.

$$HO - CH_2 - CH_2 - NH_2$$

Ethanolamine

$$HO - CH_2 - \underset{\displaystyle COOH}{\underset{|}{CH}} - NH_2$$

Serine

The preferred terminology for a cephalin accordingly is phosphatidyl ethanolamine or phosphatidyl serine. Sphingomyelin contains sphingosine rather than glycerol and is less abundant than the other phospholipids.

The phosphate group in a phospholipid molecule confers some acidity, but the quaternary ammonium group of choline is strongly basic. The cephalins are acidic. The free amino group in ethanolamine or serine is not as strongly basic as is the quaternary ammonium group of choline. Because of the free carboxyl group of serine, phosphatidyl serine is particularly acidic. The basic and acidic groups can be charged, and phospholipids thus are referred to as polar lipids.

Phospholipids, like triglycerides, are variable as to their fatty acid components. In general, their fatty acids are less saturated than are those of triglycerides.

Phospholipids occur in egg yolk, the fat globule membrane of milk, and the intramuscular lipid of meat, as well as in plant seeds. Egg yolk and soybeans are major commercial sources. Phospholipids usually occur, along with proteins, as lipoproteins in the tissues in which they occur. Lipoproteins have structural and functional significance in the tissues that contain them.

Glycolipids and Sulfolipids

Like phospholipids, most glycolipids contain either glycerol or sphingosine. Those containing glycerol usually are diglycerides with a sugar unit in glycosidic linkage with glycerol at the other position. A second sugar moiety may be in glycosidic linkage with the first. The sugar units most frequently are galactose and glucose. Glycolipids occur in small amounts in both plant and animal tissues, but the glycosyl glycerides are more prominent in plants than in animals. Galactosyl glycerides are of functional importance in breads made from wheat flour, as will be discussed in Chapter 11.

Sulfolipids contain sulfur and occur in several forms, which will not be discussed in this book. A lipid may be both a glycolipid and a sulfolipid.

A more detailed discussion of lipid structure and nomenclature is beyond the scope of this book. The reader is referred to Aurand and Woods (1973), Gurr and James (1971), and Swern (1964a) for further information.

PHYSICAL STRUCTURE AND PROPERTIES OF FATS

The physical structure and properties of fats will be discussed before the chemical reactions of lipids in order to facilitate relating the chemical reactions to physical effects.

Physical Structure

The physical structure of lipids is important to the properties of shortenings, butter, and margarine, as well as such foods as peanut butter and candies that have lipid-containing coatings. The straight chains of fatty acids are able to align themselves into crystalline structures. A solid fat consists of crystals suspended in oil. In general, the higher the proportion of crystals, the more solid is the fat at room temperature. However, the relationship between crystallinity and consistency is not that simple because a solid fat is polymorphic; it has the ability to exist in more than one crystalline form. In the β-form, representing a high degree of order, crystals not only have relatively high melting points but also are relatively large and stable. A fat with a large proportion of crystals in the β-form tends to be grainy. At the opposite extreme is the α-crystalline form, in which chain alignments are somewhat more random, the melting point is relatively low, and the crystals are relatively small. A fat with a high proportion of α-crystals is smooth but the α-crystals tend to be unstable. Whether a fat with small crystals remains smooth is dependent on the variety of molecular species present. If only a few molecular species are present, the crystals are relatively unstable because further orientation to form larger crystals is relatively easy. A large variety of molecular species favors maintenance of small crystals. The β'-crystalline form is intermediate between the α- and β-forms as to crystal properties; its crystals are smaller than those of the β-form and quite stable. This brief discussion of polymorphism is an oversimplification of a complex subject. The reader is referred to reviews by Hoerr and Paulicka (1968) and Lutton (1972).

Physical Properties

Melting point *Melting point* is the temperature at which a solid is changed to a liquid. In the case of a fat, melting point depends to a large extent on the proportion of the fat that is present in the crystalline state. This in turn depends on several internal factors, including chain length and degree of saturation of the component fatty acids, configuration at the double bonds of unsaturated fatty acids, and variety of molecular species. Previous history of a fat also affects melting point.

The greater the ease of association between the fatty acid chains, the greater is the extent of crystallization and the higher is the temperature required to disrupt the associative forces between chains; or, in other words, the higher is the melting point of the fat. Long, straight chains of the component fatty acids contribute to high melting points of fats.

Melting point increases with increasing chain length, providing only chain length, either of free fatty acids or of fatty acids in glycerides, differs. For free saturated fatty acids, melting point increases from −4°C for the 4-carbon acid to 80°C for the 22-carbon fatty acid (CRC, 1977).

Melting point increases with increasing degree of saturation if only degree of saturation differs. The reason may not be as obvious as in the case of the chain length effect. The presence of a double bond results in a bend in the chain, as seen in the following chain segment:

$$
\begin{array}{l}
-CH_2-CH \\
\qquad\quad\;\; \| \\
-CH_2-CH
\end{array}
$$

"Bent-chain" segment

The chain thus is more bulky than a saturated chain and association between chains is more difficult. Other factors being equal, the larger the number of double bonds in a chain, the less extended is the chain, the more difficult is crystallization, and the less is the amount of energy required for melting once crystallization has occurred. Among the 18-carbon fatty acids, melting point of the free acids ranges from 72°C for stearic acid (18:0) down to −11°C for linolenic acid (18:3) (CRC, 1977). Most oils are liquid at room temperature because of their relatively high concentrations of unsaturated fatty acids. An exception is coconut oil, which is highly saturated but is liquid because of a preponderance of relatively short-chained fatty acids. Among the highly unsaturated oils, some have a higher degree of polyunsaturation than others. For example, olive oil is highly monounsaturated, and many other oils, such as soybean oil and corn oil, are highly polyunsaturated.

The third molecular characteristic that affects melting point is not a variable in fats as they naturally occur, but becomes operative as an effect of processing. Naturally occurring unsaturated fatty acids, either free or esterified, have the *cis* configuration at the double bond, resulting in a bend in the chain as shown previously. With the *trans* configuration, often produced during processing, the double bond results in a small "kink" rather than a complete turnaround:

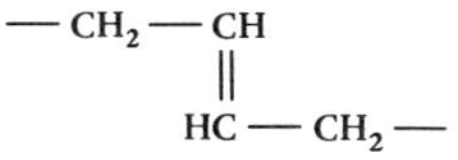

$$
\begin{array}{ll}
-CH_2-CH & \\
\quad\quad\quad\ \| & \\
\quad\quad\quad HC-CH_2- &
\end{array}
$$

"Kinked-chain" segment

The *trans* configuration thus permits the chain to be considerably more extended than an equally unsaturated chain having the *cis* configuration at the double bond(s). This results in a higher melting point of the fatty acid or glyceride than is characteristic of the corresponding lipid with the *cis* configuration. It is possible for a fat to be solid at room temperature and yet identical to an oil with regard to fatty acid chain lengths and extent of unsaturation. Whereas the melting point of oleic acid, *cis* 18:1, is 16°C, that of elaidic acid, *trans* 18:1, is 45°C (CRC, 1977). When *trans* isomers are formed, not all double bonds present are affected. The formation of *trans* isomers will be discussed later in this chapter.

Melting point also is affected by the previous history of a fat. Factors such as rate of cooling and amount of agitation during cooling affect the crystallinity and thus the melting point of a fat. This effect is based on a factor disregarded in the above discussion of melting point but mentioned earlier, the type of crystalline structure. Even a pure sample of a triglyceride containing only one fatty acid has different possible melting points, depending on the polymorphic state. For tristearin, these are 54.7, 64.0, and 73.3°C for the α-, β'-, and β-polymorphs, respectively. The rate and extent to which the transition from α to β' to β progresses depends on the details of the cooling process. Either very rapid or very slow cooling is more likely to result in a relatively large proportion of the β'-polymorph rather than the less desired β-form (Hvolby, 1974). The subject is further complicated by the differing behavior of complex lipid mixtures and will not be discussed further here.

Solidification temperature Solidification temperature of a given fat is lower than the melting point; in other words, if a solid fat is melted, it must be cooled to a temperature lower than that at which it was completely melted before it will regain its original firmness. If an oil is cooled until it solidifies, it must be warmed to a temperature higher than the temperature of solidification before it will regain its original fluidity.

Solubility Glycerol, a viscous, odorless, somewhat sweet liquid, is extremely water-soluble. The three hydroxyl groups, which constitute a large proportion of the molecule, are responsible for the water-solubility. The hydrocarbon chain of a fatty acid does not have an affinity for water, whereas the carboxyl group does. Only the fatty acids with the shortest chains are truly water-soluble and solubility decreases as the chain length increases. In a triglyceride, the hydrophilic portions of glycerol and fatty acids are "tied-up," leaving the molecule largely hydrophobic. The molecule still has some affinity for water at the ester linkage, and again, the solubility depends on the lengths of hydrocarbon chains. As natural fats contain triglycerides having a preponderance of medium to long chain fatty acids, fats are insoluble in water, slightly soluble in lower alcohols, and readily soluble in the nonpolar solvents, chloroform, ether, petroleum ether, benzene, and carbon tetrachloride.

Phospholipid and glycolipid molecules are more soluble in polar solvents than are triglycerides because of the hydrophilic groups they contain. These are phosphate and hydroxyl groups in phospholipids, carboxyl groups specifically in phosphatidyl serine, and the hydroxyl groups of the sugar moiety in glycolipids.

Density Fats have a lower density than water. The density of most fats is in the range of 0.90–0.92 g/cm^3. This results in the well-known floating of unemulsified oil on the surface of water.

Refractive index The refractive index of fats is affected by several factors. It decreases with increasing temperature, increases with increasing chain length, and increases with increasing unsaturation. The significance of the temperature effect lies in the importance of temperature control in the measurement of refractive index. Refractive index is not a property that is important in the everyday use of fats, but the effect of unsaturation on refractive index provides the food industry with a quick means of estimating the extent of hydrogenation of an oil.

Surfactant properties Surfactants were discussed briefly in Chapter 2. Free fatty acids and monoglycerides have the ability to form monolayers at water-fat interfaces. This property is most easily illustrated with fatty acids. If oil containing free fatty acids is added to water without agitation, the hydrophilic carboxyl end of a fatty acid molecule is pulled into the water and the hydrocarbon chain, being hydrophobic, remains in the oil. The result is an alignment of fatty acid molecules to form a monomolecular film at the interface:

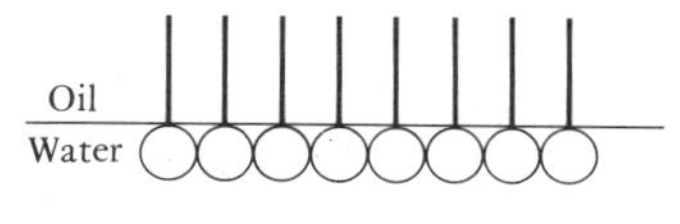

Monomolecular layer

If the mixture is agitated, the same type of film may form around oil globules, resulting in an emulsion in which each oil globule is coated by a monomolecular film:

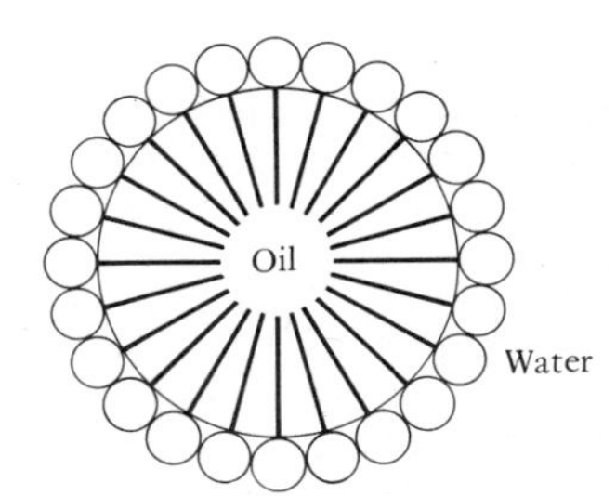

Oil globule with monomolecular layer

Monoglycerides function similarly and, therefore, are frequently added to shortenings that are likely to be used in cake batters. Their presence in shortening improves the emulsification of the shortening when batters are prepared.

Additional information concerning the physical properties of fats is available in many references, including Swern (1964b) and Weiss (1970).

CHEMICAL REACTIONS OF LIPIDS

In a discussion of the reactions to which lipids are susceptible, it is convenient to think in terms of triglycerides, though most of the reactions are applicable also to other lipids, as may be surmised from the specific molecular portions involved. Reactions of triglycerides can be classified in several ways. One basis for classification is the portion of the molecule that is involved:

Reactions involving the ester linkage
- Hydrolysis
- Saponification
- Interesterification
- Rearrangement
- Acetylation

Reactions involving double bonds
- Oxidation

Reversion
Hydrogenation
Isomerization
Halogenation

The same reactions can be classified on the basis of the general conditions under which they are most likely to occur:

Reactions occurring in commercial processing as primary reactions
Interesterification
Hydrogenation
Rearrangement
Acetylation
Reactions occurring in commercial processing as side reactions
Isomerization (*cis-trans* isomerization also may be the primary reaction)
Reactions occurring in analytical work
Saponification
Halogenation
Reactions occurring in storage and use of fats and foods containing them
Hydrolysis
Oxidation
Reversion

The reactions will be discussed in the sequence of the first classification.

Reactions Involving the Ester Linkage

Hydrolysis The ester linkage between glycerol and fatty acid is broken in hydrolysis. Complete hydrolysis of a triglyceride results in one molecule of glycerol and three molecules of fatty acid per molecule of triglyceride hydrolyzed, as illustrated with tristearin on page 260. Note that the reaction is a double decomposition reaction in which water is the other reactant. One molecule of water is required for each ester linkage broken.

Lipases, enzymes that are naturally present in butterfat, nuts, whole grains, and other fatty foods that might be stored without prior heat

$$\begin{array}{l} \overset{H}{HC}-O-\overset{\overset{O}{\|}}{C}(CH_2)_{16}CH_3 \\ \quad | \\ HC-O-\overset{\overset{O}{\|}}{C}(CH_2)_{16}CH_3 \\ \quad | \\ \underset{H}{HC}-O-\overset{\overset{O}{\|}}{C}(CH_2)_{16}CH_3 \end{array} + 3\,H_2O \rightarrow 3\,CH_3(CH_2)_{16}C\begin{array}{l}\nearrow O \\ \searrow OH\end{array} + \begin{array}{c} H \\ HCOH \\ | \\ HCOH \\ | \\ HCOH \\ H \end{array}$$

Tristearin Stearic acid Glycerol

Tristearin hydrolysis

treatment, catalyze lipid hydrolysis during storage. Most lipases preferentially split off the fatty acids in the α-position on glycerol (Aurand and Woods, 1973). The accumulation of free fatty acids results in an off flavor, rancidity. The effect is particularly noticeable in foods containing relatively high concentrations of short-chain fatty acids, such as butter, because the lower fatty acids are volatile at room temperature. The rancid condition is a symptom that may be caused by more than one mechanism, and another mechanism will be discussed later. Rancidity caused by hydrolysis is labeled *hydrolytic rancidity.*

Hydrolysis also may be catalyzed by acid in a heated system. This probably is not an important mechanism of troublesome lipid hydrolysis in food, however.

When a frying fat is used repeatedly, free fatty acids accumulate as a result of hydrolysis. In this case, heat is the catalyst. The combination of ready availability of water (in the wet food) and the high temperatures used in frying facilitates hydrolysis.

Another reaction closely follows hydrolysis when fat is overheated. Glycerol, freed by lipid hydrolysis, is decomposed through dehydration:

$$\begin{array}{c} H \\ HCOH \\ | \\ HCOH \\ | \\ HCOH \\ H \end{array} \xrightarrow{\text{heat}} \begin{array}{c} H \\ HC \\ \| \\ HC \\ | \\ HC{=}O \end{array} + 2\,H_2O$$

Glycerol Acrolein

Acrolein formation

Acrolein, the decomposition product, is a volatile, irritating compound. Its formation is evidenced by the acrid fumes emanating from overheated fat. This reaction occurs also if glycerol itself is heated as a pure liquid, though this is not significant from the standpoint of food.

Saponification *Saponification* is the formation of a metallic salt of a fatty acid; such a salt is called a soap. The reaction involves treatment of free fatty acids and/or glycerides with a base and may be considered a special case of hydrolysis when a glyceride is reacted with a base. Because only one atom of monovalent metal is taken up per fatty acid chain, regardless of chain length, the reaction provides a basis for estimating average fatty acid chain length in a fat sample. The greater the average chain length of a sample, the less sodium or potassium will a given weight of the sample take up. The amount of sodium or potassium taken up can be determined by providing a known, excessive amount of sodium hydroxide or potassium hydroxide and titrating the excess after permitting saponification to occur. Saponification is not a reaction that normally occurs in food. An exception involves excessively alkaline cakes and cake mixes.

Interesterification *Interesterification,* or alcoholysis, involves the transfer of fatty acid from the original glycerol to another alcohol. The other alcohol may be free glycerol added for this purpose. In this case, the reaction sometimes is referred to as *superglycerination.* This is the process by which monoglycerides and diglycerides are prepared commercially. The reaction is brought about under nitrogen, with a suitable catalyst and heat. Monoglycerides are useful as emulsifiers, as previously mentioned. The monoglycerides resulting from interesterification can be separated from diglycerides by distillation, but it is more practical to simply add the monoglyceride and diglyceride mixture to shortenings.

A second example of interesterification is an analytical procedure. For studying the fatty acid content of lipids by gas liquid chromatography, the fatty acids must be converted to a more readily vaporized form. This is accomplished through preparation of methyl esters by transfer of fatty acids from glycerol to methanol.

Rearrangement Another reaction brought about commercially involves the use of special processing conditions so that fatty acids become randomly distributed on the glycerol molecules. To a certain extent, this may be thought of as a variant of interesterification and has come to be used nearly as commonly as hydrogenation (Hustedt, 1976). The result of rearrangement is a greatly increased number of molecular species and, therefore, greater difficulty of crystallization and greater likelihood of the formation and maintenance of small crystals. In addition, the increased number of molecular species results in a wider melting-point range and a greater degree of plasticity of the solid fat.

Acetylation *Acetylation* is replacement of fatty acids in glycerides with acetate. The most common application probably is the acetylation of monoglycerides. Acetylated monoglycerides may be effective emulsifiers (Weiss, 1970). Because their melting points can be controlled by the reaction conditions, acetylated monoglycerides are useful as ingredients in food coatings (Luce, 1967).

Reactions Involving Double Bonds

Oxidation The type of oxidation discussed here is different from the β-oxidation that occurs in lipid metabolism. The oxidation under discussion involves reaction with molecular oxygen. Because, once started, it is self-perpetuating, it is referred to as *autoxidation.* Autoxidation actually involves a series of reactions. It begins with removal of a hydrogen atom from a carbon adjacent to a double-bond carbon, resulting in a free radical:

$$\begin{array}{ccccccccc} & H & & H & & H & & H & \\ - & C & - & C & = & C & - & C & - \\ & \cdot & & & & & & H & \end{array}$$

Free radical

The initial removal of hydrogen requires energy. Therefore, it is influenced by storage temperature. It also is catalyzed by light and by metals, which may be present as contaminants from processing equipment. Autoxidation of meat lipids is catalyzed by heme pigments and the reaction occurs more rapidly in cooked than raw meat (Chang et al., 1961). Although it frequently is stated that the conversion to ferric iron is responsible for the effect of cooking, Labuza (1971) presented arguments against the theory that iron must be in the oxidized form in order to catalyze lipid autoxidation.

The second step is the addition of molecular oxygen to give an activated peroxide:

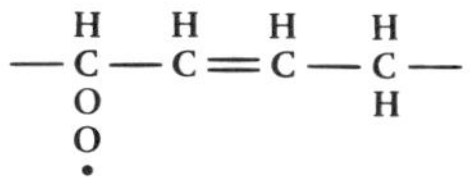

$$
\begin{array}{ccccccccc}
 & H & & H & & H & & H & \\
- & C & - & C & = & C & - & C & - \\
 & O & & & & & & H & \\
 & O & & & & & & & \\
 & \bullet & & & & & & &
\end{array}
$$

Activated peroxide

The activated peroxide is very reactive and relatively little energy is required for removal of a hydrogen from another carbon adjacent to a double-bond carbon (in another chain), resulting in the formation of a hydroperoxide on the initial chain:

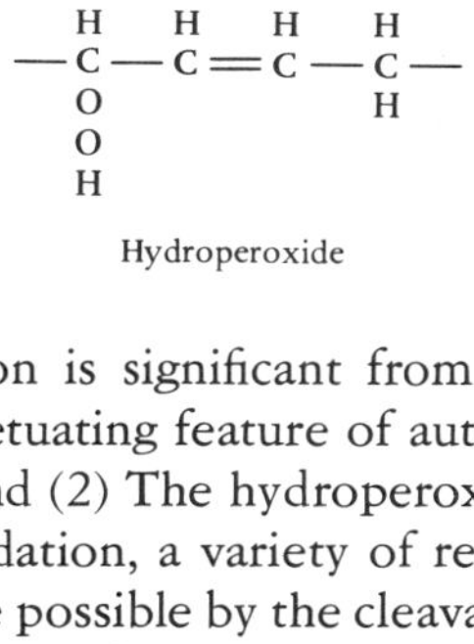

$$
\begin{array}{ccccccccc}
 & H & & H & & H & & H & \\
- & C & - & C & = & C & - & C & - \\
 & O & & & & & & H & \\
 & O & & & & & & & \\
 & H & & & & & & &
\end{array}
$$

Hydroperoxide

Hydroperoxide formation is significant from two standpoints: (1) It represents the self-perpetuating feature of autoxidation because a new free radical is formed; and (2) The hydroperoxide itself is unstable and is subject to further oxidation, a variety of rearrangements and cleavages, new reactions made possible by the cleavages, and polymerization. Among the products formed are short-chained aldehydes, ketones, acids, and hydroxyl compounds.

As with hydrolysis, an off flavor results and is referred to as rancidity. In some foods, the off-flavor development apparently parallels the development of the breakdown product hexanal. The existence of a rancid flavor in itself does not indicate whether the fat has undergone hydrolysis or autoxidation or some of each.

Several factors affect the occurrence of lipid autoxidation: (1) Extent of unsaturation of the lipid. Polyunsaturates are oxidized before monounsaturates. Linoleic acid is oxidized 15 times and linolenic acid about 30 times as fast as oleic acid. (2) Configuration at double bonds. *Trans* isomers are less susceptible than are *cis* isomers. (3) Degree of esterification. Glycerides are less susceptible than are free fatty acids. (4)

Availability of catalysts to initiate the reaction. Exposure to light is readily controlled. The catalytic activity of metal contaminants can be curtailed by the presence of chelating substances, or sequestrants. These include citrate, ascorbate, tartrate, phosphate, and ethylenediaminetetracetic acid (EDTA), which form complexes with metals and thus make the metals unavailable. Citrates and EDTA in particular are frequently added to fatty foods that are to be stored. (5) Availability of oxygen. Either vacuum packaging or replacement of air with a nitrogen atmosphere has a protective effect on food to be stored. (6) The presence of antioxidants. Antioxidants occur naturally along with some food lipids; for example, tocopherols are associated with vegetable oils. The phenolic compounds butylated hydroxyanisole (BHA), butylated hydroxytoluene (BHT), tertiary butylhydroquinone (TBHQ), and propyl gallate (PG) are antioxidants that are added commonly to commercially processed and manufactured foods, though their use is controversial in some cases. Sherwin (1976) reviewed the action of antioxidants and the properties of specific antioxidants. The phenolic antioxidants differ in their solubility and their sensitivity to heat and pH and most frequently two antioxidants are used in combination. Chelating agents do not have a direct antioxidant effect; they enhance the protective effect of the antioxidants and thus are referred to as synergists. (7) Storage temperature. Both the initiation of autoxidation and the reactions undergone by hydroperoxides are temperature dependent.

The progress of autoxidation may be followed by measuring oxygen uptake over a period of time. In the early stages of autoxidation the oxygen uptake curve is essentially flat. The period of time represented by the flat portion of the curve is referred to as the *induction period,* and may be quite long. During the induction period, free radicals form at a rate that depends on such factors as temperature, exposure to light, and presence of metals and of chelating agents. A sharp rise in the curve (the end of the induction period) tends to coincide with exhaustion of the supply of antioxidant. Rancidity develops primarily during the post-induction period because free radicals are not responsible for the off flavor. In fast-food operations, where so much frying is done that addition of new fat is almost continuous, most of the frying fat at any one time probably is in the induction period with respect to oxidation. Rancidity that develops in such fat is more likely to be hydrolytic than oxidative in origin. Accelerated storage studies, conducted routinely in the food industry for assessing shelf life of foods, are conducted at elevated temperatures in order to shorten the induction period and thus the time required for rancidity to develop.

Chemical tests used in the study of lipid autoxidation are the determination of peroxide number and the TBA test. Each method has limitations. Although peroxides are important as precursors of the off-flavor compounds, the increase in their concentration does not necessarily parallel the development of rancidity. The TBA test, which involves the formation of a measurable red condensation product between 2-thio-barbituric acid (TBA) and malonic dialdehyde, still is empirical in the sense that what is being measured is not what causes the off flavor. However, because it does measure a product rather than an intermediate, it correlates quite well with flavor change in some foods (Tarladgis et al., 1960). Recently gas liquid chromatographic analysis of specific volatile carbonyl compounds has provided a more direct means of studying rancidity development through autoxidation.

Reversion Reversion represents a specialized type of oxidative degradation. Certain oils—for example, soybean and rapeseed oils—are susceptible to a relatively limited degree of oxidation involving linolenic acid, which is contained in appreciable amounts by those oils that are susceptible (Sherwin, 1976). The term *reversion* is based on the flavor change because it occurs in oil from which a distinctive, though different, flavor has been removed in processing.

Hydrogenation Treatment of oils, which contain highly unsaturated lipids, with hydrogen under pressure in the presence of a suitable catalyst results in saturation of double-bond carbons. This chemical change in turn raises the melting point. The reaction is selective in that polyunsaturated fatty acids are hydrogenated first. Double bonds farthest from the carboxyl group of free fatty acids, or from the ester linkages of glycerides, are saturated first. Thus if the major fatty acid in an oil is linoleic acid, as in cottonseed oil, hydrogenation results in a fat in which oleic acid predominates. A detailed discussion of selectivity of hydrogenation and the effects of conditions and type of catalyst is found in Coenen's review (1976).

Hydrogenation is the basis for industrial manufacture of solid shortenings from oils. Complete hydrogenation would result in excessive hardness. The process is carried just far enough to produce the physical properties desired.

An interesting example of biological hydrogenation occurs in ruminant animals. For many years it was considered impossible to produce beef and lamb having elevated levels of dienoic fatty acids, because rumen bacteria efficiently hydrogenate dienes in unprotected rations. Recent developments, however, have resulted in the production of high-linoleate beef and lamb through feeding them protected supplements. A high-linoleate oil, encased in denatured protein, is able to pass the rumen unaltered, and the linoleic acid eventually is incorporated into body fat. Scott et al. (1971) reported linoleic acid concentrations of 3% and 28% in the subcutaneous fat of control sheep and sheep fed a protected supplement, respectively. Not surprisingly, freezer storage life of high-linoleate lamb and beef is relatively short because of susceptibility to lipid autoxidation (Bremner et al., 1976).

Isomerization Isomerization, *cis-trans* and positional, occurs as a side reaction of hydrogenation. Conversion from *cis* to *trans* configurations makes a substantial contribution to the change in melting point during commercial hydrogenation. Ottenstein et al. (1977), who studied the fatty acid content of seven margarines purchased at retail stores, reported that the samples contained total 18:1 fatty acid averaging about 45% of the total fatty acid content. Nearly half of the 18:1, about 20% of the total fatty acid content, was in the *trans* form on the average, though the samples differed considerably. With a different catalyst and a lower hydrogen pressure, *cis-trans* isomerization may be largely responsible for the change in physical state. "Hard butters" appropriate for confectionery coatings may be prepared by hydrogenation of soybean oil under optimum *trans* conditions, resulting in a greater than 50% *trans* fatty acid content (Paulicka, 1976).

Positional isomerization is a migration of double bonds in fatty acid chains. Different types of isomerization actually are different reactions but tend to occur simultaneously. Carpenter and Slover (1973) studied both *trans* and positional isomers in purchased margarines and reported that positional isomerization more frequently was associated with the *trans* configuration than with the *cis*.

Halogenation The ability of a halogen to be added to double bond carbons in unsaturated fatty acids is the basis for analytical estimation of the extent of unsaturation of a fat. Iodine value is the number of grams of iodine absorbed by 100 g of fat and ranges from 30–40 for butterfat to more than 130 for some seed oils.

PROCESSING OF FATS

Fats do not occur free in nature but are isolated and purified from foods high in fat, such as cream, animal tissues, fruits, and seeds. Recent technological advances make possible the production of fats that are tailor-made for specific uses. The variety of fats available to food manufacturers far exceeds that on the retail market; however, most consumers probably use two or three types of fat. Discussion of processing methods will be limited to principles that are related to chemical and physical properties, already covered, and to functional properties, yet to be discussed.

Animal Fats

Butter Cream, separated from milk by centrifugation, is pasteurized to inactivate lipase and destroy most of the organisms present. It usually is ripened by means of a mixed bacterial culture containing both lactose- and citrate-fermenting species and is churned. The churning process results in clumping of fat globules and transformation of the oil-in-water emulsion to a water-in-oil emulsion. Buttermilk is separated from the mass of fat, which is washed with water, colored for uniformity, and usually salted. The final product contains 80–81% fat. Short-chain fatty acids and diacetyl are largely responsible for the characteristic flavor of butter. Diacetyl is an oxidation product of acetylmethylcarbinol, which is produced as the first product of citrate fermentation.

$$CH_3-\overset{\overset{O}{\|}}{C}-\underset{H}{\overset{OH}{\overset{|}{C}}}-CH_3 \xrightarrow{-2\,H} CH_3-\overset{\overset{O}{\|}}{C}-\overset{\overset{O}{\|}}{C}-CH_3$$

Acetylmethylcarbinol → Diacetyl

Diacetyl formation

Consistency and texture depend on the fatty acid content of the cream and on the specific chilling and churning treatments (Mattill, 1964).

Lard Lard is heat-rendered from fatty tissues of the hog. The fatty tissues, containing connective tissue as well as fat, are chopped and heated, with or without added water. Wet-rendering is the more common procedure. The heating may be done in a steam-jacketed kettle or may consist of direct treatment with steam. Once the fat has been separated out, it may be modified in any of several ways, including bleaching, hydrogenation, deodorization, rearrangement, addition of emulsifier, and addition of antioxidant.

Lards produced by current technology are quite different in their properties from those represented by data presented in early research literature. Present-day lards also are tailor-made for special purposes. Lard's normal β-crystal structure is modified (by rearrangement) to give the β'-structure in lard that is to be used in cakes and icings. Rearrangement is omitted and the β-type structure is retained in lard that is to be used in pastry (Weiss, 1970).

Beef fat Several shortenings on the retail market consist of a mixture of meat fat and vegetable oil. Beef fat, one of the fats used in such shortenings, usually is dry-rendered, deodorized, and hydrogenated. It crystallizes in the β'-form (Weiss, 1970). Luddy et al. (1973) fractionated beef tallow into fractions with different properties for special uses.

Plant Fats

Vegetable oils Oilseeds are the major source of oil from plants and their use has increased tremendously in recent years. An entire issue of the *Journal of the American Oil Chemists' Society* (AOCS, 1976) recently was devoted to the proceedings of a 1976 world conference on processing of oilseeds and vegetable oil.

Seeds such as corn, cottonseed, soybean, and peanuts are cleaned and tempered and dehulled as needed, then flaked or crushed. Oil is removed by solvent extraction, application of pressure, or a combination of these methods. The flesh of olives is pressed for "virgin olive oil" and the residue and pit kernels are solvent-extracted for "pure olive oil." Coconut oil is pressed from dried coconut flesh. Vegetable oils undergo further processing to remove pigments by adsorption, free fatty acids by light saponification, and volatile odor-causing compounds by steam distillation under reduced pressure. Olive oil undergoes minimal processing because of the prized uniqueness of its flavor. Cottonseed oil must be freed of gossypol, a toxic polyphenolic component of cottonseed. Silicones frequently are added to frying oils to suppress foaming. Lesieur (1976) summarized trends in consumer use of oils and discussed criteria for choosing salad and cooking oils.

Some oils are winterized to permit them to remain liquid at refrigerator temperature. *Winterizing* involves chilling to crystallize the saturated high molecular weight components, then separating the solids from the cold liquid by filtration (Kreulen, 1976). Oils from some sources are so highly unsaturated that they remain uniformly liquid at refrigerator temperature.

The residual solids are rather high in protein and are used in animal feeds. Recent interest in expanding the supply of protein for humans has resulted in extensive study and considerable utilization of these products in human food. For example, the defatted soy flakes may be processed into many forms, including flour, soy protein concentrate (about 70% protein), or soy protein isolate (approx 90% protein). The isolate, in turn, may be processed into a variety of textured forms such as spun fibers for use in meat analogs and granular forms for use as meat extenders.

Vegetable shortenings Plastic solid shortenings are made by the hydrogenation of refined oils. Oils from different sources frequently are combined and treated with hydrogen in the presence of a nickel catalyst or, occasionally, a copper or nickel-silver catalyst. The principles of the hydrogenation process were discussed earlier in this chapter. When the desired extent of hydrogenation has been attained, the monoglyceride-diglyceride emulsifier is added and the shortening is chilled and whipped. Although gaps exist in the available information as to the underlying science, the state of shortening technology today is quite advanced.

Margarines Margarines, made to simulate butter, most frequently are mixtures of vegetable oils that are hydrogenated to the desired consistency and mixed with pasteurized and cultured skim milk, salt, emulsifiers, preservatives, and color. Diacetyl as such may be added for flavor or it may be developed in the milk, along with lactic acid, by the bacterial starter (Haighton, 1976). Like butter, margarine contains 80–81% fat. Various modifications have resulted in a recent proliferation of margarine types. One modification consists of hydrogenating some oil to a very hard stage and then mixing unhydrogenated oil with the hard fat. This results in the desired consistency along with a higher concentration of polyunsaturated fatty acids than is characteristic of the margarines made by hydrogenating all of the oil to the desired consistency. (Remember the selectivity of the hydrogenation reaction.)

A more recent variant on the retail market is a margarine that is similar to a conventional margarine but has a high enough concentration of unhydrogenated oil to result in a semisolid consistency even at room temperature. The cold margarine is soft enough to be easily extruded from a squeeze-type bottle.

Some margarines are physically modified by whipping to incorporate air and thus increase volume. Such products on the market represent about a 50% increase in volume over that of conventional margarines. It should be remembered that these products are lower in calories only if they are substituted for a conventional spread on a volume basis. Gram for gram, they are not lower in calories. On the other hand, there are some special-purpose spreads that are made with lower fat (about 40%) and higher water contents than those of butter and margarine.

Weiss (1970) presents additional information concerning margarine.

FUNCTIONS OF FATS IN FOOD

The contributions of fat to food properties are related to the chemical nature of lipids and the resulting physical properties of fats. The effects of fats on food are largely textural but to a lesser extent also involve color and flavor.

Tenderization

One of the most important functions of fat is to tenderize baked products that otherwise might be solid masses firmly held together by strands of gluten. This function is particularly important in pastry and breads, which have little or no sugar. Fat, being insoluble in water, interferes with gluten development during mixing. Fat adsorbed on surfaces of gluten proteins interferes with hydration and thus with the development of a cohesive gluten structure. In the mixing of pastry and biscuits, intimate mixing of the fat with the other ingredients is deliberately avoided so that the fat will cause layers of dough to form. These products are likely to be flaky as well as tender. However, flakiness and tenderness are different properties and may not be entirely compatible; as flour and fat are mixed more thoroughly, the product may become more tender but less flaky. The measurement of pastry tenderness will be discussed in Chapter 16.

Langmuir's classic study (1917) of monomolecular films of fatty acids and triglycerides on water led to the theory that the shortening power of a fat is related to the surface covered. Langmuir observed that the surface covered by a given weight of lipid varies inversely with fatty acid chain length and is greater for unsaturated fatty acids and triglycerides than for saturated compounds. The latter difference apparently reflects the attraction of double bonds, as well as the hydrophilic polar ends of the molecules, to the aqueous phase. The amount of surface covered is not proportional to the number of double bonds in the chain. Whereas oleic acid (18:1) occupies a greater interfacial area than does stearic acid (18:0), linoleic acid (18:2) does not occupy more space than oleic. Linolenic acid (18:3) does occupy more space than any of the other 18-C fatty acids. The imperfect relationship might reflect a steric factor.

Attempts to relate shortening power to degree of lipid unsaturation, physical state, melting point, plasticity, and other properties of shortenings have not had consistent results. Lowe et al. (1938) found a relationship between the shortening value and the iodine number of a series of fats, most of which were lards. Hornstein et al. (1943), studying more varied fats, did not observe such a relationship. Matthews and Dawson (1963) considered their data to show that plasticity of fat is important to tenderness of baking powder biscuits but not to that of pastry. Hornstein et al. (1943) reported a high negative correlation between breaking strength of pastry and consistency of the worked fat (plasticity) at 22°C for machine-made pastry but not under the other conditions tested. Matthews and Dawson (1963) found oil to give greater tenderness than solid shortenings at some but not all levels of fat in pastry.

One of the many problems in relating fat properties to shortening power is the dependence of the tenderness of pastry on many factors in addition to the fat. For example, Hornstein et al. (1943), who compared a number of fats in pastry mixed at different temperatures and by different methods, found that the relative shortening values of the fats depended on mixing temperature and method. Correlations between shortening value and properties of fat also depended on mixing temperature and method. Even if control of conditions is excellent and the effects of the fats on tenderness are obvious, interpretation frequently is difficult because of the heterogeneity and variability of fats and the interrelatedness of their chemical and physical properties. For example, if an oil performs differently from a solid fat, is the effect attributable to a difference in unsaturation, in physical state, or in crystallinity? It probably is impossible to compare two fats that differ from one another in a single respect. This is not an argument against comparing fats as to shortening power but an argument for exercising caution in interpreting results, as well as for control of experimental conditions.

Aeration

Fat contributes to the incorporation and retention of air in the form of small bubbles distributed throughout the batter. These bubbles serve as gas cell nuclei into which carbon dioxide and steam diffuse during baking; thus they are important to grain and volume of the baked product. The role of fat in aeration may be related to its effect on batter viscosity. Finely dispersed fat enhances batter viscosity; increased viscosity, up to a point, increases gas retention. Monoglycerides added to hydrogenated shortenings increase the extent of emulsification of shortening; increased viscosity and increased batter aeration are associated with the increased emulsification.

Heat Transfer

The temperature to which a fat can be heated is limited only by its smoke point, the temperature at which degradation is sufficient to result in evolution of smoke. Smoke point is much higher than the boiling point of water; therefore, frying temperatures ranging from 175 to 195°C are used and permit much more rapid cooking than occurs in water. In addition, surface drying and browning occur and a distinctive flavor is associated with the browning. Although air can be heated to an even higher temperature than can fat, air is far less efficient as a heating medium.

Smoke points of fats currently available are beyond the temperature range used for normal frying, and problems develop only with misuse of fat. Fat degradation with consequent lowering of smoke point occurs gradually during heating, increasing with both time and temperature. The availability of water affects the rate of degradation, as mentioned previously. Monoglycerides and diglycerides in the fat lower smoking temperature; therefore, a fat that is ideal for a cake batter is not the best frying fat.

In addition to the degradation associated with smoking, frying fat gradually oxidizes, polymerizes, and increases in tendency to foam. Protection from free metals and light and the use of fats containing silicones are helpful. The chemical and physical changes that occur in fat during heating have been reviewed by Roth and Rock (1972a, b).

Food being fried absorbs some fat, the amount of absorption varying with the food and with heating conditions (Bean et al., 1963; Rock and Roth, 1964). Composition of the food that is fried has variable effects on the lipids of the frying fats and/or those of the fried food (Bennion, 1967; Heath et al., 1971; Kilgore and Bailey, 1970; McComber and Miller, 1976).

Contribution to Emulsion Structure

Food emulsions could not exist without fats; fats constitute one of the essential phases. Food emulsions occur naturally, as in milk, and are prepared in foods such as mayonnaise and cake batters. Except in margarine and butter, oil is the dispersed, or discontinuous, phase in food emulsions, and water is the dispersion medium, or continuous phase. Oil would not remain dispersed in an aqueous medium in the absence of a third phase, the emulsifying agent. The lipoprotein of egg yolk is a particularly effective emulsifying agent and functions in that capacity in many foods. In mayonnaise, in which the oil concentration is particularly high, dilution of egg yolk with egg white, or total substitution of egg white for yolk, results in a product that is less viscous and less stable than the product with egg yolk. Starch contributes to emulsification of fat in some foods, such as gravies and sauces. Vegetable gums, which are used increasingly in formulated foods, function in many ways, including emulsification.

Evaporation of water from an emulsion can be disastrous. In an oil-in-water emulsion in which the amount of dispersed phase is about at its upper limit relative to the dispersion medium (water), the system is vulnerable. Gravy is a food in which the fat and water proportions vary greatly. Gravy also is a food that sometimes is subjected to prolonged heating and holding and thus to opportunity for evaporation to occur. It is not unusual, therefore, for fat to separate from a gravy during holding. If this happens to a gravy that initially appears homogeneous, the solution is to replace the lost dispersion medium or, in other words, to add water.

Other Contributions to Texture

Fats have additional textural effects in foods. They affect the smoothness of crystalline candies and frozen desserts through retardation of crystallization. They affect consistency of starch-thickened mixtures by influencing gelatinization. They contribute to apparent juiciness of meats. They contribute to the foam structure of whipped cream.

Contributions to Flavor

Fats influence flavor, whether added for that purpose or to serve another function. Often a specific fat is chosen for a specific use because of its unique flavor. Butter, bacon fat, and olive oil are examples. Fats also act as carriers or solvents for added food flavors.

EXPERIMENTS

I. Crystallization Behavior of Fats

A. Procedures

1. Obtain fatty tissue of pork and devise a reasonable means of wet-rendering the fat and separating the fat from the residual solids and the water.
2. Melt equal portions of the pork fat and a hydrogenated shortening, heating to the same temperature. Divide each

into four equal portions in beakers (30 ml of melted fat in a 50-ml beaker if the supply of pork fat permits).
3. Cool the four samples of each fat as follows:
   a. Undisturbed, in an ice bath
   b. With constant stirring in an ice bath
   c. Undisturbed, at room temperature
   d. With constant stirring at room temperature
4. When samples c and d have solidified, refrigerate all samples long enough to ensure that all samples will be at the same temperature when compared.
5. Observe translucency and apparent smoothness of the solidified fats.

B. *Results*

Make a chart, listing the treatments down the left side and translucency and smoothness across the top. Write brief descriptions on the chart.

C. *Questions*

What is the effect of rate of solidification on the properties of the two fats? What is the effect of agitation on the properties? What is the chemical basis for the physical differences?

II. A Comparison of Emulsifying Agents

A. *Procedures*

1. Pour 5 ml of salad oil into each of seven test tubes.
2. Make one of the following additions to each tube:
   a. 5 ml of water
   b. 5 ml of vinegar
   c. 5 ml of vinegar with 1/4 tsp added paprika
   d. 5 ml of a 1 : 1 aqueous dilution of egg white
   e. 5 ml of a 1 : 1 aqueous dilution of egg yolk
   f. 5 ml of a gelatinized 1% starch suspension
   g. 5 ml of a heated and cooled 1% gelatin dispersion
3. With the thumb over the end of the tube, shake each tube vigorously 50 times.

B. Results

Make a chart with treatments listed down the left side and minutes required for beginning of separation and for completion of separation across the top. For any sample that is so stable that no change is observed, record the time in which no apparent separation occurs. (Write an appropriate footnote.)

C. Questions

How do the added substances compare as to emulsifying effectiveness? What is a temporary emulsion? What is a permanent emulsion?

III. Shortening Power of Different Types of Margarine

In the first part of this experiment, two solid margarines that differ as to extent of polyunsaturation are compared as shortening agents in pastry. In the second part, a conventional margarine and a semiliquid margarine are compared. All margarines should be at refrigerator temperature when added to the flour.

A. Procedures

1. Basic pastry formula, mixing method, and wafer preparation method (use assembly line for control).

All-purpose flour, g	110
Salt, g	1
Fat, g	50
Water, ml	25–35

Put flour, salt, and fat into small mixing bowl. Using a pastry blender, cut fat into the flour until no particles are larger than peas. Add water, sprinkling over the flour-fat mixture and cutting in with the blender. After every five blending strokes, scrape dough from the blender with a rubber scraper. Put ball of dough on a piece of waxed paper large enough to roll it on later and let stand 5 min. Pat out with the hands until 2.5 cm thick. Place a metal guide strip on each side of dough and cover dough with another piece of waxed paper. Roll until the thickness of the dough is the same as that of the guide strips. Peel off the upper paper and invert the pastry on the baking sheet. Remove the other piece of paper and cut the wafers 4.5 × 9 cm. Do not separate the wafers from each other and do not remove the extra dough around the edges. Prick the wafers uniformly with a fork or with a meat tenderizer (a block of wood containing narrow blades). Also prick the dough around the edges in a few places. Bake at 218°C (425°F) until the edges of the dough are light brown. Cool. Measure the index to flakiness (thickness of a stack of four wafers) with a vernier caliper and the breaking strength of as many wafers as possible with a shortometer (Chapter 16). If a shortometer is not available, conduct sensory evaluation of tenderness.

2. Variations.
   a. Compare a conventional margarine and a margarine that is a solid mixture of oil and hard margarine (read the labels).
   b. Compare a conventional margarine and a semiliquid margarine, preferably of the same brand.

B. *Results*

Make separate tables for 2a and 2b. Write the treatments (types of margarine) down the left side and the index to flakiness and breaking strength across the top.

C. *Questions*

What problems of control were encountered in the experiment? How might control procedures have been improved? Summarize the results of 2a and 2b and formulate possible explanations.

IV. Batter Aeration and Fat Emulsification

A vegetable oil and a hydrogenated vegetable oil are dyed and compared in "cake" batters. Several observations on the batters and baked samples are used as criteria of aeration and emulsification.

A. *Procedures*

1. Basic formula, mixing method, and observations.

Fat, g	96 (Dye with Sudan III).
Sugar, g	200
Egg white, g	96
Flour, g	190
Milk, ml	108

Cream sugar and fat 2 min on moderate speed, stopping after 1 min to scrape bowl and beaters. Add egg white and beat 1 min on moderately high speed. Add the flour in three portions, alternating with the milk in two portions, mixing 30 sec on low speed after each addition; beat 1 min at moderately high speed. Determine average line-spread (Chapter 16) and either batter density or specific gravity (Chapter 16). Prepare slides for microscopic observation. Bake three 40-g samples of each batter in a muffin pan at 218°C (425°F). Evaluate volume, using an appropriate scale. (See the Appendix.)

2. Variation.

Compare a vegetable oil and a hydrogenated vegetable oil.

B. *Results*

Make a table listing the fats down the left side and the average line-spread, batter density or specific gravity, batter appearance (microscopic), and volume score across the top. In recording batter appearance, sketch the air cells and fat globules.

C. *Questions*

How do the batters in this experiment differ from true cake batters in addition to their containing dyed fat? Which observations are criteria of batter aeration and emulsification respectively? Does there appear to be a relationship between aeration and emulsification? What is the apparent effect of physical state of the fat on batter aeration and fat emulsification? Why can the apparent effect *not* be attributed definitely to a difference in physical state?

SUGGESTED EXERCISES

1. Obtain conventional margarine and whipped margarine, preferably of the same brand. Determine either their density or their specific gravity (Chapter 16).
2. Compare tenderness of pastry made with shortening and margarine substituted both on an equal weight and on an equal fat basis.

REFERENCES

AOCS. 1976. World conference on oilseed and vegetable oil processing technology. Complete proceedings. J. Amer. Oil Chemists' Soc. 53: 221.

Aurand, L. W. and Woods, A. E. 1973. "Food Chemistry." Avi Publ. Co., Westport, CT.

Bean, M. L., Sugihara, T. F. and Kline, L. 1963. Some characteristics of yolk solids affecting their performance in cake doughnuts. I. Effects of yolk type, level, and contamination with white. Cereal Chem. 40: 10.

Bennion, M. 1967. Effect of batter ingredients on changes in fatty acid composition of fats used for frying. Food Technol. 21: 1638.

Bremner, H. A., Ford, A. L., Macfarlane, J. J., Ratcliff, D. and Russell, N. T. 1976. Meat with high linoleic acid content: oxidative changes during frozen storage. J. Food Sci. 41: 757.

Carpenter, D. L. and Slover, H. T. 1973. Lipid composition of selected margarines. J. Amer. Oil Chemists' Soc. 50: 372.

Chang, P.-Y., Younathan, M. T. and Watts, B. M. 1961. Lipid oxidation in precooked beef preserved by refrigeration, freezing, and irradiation. Food Technol. 15: 168.

Coenen, J. W. E. 1976. Hydrogenation of edible oils. J. Amer. Oil Chemists' Soc. 53: 382.

CRC. 1977. "Handbook of Chemistry and Physics," 58th ed., Ed. Weast, R. C. CRC Press, Inc., Cleveland.

Gurr, M. I. and James, A. T. 1971. "Lipid Biochemistry: An Introduction." Cornell University Press, Ithaca, NY.

Haighton, A. J. 1976. Blending, chilling, and tempering of margarines and shortenings. J. Amer. Oil Chemists' Soc. 53: 397.

Heath, J. L., Teekell, R. A. and Watts, A. B. 1971. Fatty acid composition of batter coated chicken parts. Poultry Sci. 50: 219.

Hoerr, C. W. and Paulicka, F. R. 1968. The role of x-ray diffraction in studies of the crystallography of monoacid saturated triglycerides. J. Amer. Oil Chemists' Soc. 45: 793.

Hornstein, L. R., King, F. B. and Benedict, F. 1943. Comparative shortening value of some commercial fats. Food Res. 8: 1.

Hustedt, H. H. 1976. Interesterification of edible oils. J. Amer. Oil Chemists' Soc. 53: 390.

Hvolby, A. 1974. Expansion of solidifying saturated fats. J. Amer. Oil Chemists' Soc. 51: 50.

Kilgore, L. and Bailey, M. 1970. Degradation of linoleic acid during potato frying. J. Amer. Dietet. Assoc. 56: 130.

Kreulen, H. P. 1976. Fractionation and winterization of edible fats and oils. J. Amer. Oil Chemists' Soc. 53: 393.

Labuza, T. P. 1971. Kinetics of lipid oxidation in foods. CRC Critical Reviews in Food Technol. 2: 355.

Langmuir, I. 1917. The constitution and fundamental properties of solids and liquids. II. Liquids. J. Amer. Chem. Soc. 39: 1848.

Lesieur, B. 1976. Salad and cooking oils. J. Amer. Oil Chemists' Soc. 53: 414.

Lowe, B., Nelson, P. M. and Buchanan, J. H. 1938. The physical and chemical characteristics of lards and other fats in relation to their culinary value. I. Shortening value in pastry and cookies. Res. Bull. 242. Iowa Agric. Exp. Stn., Ames.

Luce, G. T. 1967. Acetylated monoglycerides as coatings for selected foods. Food Technol. 21: 1462.

Luddy, F. E., Hampson, J. W., Herb, S. F. and Rothbart, H. L. 1973. Development of edible tallow fractions for specialty fat uses. J. Amer. Oil Chemists' Soc. 50: 240.

Lutton, E. S. 1972. Lipid structures. J. Amer. Oil Chemists' Soc. 49: 1.

Matthews, R. H. and Dawson, E. H. 1963. Performance of fats and oils in pastry and biscuits. Cereal Chem. 40: 291.

Mattil, K. F. 1964. Butter and margarine. In "Bailey's Industrial Oil and Fat Products," 3rd ed., Ed. Swern, D. Interscience Publishers, New York.

McComber, D. and Miller, E. M. 1976. Differences in total lipid and fatty acid composition of doughnuts as influenced by lecithin, leavening agent, and use of frying fat. Cereal Chem. 53: 101.

Ottenstein, D. M., Wittings, L. A., Walker, G., Mahadevan, V. and Pelick, N. 1977. *trans* fatty acid content of commercial margarine samples determined by gas liquid chromatography on OV-275. J. Amer. Oil Chemists' Soc. 54: 207.

Paulicka, F. R. 1976. Specialty fats. J. Amer. Oil Chemists' Soc. 53: 421.

Rock, S. P. and Roth, H. 1964. Factors affecting the rate of deterioration in the frying qualities of fats. I. Exposure to air. J. Amer. Oil Chemists' Soc. 41: 228.

Roth, H. and Rock, S. P. 1972a. The chemistry and technology of frying fat. I. Chemistry. Baker's Digest 46(4): 38.
Roth, H. and Rock, S. P. 1972b. The chemistry and technology of frying fat. II. Technology. Baker's Digest 46(5): 38.
Scott, T. W., Cook, L. J. and Mills, S. C. 1971. Protection of dietary polyunsaturated fatty acids against microbial hydrogenation in ruminants. J. Amer. Oil Chemists' Soc. 48: 358.
Sherwin, E. R. 1976. Antioxidants for vegetable oils. J. Amer. Oil Chemists' Soc. 53: 430.
Swern, D. 1964a. Structure and composition of fats and oils. In "Bailey's Industrial Oil and Fat Products," 3rd ed., Ed. Swern, D. Interscience Publishers, New York.
Swern, D. 1964b. Physical properties of fats and fatty acids. In "Bailey's Industrial Oil and Fat Products," 3rd ed., Ed. Swern, D. Interscience Publishers, New York.
Tarladgis, B. G., Watts, B. M., Younathan, M. T. and Dugan, L. Jr. 1960. A distillation method for the quantitative determination of malonaldehyde in rancid foods. J. Amer. Oil Chemists' Soc. 37: 44.
Weiss, T. J. 1970. "Food Oils and Their Uses." Avi Publ. Co., Westport, CT.

9. Starch and Flour

Starch, extracted from cereals and from some other plants, is chemically rather pure. Therefore, it will be discussed in general terms, along with some consideration of differences among starches from different sources. Flours are multicomponent substances and differ considerably with source. Wheat flour will be emphasized in this chapter.

STARCH

Starch, a polysaccharide, occurs in colorless plastids (leucoplasts) of some plant cells, in which it serves a storage function. Major sources include cereal seeds and certain roots and tubers. The details of the isolation of starch from plant tissue vary with the tissue and will not be presented here.

The chemical nature of starch, involving the molecular structures and the changes they can undergo, and the physical nature of starch, involving the granular structures and the changes they can undergo, need to be distinguished from, and yet related to, one another.

Chemical Structure and Properties

Two types of starch molecule occur together in most sources. *Amylose* is an essentially linear chain with the glucose units in α1,4-glycosidic linkage (Figure 9.1). Possibly a small degree of branching exists, of the order of one branch point per 500 glucose residues (Banks et al., 1973). Free amylose in aqueous dispersion probably forms helical structures (Foster, 1965). The following is a segment of a helix:

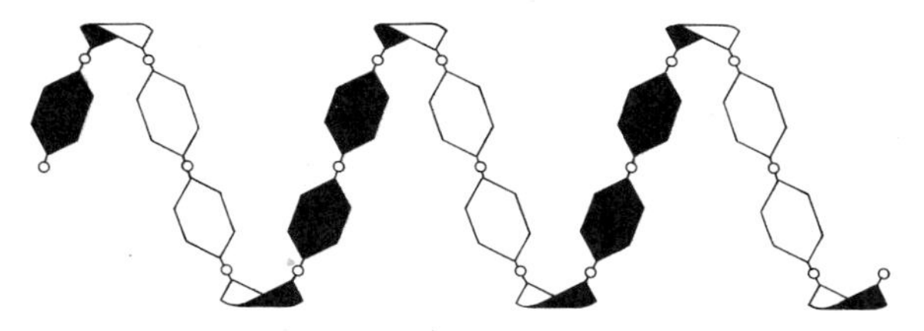

Segment of a helix

Amylopectin has a highly branched molecule, represented in Figure 9.2. The chains are formed from glucose units in α1,4-glycosidic linkages and the branch points result from α1,6-linkages (Figure 9.3). The branch points in amylopectin average one per 16–25 glucose residues (Banks et al., 1973) in different starches. Osman (1972) points out that average values for frequency of branch points can be misleading; it is likely that considerable variation occurs, perhaps even within a single amylopectin molecule.

Each starch molecule, whether linear or branched, and regardless of size, has one reducing end-group (Study Figures 9.1, 9.2, and 9.3, keeping in mind that the potential reducing group is at the number 1 carbon.). Reducing power, therefore, has been used as an indication of

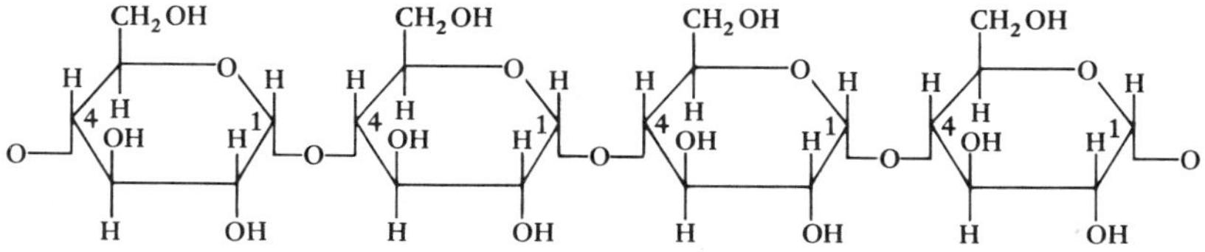

Figure 9.1. Segment of amylose chain.

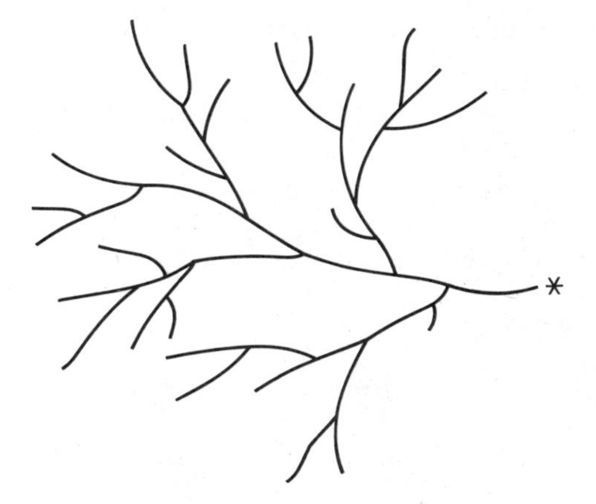

Figure 9.2. Diagrammatic representation of amylopectin molecule. The asterisk marks the location of the reducing group.

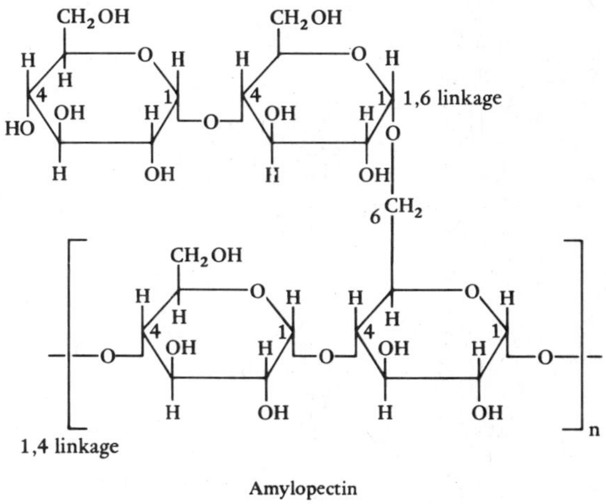

1,6 linkage

1,4 linkage

Amylopectin

Figure 9.3. Segment of amylopectin molecule.

changing molecular size during hydrolysis. For every glycosidic bond hydrolyzed, a number 1 carbon is exposed.

Molecular weights of the amylose and amylopectin components of starch vary greatly between sources, between specific samples, and undoubtedly even between molecular species in a given sample. Averages thus are not very meaningful, but amylopectin is recognized as having a much greater molecular size than amylose; molecular weight values that have been reported for amylopectin run into the millions (Greenwood, 1976).

The ability of amylose to form a long helix and of amylopectin branches to form only short helices is responsible for a difference in the color change resulting from addition of iodine to these starch fractions separately. The structure of the helix in either an amylose molecule or an amylopectin branch is such that certain substances readily complex with the chain. Iodine, for example, is visualized as a core running through the center of helices when an iodine solution is added to a starch dispersion. One molecule of iodine complexes with six glucose units in one turn of the helix. If the helix is long enough, as in amylose, the complex is blue. The branches in amylopectin are not long enough to give the blue color with iodine, so free amylopectin treated with iodine has a reddish purple color. Holló and Szeitli (1968) discuss the reaction of starch with iodine in detail. Fatty acids also can form complexes with long chains such as those of amylose.

Starch molecules are susceptible to the action of amylases. The α-amylases are endo enzymes, hydrolyzing α1,4-glycosidic linkages in the inner regions of molecules. Products are of varying size and include oligosaccharides and low molecular weight dextrins. The β-amylases are exo enzymes, splitting unmodified amylose chains into maltose units, beginning at the nonreducing end. Acting on α1,4 bonds, they hydrolyze maltose molecules from the branches of amylopectin until branch points are reached, resulting in a single large dextrin per amylopectin molecule. Dextrin so formed is referred to as *limit dextrin.* Flour contains appreciable amounts of β-amylase but only traces of α-amylase.

Two additional types of amylases should be mentioned. Amyloglucosidases are of fungal origin and act similarly to β-amylase except that they release glucose rather than maltose. The α1,6-glucosidases can hydrolyze the α1,6-branching linkages. Once this has occurred, β-amylase catalysis of maltose production can proceed. Detailed discussion of the properties and actions of amylases is found in the reviews of Robyt and Whelan (1968a, b), Fleming (1968), and Price (1968). Amylolytic enzymes are useful in the conversion of cornstarch to corn sirups.

Chemical modification of starch will be discussed briefly later in the chapter.

Physical Structure and Properties

The physical structural unit of starch is the granule, which has a distinctive microscopic appearance for each plant source (Figure 9.4). Granules of some starches, such as potato starch, appear to have concentric rings, which were thought to result from different rates of molecular deposition during alternating light and dark cycles during plant growth. However, this interpretation is subject to question (Banks et al., 1973). Granules from different sources vary as to size, ranging from 3–8 μm for rice starch to 15–100 μm for potato starch. Wheat-starch granules in a single sample seem to fall into two size ranges: 2–10 μm and 20–35 μm (Heckman, 1977).

Large numbers of amylose and amylopectin molecules are associated in a single granule through hydrogen bonding made possible by the many hydroxyl groups in their molecules. Much conjecture and study have been devoted to the amylose-amylopectin arrangement within granules; although theories have been advanced, along with some supporting evidence, much is yet to be learned and this aspect of granule structure will not be discussed here.

Amylose constitutes 20–30% of the total starch in the nonwaxy cereal starches and in potato starch. Waxy starches, such as waxy maize, have essentially no amylose. Some sources of relatively high amylose starch exist. An example is a corn mutant, amylomaize, which has starch with more than 50% amylose (Banks et al., 1973). These differences contribute to differences in functional performance of different starches, as will be seen later.

Some portions of the starch granule are amorphous, apparently because of a high degree of branching and consequent difficulty of orientation. In other portions of the granule, crystallinity exists as a result of a high degree of orientation of linear chains. Because of these crystalline areas, a distinct pattern of birefringence is observed under a polarizing microscope. This property is of practical significance in that

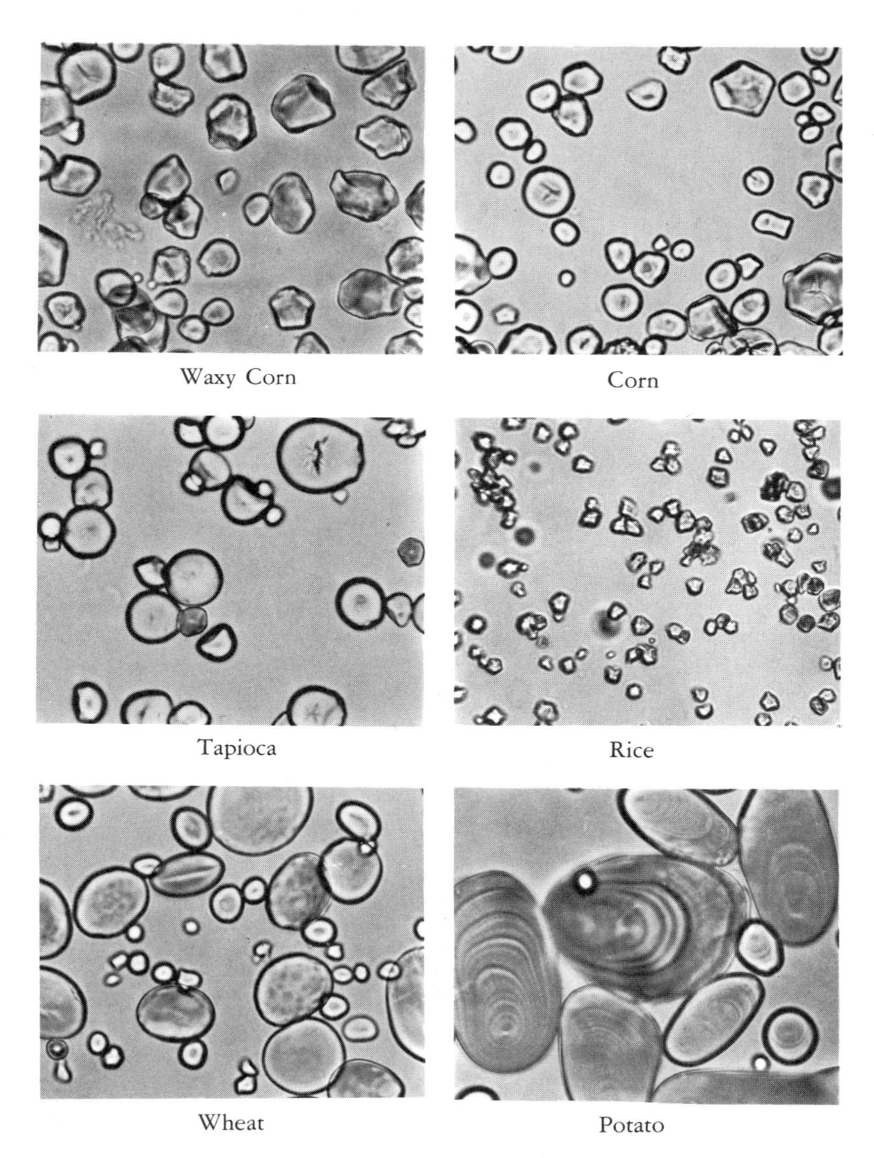

Figure 9.4. Microscopic appearance of ungelatinized starch granules. (Courtesy of Northern Regional Research Center, ARS, U.S. Department of Agriculture, Peoria, Illinois.)

loss of birefringence is a useful parameter of starch gelatinization; it is used in the assessment of effects of various treatments on starch gelatinization. Granule crystallinity is discussed at length in Sterling's (1968) review of granule structure.

Starch particles, as they exist in powdered starch, are agglomerates of granules. The agglomerates readily break up into a suspension of granules when the starch is combined with water. Starch is not water-soluble because the granules are too large to form a solution. Molecular starch does form a colloidal dispersion, but a molecular dispersion is not attained simply by combining raw starch with water. Raw granular starch is considerably denser than water and settles out of a suspension that is allowed to stand undisturbed.

Gelatinization

The basis for much of the current understanding of the physical process of gelatinization was established by Meyer (1952), who pioneered in many aspects of the chemical and physical nature of starch. Gelatinization consists of the changes that occur when starch is heated in water. Not enough water is taken up by granules at room temperature to cause noticeable swelling, except in the case of mechanically damaged starch. As heat is applied, the thermal energy permits some water to pass through the molecular network on the granule surface. Water penetrates first into the relatively open amorphous regions. With continued heating, the energy level becomes high enough to disrupt hydrogen bonding in the crystalline areas. This effect may be observed, with the aid of a polarizing microscope, as loss of birefringence. With the entire granule structure now more "loose," water uptake proceeds readily as heating continues, resulting in rapid swelling of granules. The temperature at which rapid swelling of granules begins depends on the type of starch and on other ingredients but tends to fall in the range of 64–72°C for cornstarch.

As starch granules swell, their density decreases and they eventually are able to remain suspended. The increasing size of granules also results in increasing internal friction and thus increasing viscosity of the suspension. Furthermore, granule surfaces become increasingly "open" with swelling so that some of the straight chains of amylose are able to leave the granules. Amylose that has left the granules is in colloidal dispersion, and the dispersion thus is a sol in which the intact granules are in suspension. The loss of some amylose results in implosion. The granules, though still large relative to their ungelatinized state, appear somewhat shrunken and frequently appear folded when observed under the microscope. Implosion need not result in appreciable thinning of the suspension because the long chains of free amylose, as well as the intact imploded granules, contribute to viscosity. The temperature at which maximum viscosity is achieved depends on the starch and on other ingredients, but usually is at least 90°C. Starch-thickened mixtures are commonly heated to the boiling point to ensure maximum gelatinization. Fragmentation of granules is not extensive unless acid is present during a long heating period or unless starch granules are relatively fragile and agitation is excessive. When fragmentation does occur, it does not involve bursting of granules.

The heated starch paste is referred to as a *paste* and the process of preparing a paste, or gelatinizing starch, is referred to in some of the literature as *pasting*. Two additional points should be emphasized with respect to gelatinization. It has nothing to do with gelatin, and the process is neither the same as nor inclusive of gelation.

Gelation

Gelation is the formation of a gel, regardless of the gel-forming agent. In the case of starch, gelation does not occur until a hot paste is cooled; in other words, gelatinization must precede gelation. If conditions favor gelation of a starch paste, extensive hydrogen bonding occurs during cooling. Probably free amylose molecules form hydrogen bonds not only with one another but also with amylopectin branches extending from granule surfaces. The net effect, as described by Collison (1968), is a continuous three-dimensional network of swollen granules. Intact (unfragmented) granules and bonding between them are important. Free amylose apparently contributes to intergranular bonding through the orienting ability of the straight chains. As is the case with any type of gel, water is held enmeshed in the continuous solid network.

Factors Affecting Paste Viscosity and/or Gel Properties

Hot paste viscosity is not necessarily predictive of gelling ability. Starches from different sources have different properties; for example, root and tuber starches may form pastes that are at least as viscous as pastes of cereal starches and yet show less gel-forming ability than cereal starches. In a series of gelatinized samples of a given cereal starch, hot paste viscosity might correlate quite well with gel strength.

Concentration and kind of starch Increasing concentration of the thickening and gelling agent results in increased paste viscosity and, in the case of gelling starches, increased gel strength.

Osman and Mootse (1958) determined the amounts of different starches required to give a specified paste viscosity. Smaller amounts of waxy cereal starches were required than of the corresponding nonwaxy starches; concentrations of 2.98, 3.13, and 3.42% waxy corn, rice, and sorghum starches respectively were equivalent in thickening power to 4.90, 5.49, and 4.66% concentrations of nonwaxy corn, rice, and sorghum starches respectively. On the other hand, the nonwaxy starch pastes gelled, whereas the waxy starch pastes did not. Potato and tapioca starch pastes also failed to gel.

Ott and Hester (1965) added pure amylose to suspensions of non-gelling waxy cornstarch and in doing so brought about gelation after heating and cooling. As the amylose concentration was increased beyond that required for gelation, gel strength increased.

Chemically modified starches show different gelatinizing and gelling behavior from that of the parent starches. Chemical modification involves holding starch in a chemical solution at a temperature that is high enough to permit the chemical to react with molecules within the granules but not high enough to cause appreciable granule swelling. The starch then is dried. The effects of modification that become apparent at the time of use depend on the specific modification. A huge variety of modified starches, having nearly any desired properties, is available to food manufacturers. Of greatest theoretical interest here are the cross-linked starches. *Cross-linking* involves treating starch with chemicals that can react with hydroxyl groups of two different starch chains, forming links between them. If the treated starch has granules that tend to fall apart during gelatinization, the cross-linking treatment can help maintain granule integrity during gelatinization. Cross-linking also can prevent gelation by inhibiting loss of amylose from granules.

Extent of heating Maximum viscosity depends on sufficient heating to achieve maximum swelling of starch granules. Maximum gel strength depends on sufficient heating to free some amylose, with minimal fragmentation of granules. If heating conditions cause excessive granule fragmentation, not even a large amount of amylose freed from the granules can cause gelation.

Other ingredients Interactions among ingredients complicate generalizations as to their individual effects. Sucrose, a common ingredient of starch-thickened mixtures, and citric acid, a component of lemon juice, which is another common ingredient, should be considered separately and together.

Sucrose Hester et al. (1956) observed that sucrose present during heating of starch suspensions retarded the hydration of starch granules by competing for water. In such a case, a given heat treatment results in a less advanced stage of gelatinization than occurs in a control paste containing no sugar. Depending on the extent of gelatinization, the paste containing sucrose may be somewhat less viscous than the control, or it may be as viscous but show less gel-forming ability because of insufficient implosion to provide free amylose to the same extent as in samples without sugar. Unless the concentration of sucrose is very high, the retarding effect can be at least partially overcome by increasing heating time.

Citric acid Hansuld and Briant (1954) heated cornstarch and wheat starch in water with several levels of citric acid and observed decreased paste viscosity and decreased gel strength. Granule fragmentation occurred. In the presence of acid, gelatinization is accelerated because hydrolysis of some molecules on granule surfaces occurs, resulting in increased permeability of the surface and increased rate of swelling. As swelling proceeds, the acid gains access to the interior; hydrolysis of molecules results in fragmentation of granules and reduced gel-forming ability.

The potential effect of acid on the thickening power of starch is recognized in recipes that call for late addition of lemon juice. However, it should be remembered that adding lemon juice after completion of heating involves addition of water as well as acid. Water added after completion of heating has little opportunity to be taken up by starch granules. In a situation such as the thickening of the cherry juice for a cherry pie, the acid is present in the juice that is thickened and there is no way to add the acid after completion of thickening. The amounts of acid used in starch-thickened mixtures need not cause undue thinning if heating is carried out rapidly. The effect of heating on the flavor of lemon juice sometimes is more apparent than a thinning effect.

Sucrose and citric acid Campbell and Briant (1957), studying the effects of sucrose and citric acid in a factorial study of starch pastes and gels, observed that the effect of sucrose depended on the level of citric acid. They theorized that this interaction reflects the opposing effects of sucrose and citric acid on the rate of gelatinization, as well as the apparent dependence of gelation on both intact granules and free amylose. Sucrose, by retarding gelatinization, may result in a high proportion of intact granules but little free amylose; citric acid, by accelerating gelatinization, may result in a high proportion of free amylose but few intact granules. Together, in certain proportions, they can tend to counteract one another's effects.

Monoglycerides Longley and Miller (1971) heated dilute wheat-starch suspensions in the presence of a homologous series of monoglycerides with fatty acid chain lengths of 4–20 carbons. The monoglycerides reduced the degree of gelatinization, the effect tending to increase with fatty acid chain length and with monoglyceride concentration.

Ingredients of baked products Completely different approaches were used in two laboratories in studies of the effects of dough ingredients on starch gelatinization. d'Appolonia (1972) used the Brabender amylograph for studying the individual effects of bread ingredients on starch gelatinization. In the amylograph (see Chapter 16), a slurry is heated at the rate of 1.5°C per min, with constant stirring, and a continuous record is made of the change in viscosity. Starch slurries must be quite dilute (not more than 10% starch) and the conditions in the actual product, therefore, are not simulated. Bread ingredients studied by d'Appolonia in 10% wheat starch suspensions were sucrose, sodium chloride, nonfat dry milk, shortening, oxidizing agents, reducing agents, and sodium propionate. One set of curves, showing the effect of added sucrose, is shown in Figure 9.5. The effect of sucrose in retarding gelatinization is apparent.

Hoseney et al. (1977) extracted starch from a variety of systems,

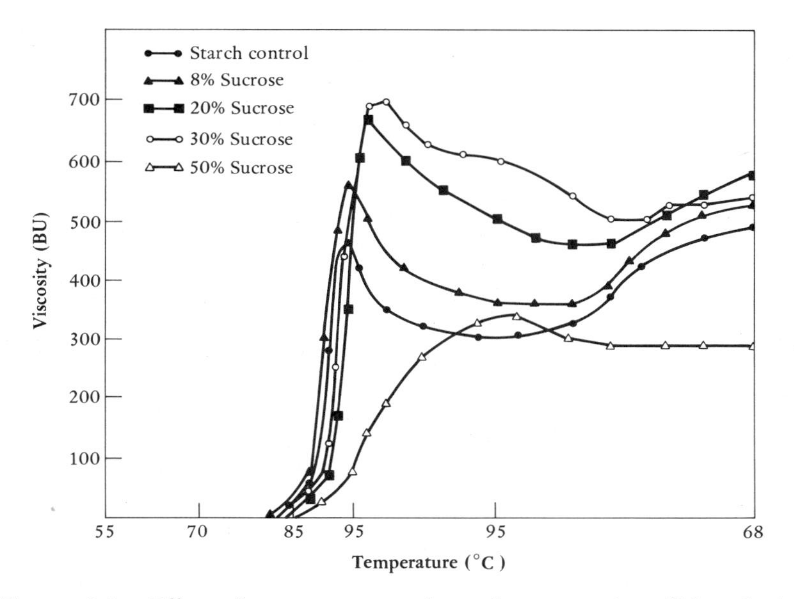

Figure 9.5. Effect of sucrose on starch pasting properties. (d'Appolonia, 1972)

including baked products, and assessed the effects of heating from the appearance of the granules. Selected electron micrographs are shown in Figures 9.6, 9.7, and 9.8. In Figure 9.6, it is apparent that the granules were changed somewhat more in 10% starch slurries than in doughs with lower moisture levels. The gelatinization-retarding effect of sucrose is seen in Figure 9.7. Increased concentration of sucrose resulted in decreased granule deformation. In Figure 9.8, the small extent of gelatinization in pastry is compared with the large extent of gelatinization in angel food cake, reflecting a difference in the moisture levels of the mixtures.

Low temperature storage Hydrogen bonding increases at decreasing energy levels. Therefore, a paste that is capable of gelling at room temperature is subject to extensive "tightening" of structure and decreased water-holding capacity when held at low temperatures. What

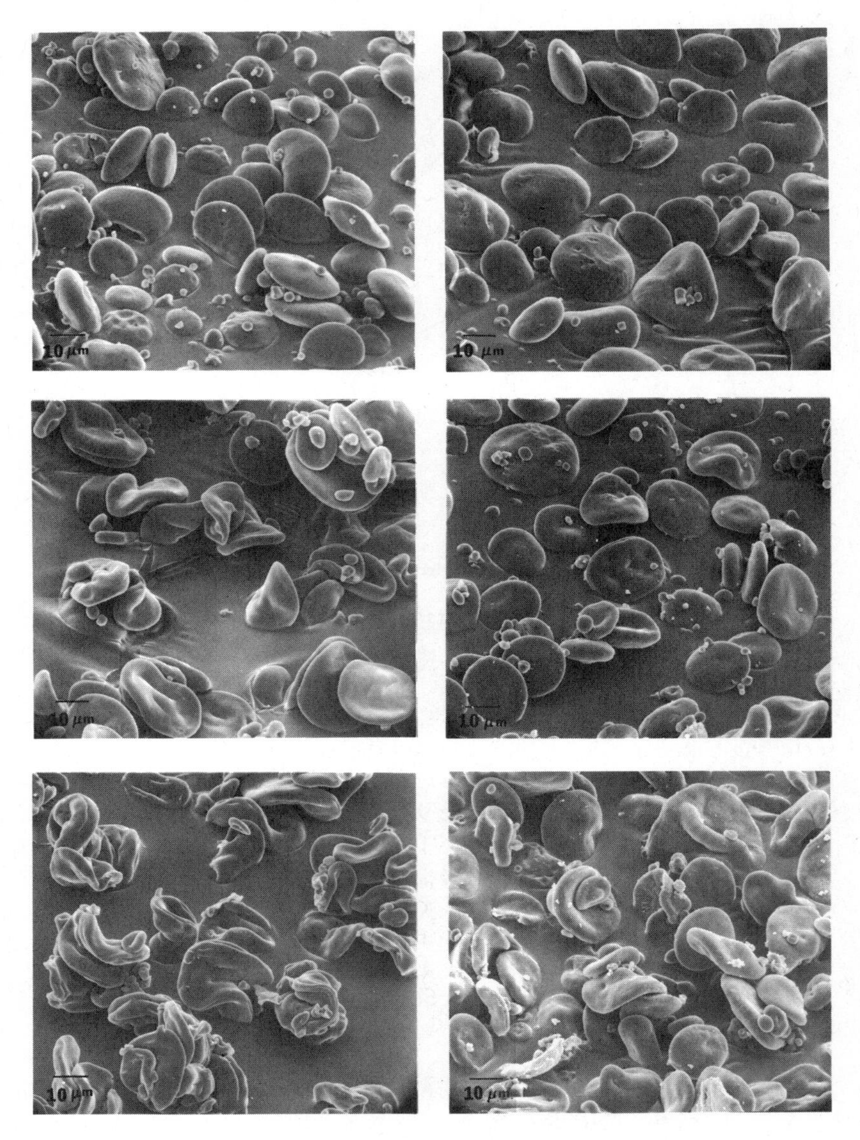

Figure 9.6. Starch extracted from a 10% wheat starch-water suspension (left column) and a 65% moisture flour-water dough (right column); temperatures from top to bottom: 25, 70, and 90°C. (Hoseney et al., 1977)

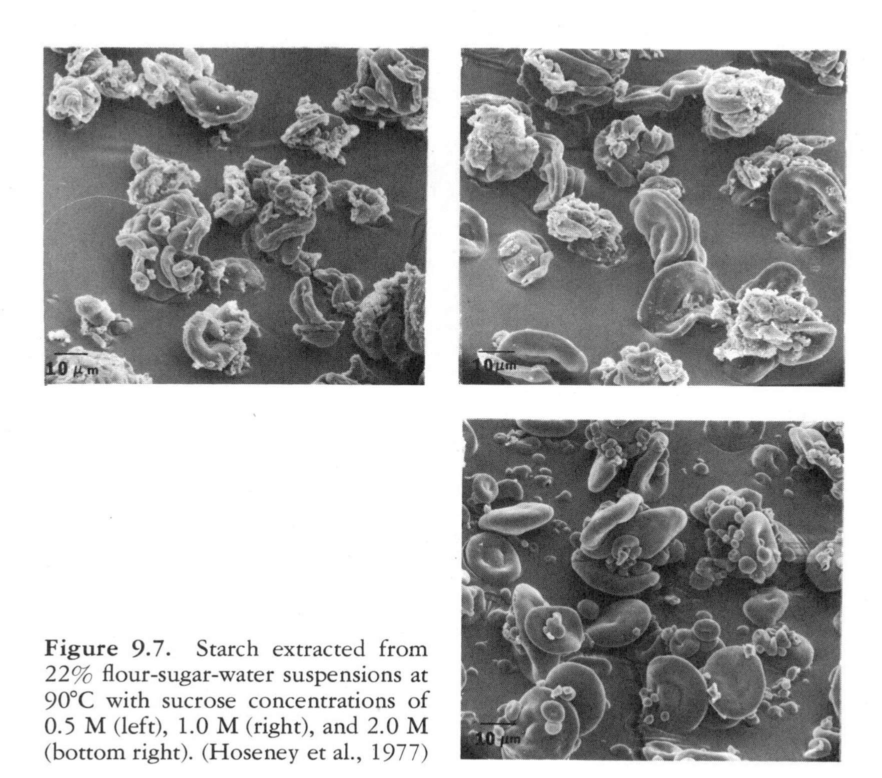

Figure 9.7. Starch extracted from 22% flour-sugar-water suspensions at 90°C with sucrose concentrations of 0.5 M (left), 1.0 M (right), and 2.0 M (bottom right). (Hoseney et al., 1977)

occurs is an extension of the gelation process and is referred to as *retrogradation.* A conventional starch-thickened pudding or pie filling that has been frozen and thawed has a rough, spongy, cottony texture and loses water readily. Gravies and sauces, in which the starch concentration is relatively low, tend to show separation. Starches that do not have gel-forming ability do not retrograde. Therefore, waxy starches are used more successfully in such foods that are to be frozen than are nonwaxy starches.

Some chemically modified starches are even more satisfactory than waxy starches for starch-thickened foods that are to be frozen. Increased bulkiness of molecules results from cross-linking and some other types of chemical modification; therefore, the difficulty of releas-

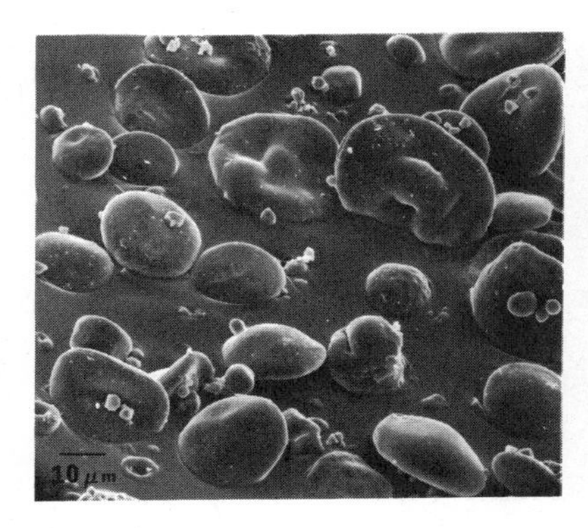

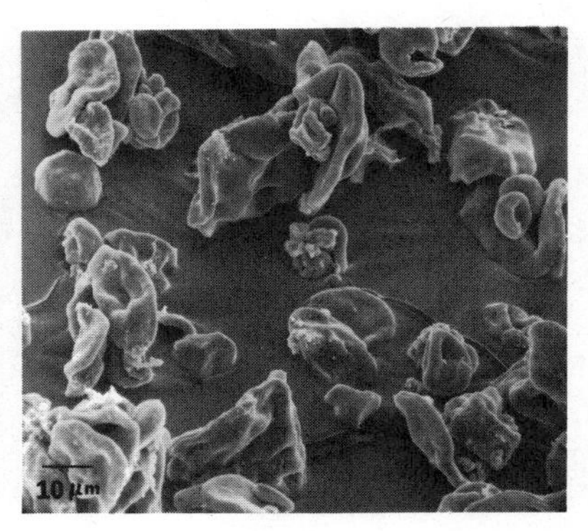

Figure 9.8. Starch extracted from commercially obtained baked products: left, fried pie crust; right, angel-food cake. (Hoseney et al., 1977)

ing amylose from the granules during gelatinization increases. High viscosity is possible, but actual gelation does not occur and neither does retrogradation. Chemically modified starches are not on the retail market but are available within the food industry and are ingredients in many consumer products.

Methods of Studying Starch

It should be recognized that much of the reported research concerning starch gelatinization and gelation has involved the use of suspensions containing about 5–6% starch, heated slowly to 90–95°C, and held at the maximum temperature for 10–15 min. These conditions are not characteristic of those of most food systems. However, the methods and equipment used in research provide for control of heating conditions and permit comparisons of treatment effects under those controlled conditions.

Many methods that have not been mentioned are used in the study of starch. The reader is referred to the approved methods of the American Association of Cereal Chemists (AACC, 1962) for details of many official tests that are applied to starch. In addition, two research reports (Medcalf and Gilles, 1965; Kulp, 1972) are particularly recommended for reading because of their inclusion of a large number of representative procedures.

FLOUR

The major portion of this section of the chapter will be devoted to discussion of wheat flour, more specifically white wheat flour. The properties of flour are influenced by the raw material, wheat, as well as by milling.

Wheat

Wheat often is classified as hard or soft, in reference to the resistance of the endosperm to grinding. Hardness is a genetically determined quality. Stenvert and Kingswood (1977) reported that the hardness of the endosperm is a function of the continuity of the protein matrix and the degree of order in the endosperm structure. Hoseney and Seib (1973) presented electron photomicrographic evidence that the bonding between starch and protein is stronger in hard wheats than in soft wheats. The greatest significance of the differences between flours of hard and soft wheats is in their baking qualities, since flour made of hard wheat is especially suited to making bread, whereas that made of soft wheat is best for cakes, pastry, and crackers. The superiority of hard wheat flour for bread reflects the large amount and good quality of gluten it forms when mixed with water. Durum wheat is an extremely hard high-protein wheat, which belongs to a botanical species different from that of the common wheats. It is used to make macaroni, spaghetti, and similar products. Geneticists continually work at developing wheat varieties that combine the qualities of high yield, high disease and pest resistance, high protein content, good milling qualities, and high baking potential. It takes as many as 10–15 yr to produce and propagate a new variety of winter wheat or to introduce a new trait into a variety (Mattern, 1973).

Wheat is classified also as spring or winter wheat. Spring wheat is planted in the spring and harvested in late summer, whereas winter wheat is planted in the fall so that it can develop a root system before cold weather. It grows rapidly in the spring and is harvested in early summer. Winter wheat can be grown only in areas where the root system can survive the winter. Of these two classes, winter wheat has the advantage of higher yields, while spring wheat usually is harder. However, winter wheats may be either hard or soft depending on the variety and the growing conditions. The terms red wheat and white wheat are applied to wheat varieties and refer to the color of the kernels.

Most of the hard wheat grown in the United States is produced in the Midwest and upper Midwest. The largest producing areas of soft wheat are in the Pacific Northwest and in Ohio and its neighboring states. Most of the durum wheat is grown in North Dakota.

The protein content of wheat can be influenced by the fertility of the soil in which it is grown. Climatic factors such as temperature, rainfall, and sunshine also affect protein content and, therefore, baking quality.

Milling

Structure of wheat kernel The milling process is designed to produce white flour of good baking quality by separating the endosperm from the bran and germ that surround it. This separation is possible because endosperm is more easily crushed than are the bran and germ. Longitudinal sections and cross sections of the wheat kernel are shown in Figure 9.9. The crease, which is seen best in the cross section, penetrates nearly to the center of the kernel of the commonly grown varieties of wheat. The crease and the hairs of the brush cause problems in cleaning the grain. The bran is made up of an outer pericarp, called beeswing by millers because it is thin and papery, an inner pericarp, and a thin seed coat that is fused to the pericarp. In each of these three parts, several layers can be distinguished microscopically. Botanically, the aleurone layer is an outer row of thick walled cells of the endosperm, but it is removed with the bran during the milling process. It contains no starch or gluten protein but has reserve foods in the form of oil and nongluten protein. The endosperm, sometimes called starchy endosperm to distinguish it from the aleurone layer, occupies the largest part of the kernel, about 83%. It is composed of cells containing many starch granules embedded in a matrix of protein. It can be seen in Figure 9.9 that the germ occupies only a small portion of the kernel. It

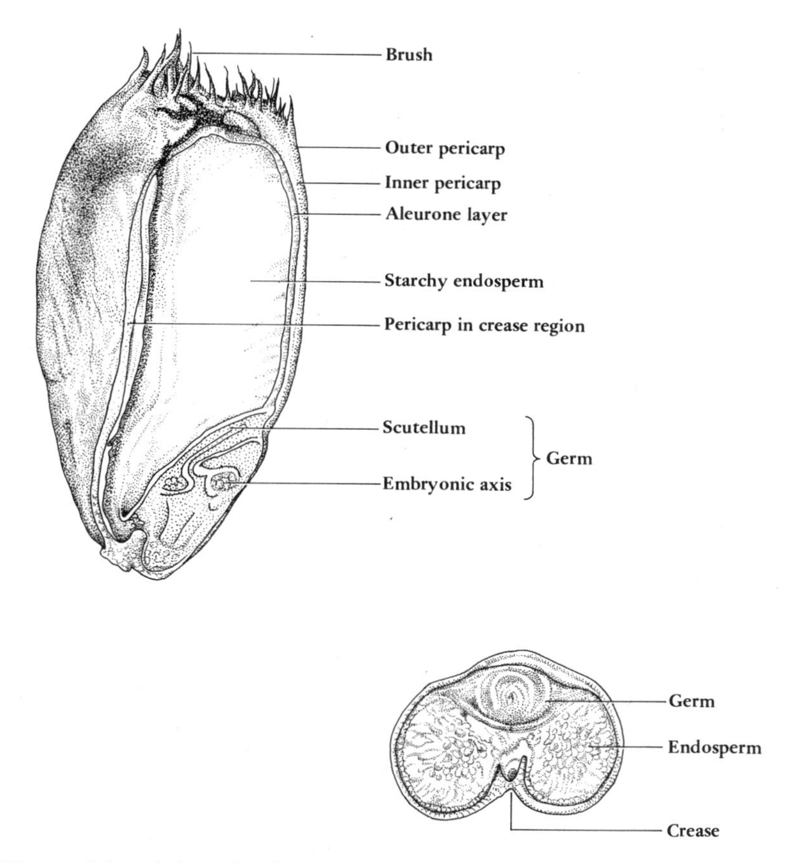

Figure 9.9. A kernel of wheat. Longitudinal section bisected through the crease, and lateral cross section. (Bradbury et al., 1956)

is composed of two parts: (1) the embryonic axis, which develops into the seedling; and (2) the scutellum, which contains most of the thiamin in the wheat kernel. The entire germ is rich in oil and protein. Much more detail concerning kernel structure is found in the review of MacMasters et al. (1971).

Basic milling processes The milling processes are diagrammed in Figure 9.10. Either before or after the wheat is selected and blended for

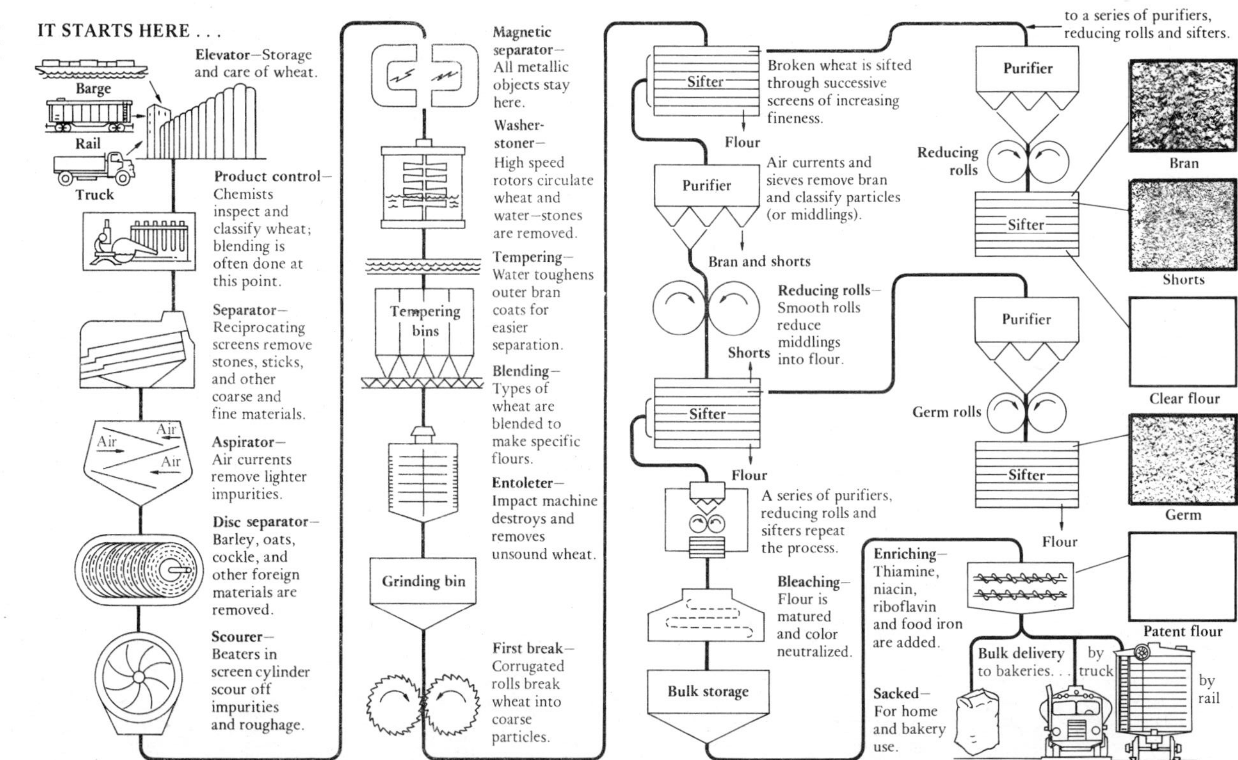

Figure 9.10. The total process of making flour. (Courtesy of Wheat Flour Institute.)

the type of flour desired, it is cleaned and tempered. During cleaning, the brush and the dirt in the crease are removed, as well as seeds of other plants, stones, chaff, straw, and dirt. Tempering, also called conditioning, facilitates milling by toughening the bran and mellowing the endosperm, making their separation easier. This is accomplished by adding water and allowing the grain to stand, sometimes at elevated temperatures.

Endosperm is separated from the bran and germ by a complex process, which varies in its details with the mill and the products desired. Grinding wheat finely in one operation would result in flour containing appreciable amounts of bran and germ. Therefore, grinding is gradual and the particles produced by each grinding are sized by sifting to permit efficient separation of the endosperm. The first grinding, called the first break, is done between corrugated steel rollers set far apart so that large particles are produced. The sheared particles are passed through sieves, progressing from the coarsest to the finest. At each sieve level, those particles that do not pass are diverted to the second break rolls, to a middlings purifier, or to the reduction rolls. Particles of bran and germ are removed from the middlings in the purifier. The material that goes to the smooth reduction rolls is relatively pure endosperm and is crushed further and sifted as flour. The total process is repeated several times. The flour streams are combined into flour(s) as desired and the bran and germ are used in animal feed.

The milling process is carried out automatically and continuously, material being conveyed by such devices as elevators and tubes from one operation to the next. Although over a hundred separations may be involved in converting wheat to flour, the entire process takes only about 30 min. Each grinding step produces a certain amount of material as fine as flour, or a mill stream. A large mill may have 30–40 mill streams, each with different characteristics, which are blended to produce the grades of flour and feed desired, as indicated in Figure 9.11. The figure shows that, from 100 lb of wheat, about 28 lb of feed and 72 lb of white flour can be expected. This extent of recovery of flour is referred to as a 72% rate of extraction. Straight flour is the result of combining all of the mill streams, that is, the entire 72 lb of flour. Frequently the mill streams from the starchier end of the mill-stream spectrum are combined to make patent flour. The remainder becomes clear flour. If only 40–60% of the streams are combined, an extra short patent flour is produced, leaving a fancy clear. As more mill streams are

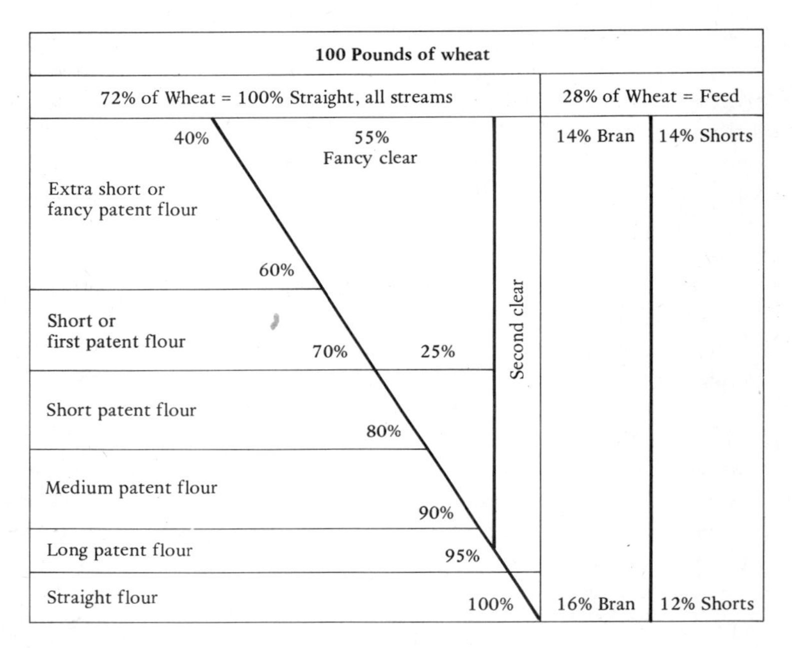

Figure 9.11. Chart showing the amounts of various grades of flour and feed produced from 100 lb of wheat. (Courtesy of Wheat Flour Institute.)

combined, the patent is longer and the amount and quality of clear are reduced. Patent flours are commonly used for household and bakery needs, whereas clear flour is used in products for which its creamy to grayish color is not a disadvantage, such as rye or whole wheat bread and pancake flour. The milling process is described in detail in the review of Ziegler and Greer (1971).

Additional physical and chemical treatments *Impact milling,* of which pin-milling is an example, further reduces particle size by forcing conventionally milled flour against pins or baffles with great force. It might be done if there is a reason to increase the level of damaged starch in a flour. Tipples and Kilborn (1968) tested the theory that flour absorption (water uptake during mixing) would be increased by increasing the level of damaged starch by pin-milling. They obtained increased absorption, and therefore greater yield, with certain bread-making methods, but the results were not consistent. Impact milling also might be carried out in preparation for air classification. *Air classification* is a means of fractionating a flour, on the basis of particle size and density, by application of centrifugal force. The fractions differ in composition as well as in size and density, and the process provides the possibility of making flours that are highly specialized for different uses. Hayashi et al. (1976) compared fractions obtained through pin-milling and air classification in several baked products and found that the fractions differed considerably as to their performance.

Instantizing or *agglomeration,* another physical treatment, involves moistening of conventionally milled flour and drying it, all with constant agitation. The particles, made sticky by the moisture, adhere to one another to form agglomerates during drying. The agglomerates are porous and, therefore, are readily dispersed without lumping, and disintegrated in water. Miller et al. (1969) reviewed the properties of instantized flour and the factors that need to be considered in using it in food preparation.

Chemical treatments of flour consist of the possible addition of enrichment mixtures (Schmidt, 1973) and of bleaching and improving chemicals. The carotenoid pigments in flour give it a creamy to yellowish color when it is first milled. During storage for several months, its color becomes lighter because of oxidation of the pigments. Its baking quality improves also, because of oxidative effects on gluten proteins. These changes can be accelerated by the use of certain chemicals at the mill. These include chlorine dioxide, acetone peroxide, and azodicarbonamide in bread and all-purpose flours. Cake and pastry flours are treated with chlorine. Chlorination is especially important for cake flours. Both starch and flour proteins are known to be affected by chlorination, but the mechanism of the beneficial effect of chlorination on cake flour is not yet clear.

Types of flour Milling of wheat produces several major types of flour, including bread flour, a hard wheat flour having about 14% protein; all-purpose, or family, flour, made from either hard or soft wheat and having about 10.5% protein; and cake flour, a soft wheat flour having about 7.5% protein. Special purpose flours include pastry flour, which is a longer patent soft wheat flour than cake flour, and self-rising flour, to which salt and the leavening agent have been added for convenient use in quickbreads.

Flours from other sources, such as soy, rye, triticale, and others, are quite different from wheat flours in their composition and properties. They will be mentioned in Chapter 11 as possible substitutes for wheat flour in breads.

Flour Quality

Determinants The gluten proteins are vital to the dough structure that develops when flour is hydrated and made into a dough. They also are probably responsible for the differences in baking quality of flours from different wheats. Although the gluten proteins, glutenin and gliadin, are separate components in flour, they complex to form gluten during dough formation. Glutenin has a much higher molecular weight than gliadin. Glutenin has a large amount of intermolecular disulfide bonding, as well as intramolecular bonding. Gliadin has only intramolecular disulfide bonding. Neither component has the ability to form a satisfactory dough structure alone. Their complexing to form gluten involves breaking of some disulfide bonds and formation of new ones. Oxidizing and reducing agents thus affect their interaction. Lipids, specifically the polar galactosyl glycerides that are present in flour, are involved in the gluten complex and thus are important to the baking quality of flour. Reviews of gluten proteins include those of Kasarda et al. (1976) and Mecham (1972).

Tests for flour quality Tests for flour quality include those on the flour itself, such as moisture, ash, nitrogen, color, sedimentation, water retention capacity, and amylograph viscosity. Dough tests include the farinograph test and measurement of dough extensibility. The farinograph test provides information as to dough development time and dough stability. Baking tests include the cookie baking test for long patent or straight soft wheat flours, the cake baking test for cake flour, and the bread-baking test for bread flour. Most of these tests and more are among the approved methods of the American Association of Cereal Chemists (AACC, 1962).

A useful technique for the study of flour involves fractionation of flours that differ as to baking quality into their separate components—gluten, starch, "tailings," and water-solubles. The fractions then are recombined for baking tests. This is done first without interchange in order to confirm that fractionation did not alter the functional properties of the components, and then with systematic interchange in order to determine the contribution of each fraction to the quality difference between the parent flours. Flour fractionation has been used particularly extensively and effectively by Hoseney and Finney (1971) and their coworkers.

EXPERIMENTS

I. Endpoint Temperature of Starch Pastes

A single cornstarch suspension is sampled at several endpoint temperatures for observation of microscopic appearance of granules, average line-spread at 50°C, and penetrability or percent sag at room temperature (see Chapter 16).

A. *Procedures*

1. Disperse 72 g of cornstarch in 1200 ml of distilled water in a 2-liter pan. Prepare a slide with a small drop of the suspension and observe under the microscope.
2. Suspend a thermometer in the suspension and heat with constant stirring to a temperature of 70°C.
   a. Prepare a slide with a small drop of the suspension. Observe under the microscope.
   b. Fill a line-spread ring and measure average line-spread at 50°C.
   c. Fill a custard cup to the top line, cover with foil, cool at room temperature, and measure penetrability or percent sag. This sample probably will have to be held until the next class period unless it is possible to make the measurement between class periods.
3. Return the suspension to the burner and heat with constant stirring to 80°C. Repeat steps 2a, b, and c.
4. Return the suspension to the burner and heat with constant stirring to 85°C. Repeat steps 2a, b, and c.

5. Return the suspension to the burner and heat with constant stirring to 90°C. Repeat steps 2a, b, and c.

B. Results

1. Make a chart listing endpoint temperatures down the left side (with adequate space for drawing starch granules) and granule appearance and the measurements across the top.
2. Write summarizing statements as to the relationships among granule appearance, apparent viscosity of the warm suspension, and gel-forming ability.

C. Questions

What was the appearance of the granules when line-spread first could be measured? What was the effect of continued heating on apparent viscosity? What was the relationship between apparent viscosity and gel strength?

II. Sucrose and Citric Acid in Cornstarch Pastes

Gel-forming ability is observed in a factorial experiment in which two levels of sucrose and two levels of citric acid are used. Samples will have to be held for measurement of percent sag or penetrability at room temperature.

A. Procedures

1. Prepare the following starch suspensions:

	Treatments			
	1-a	1-b	1-c	1-d
Cornstarch, g	10	10	10	10
Sucrose, g	—	25	—	25
Distilled water, ml	200	200	—	—
0.05 N citric acid soln, ml	—	—	200	200

a. Heat suspensions, at a moderate rate, in 400-ml beakers

into which are suspended calibrated thermometers so that their bulbs are covered with suspension. Carefully and continuously stir with a glass rod to prevent settling.

b. Remove about 3 ml at each of the temperatures 70 and 90°C and dilute each sample with 10 ml of water.

c. Fill a custard cup to the top line with each 90° paste. Hold at room temperature until cool. Measure percent sag or penetrability.

2. Prepare slides from the diluted samples and examine immediately with a microscope. Sketch a typical granule from each subsample.
3. Measure percent sag or penetrability of cooled samples in custard cups.

B. *Results*

1. Make a chart listing sample codes down the left side and granule appearance and percent sag or penetrability at the top.
2. Write statements summarizing the single and combined effects of sucrose and citric acid.

C. *Questions*

What was the effect of sucrose in the absence of citric acid? What was the effect of citric acid in the absence of sucrose? What was the effect of the combined presence of sucrose and citric acid? Did granule appearance contribute to an explanation of the results with regard to gel strength?

III. Freeze-thaw Stability of Mixtures Thickened with Unmodified and Chemically Modified Starches

Lemon fillings are prepared from unmodified cornstarch and from starch declared by the manufacturer to have freeze-thaw stability. Ideally the unmodified starch should be the parent starch of the modified starch, but this is not feasible. The fillings are subjected to low-temperature storage.

A. Procedures

1. Prepare two lemon fillings, one with unmodified cornstarch and one with chemically modified cornstarch from the formula:

Sugar, g	160
Cornstarch, g	36
Water, ml	460
Egg (beaten), g	48
Lemon juice, ml	40
Margarine, g	21

   a. Combine sugar and cornstarch in a 1-liter pan. Add boiling water gradually with stirring.
   b. Cook over low heat to boiling, stirring constantly.
   c. Combine egg and juice. Add a little hot starch paste and blend; add to the starch in the pan.
   d. Cook over low-moderate heat for 2 min, stirring constantly.
   e. Add margarine.
2. Make line-spread readings on the pastes at about 70°C. The exact temperature is not as important as its being the same for all samples.
3. Fill custard cups to the top line. Two will be needed for each mixture. It will be necessary to flatten and smooth the surfaces as much as possible. Cover immediately with foil. Store one sample of each mixture in the refrigerator and one in the freezer for 24–48 hr. Thaw the frozen samples in a refrigerator and take penetrometer readings on all samples. Observe smoothness of samples. If water separates, decant and measure it.

B. Results

1. Make a chart listing treatment combinations down the left side and line-spread, penetrability, smoothness, and water separation across the top.
2. Summarize the effect of low-temperature storage on each starch-thickened mixture.

C. *Questions*

Why should the two mixtures not be compared directly as to their properties? What were the effects of storage on the properties of each mixture? Recognizing the unknowns concerning the starch used, formulate a possible explanation for the differing responses to storage treatment.

IV. Gluten Development as Affected by Kind of Flour and Added Ingredients

Doughs are made from flour and water plus individually added ingredients. Gluten is separated from the other dough constituents, and the yields, reflecting the effects of flour type and added ingredients, are compared.

A. *Procedures*

1. Preparation of doughs and gluten balls
   a. To 100 g of flour (hard wheat all-purpose unless otherwise specified), add any other ingredient specified below, blend thoroughly, and add distilled water gradually from a graduated cylinder to produce a dough that can be kneaded with the hands. Record the volume of water used.
   b. Knead the dough for 10 min and immerse in a beaker of distilled water for 5 min.
   c. Knead the dough gently in the beaker until the water becomes "milky" in appearance, being careful not to let the ball of dough fall apart. Decant water through a strainer lined with finely woven cheesecloth.
   d. Add a fresh supply of water and repeat c.
   e. Repeat d until the water is clear. Retrieve bits of gluten from the cheesecloth and add to the ball.
   f. Squeeze as much water as possible from each gluten ball and weigh. If any of the gluten balls for a replication are not ready, return the ball to water until all are ready.
   g. Weigh from each ball an amount equal to the weight of the smallest ball in each replication.

h. Bake all samples for a replication together on a baking sheet at 232°C (450°F) for 15 min. Lower oven to 149°C (300°F) and bake 40 min longer.

2. Variations
   a. —— (control)
   b. 2 g sodium bicarbonate added
   c. 1.5 g cream of tartar added
   d. 24 g shortening added (cut into flour)
   e. 24 g sugar added
   f. 2.5 g sodium chloride added
   g. Soft wheat all-purpose flour substituted if available
   h. Bread flour substituted if available

B. *Results*

1. Rank the samples in order of increasing yield of gluten. If different flours are used, make separate rankings for added ingredients and kinds of flour, using the control in each ranking.
2. Rank the samples in order of increasing size of baked gluten ball, again making separate rankings for added ingredients and kinds of flour.

C. *Questions*

To what extent did results between replications (if more than one done) agree? How can disagreement between replications be explained? What were the effects of the various additives and of different flours and why?

SUGGESTED EXERCISES

1. Obtain several chemically modified starches for which various claims are made by their manufacturers, such as acid stability, high temperature stability, and freeze-thaw stability. Advertisements in a recent issue of *Food Technology* provide information regarding sources. Plan and conduct experiments to test the claims.

2. Compare starches from several sources, such as potato, corn, wheat, and rice. Use the starch and water proportions and the heating procedure described in Experiment II. Prepare slides and observe granule appearance before and after heating. Measure line-spread and penetrability or percent sag.
3. Plan and conduct an experiment to show the effects of holding temperature and temperature of measurement on apparent strength of cornstarch gels.
4. Substitute different types of flour in baked products.
5. Compare cake flour, all-purpose flour, and bread flour if available as to:
   a. rate of settling from a suspension in a graduated cylinder
   b. microscopic appearance
   c. amount of water required to give dough of a certain consistency
   d. elasticity of dough, as measured by area covered by a weighed amount of dough rolled on an oiled surface to a controlled thickness and permitted to stand for 1 min.

REFERENCES

AACC. 1962. "Approved Methods of the American Association of Cereal Chemists," 7th ed. American Association of Cereal Chemists, St. Paul, MN.

Banks, W., Greenwood, C. T. and Muir, D. D. 1973. The structure of starch. In "Molecular Structure and Function of Food Carbohydrate," Eds. Birch, G. G. and Green, L. F. Applied Sci. Publishers Ltd., London.

Bradbury, D., Cull, I. M. and MacMasters, M. M. 1956. Structure of the mature wheat kernel. I. Gross anatomy and relationships of parts. Cereal Chem. 33: 329.

Campbell, A. M. and Briant, A. M. 1957. Wheat starch pastes and gels containing citric acid and sucrose. Food Res. 22: 358.

Collison, R. 1968. Swelling and gelation of starch. In "Starch and Its Derivatives," 4th ed., Ed. Radley, J. A. Chapman and Hall, Ltd., London.

d'Appolonia, B. L. 1972. Effect of bread ingredients on starch gelatinization properties as measured by the amylograph. Cereal Chem. 49: 532.

Fleming, I. D. 1968. Amyloglucosidase: α-1,4-glucan glucohydrolase (E.C. 3.2.1.3). In "Starch and Its Derivatives," 4th ed., Ed. Radley, J. A. Chapman and Hall, Ltd., London.

Foster, J. F. 1965. Physical properties of amylose and amylopectin in solution.

In "Starch: Chemistry and Technology, Vol. 1. Fundamental Aspects," Eds. Whistler, R. L. and Paschall, E. F. Academic Press, New York.

Greenwood, C. T. 1976. Starch. In "Advances in Cereal Science and Technology," Ed. Pomeranz, Y. American Association of Cereal Chemists, St. Paul, MN.

Hansuld, M. K. and Briant, A. M. 1954. The effect of citric acid on selected edible starches and flours. Food Res. 19: 581.

Hayashi, M., d'Appolonia, B. L. and Shuey, W. C. 1976. Baking studies on the pin-milled and air-classified flour from four hard red spring wheat varieties. Cereal Chem. 53: 525.

Heckman, E. 1977. Starch and its modifications for the food industry. In "Food Colloids," Ed. Graham, H. D. Avi Publ. Co., Westport, CT.

Hester, E. E., Briant, A. M. and Personius, C. J. 1956. The effect of sucrose on the properties of some starches and flours. Cereal Chem. 33: 91.

Holló, J. and Szeitli, J. 1968. The reaction of starch with iodine. In "Starch and Its Derivatives," 4th ed., Ed. Radley, J. A. Chapman and Hall, Ltd., London

Hoseney, R. C. and Finney, K. F. 1971. Functional (breadmaking) and biochemical properties of wheat flour components. XI. A review. Baker's Digest 45(4): 30.

Hoseney, R. C. and Seib, P. A. 1973. Structural differences in hard and soft wheat. Baker's Digest 47(6): 26.

Hoseney, R. C., Atwell, W. A. and Lineback, D. R. 1977. Scanning electron microscopy of starch isolated from baked products. Cereal Foods World 22: 56.

Kasarda, D. D., Bernardin, J. E. and Nimmo, C. C. 1976. Wheat proteins. In "Advances in Cereal Science and Technology," Ed. Pomeranz, Y. American Association of Cereal Chemists, St. Paul, MN.

Kulp, K. 1972. Physicochemical properties of starches of wheats and flours. Cereal Chem. 49: 697.

Longley, R. W. and Miller, B. S. 1971. Note on the relative effects of monoglycerides on the gelatinization of wheat starch. Cereal Chem. 48: 81.

MacMasters, M. M., Hinton, J. J. C., Bradbury, D. 1971. Microscopic structure and composition of the wheat kernel. In "Wheat Chemistry and Technology," Ed. Pomeranz, Y. American Association of Cereal Chemists, St. Paul, MN.

Mattern, P. J. 1973. Potential benefits for bakers from improved protein contents in hard winter wheats. Baker's Digest 47(5): 62.

Mecham, D. K. 1972. Flour proteins and their behavior in doughs. Cereal Sci. Today 17: 208.

Medcalf, D. G. and Gilles, K. A. 1965. Wheat starches. I. Comparison of physicochemical properties. Cereal Chem. 42: 558.

Meyer, K. H. 1952. The past and present of starch chemistry. Experientia 8: 405.

Miller, B. S., Trimbo, H. B. and Derby, R. I. 1969. Instantized flour—physical properties. Baker's Digest 43(6): 49.

Osman, E. 1972. Starch and other polysaccharides. In "Food Theory and Applications," Eds. Paul, P. C. and Palmer, H. H. John Wiley and Sons, Inc., New York.

Osman, E. M. and Mootse, G. 1958. Behavior of starch during food preparation. I. Some properties of starch-water systems. Food Res. 23: 554.

Ott, M. and Hester, E. E. 1965. Gel formation as related to concentration of amylose and degree of starch swelling. Cereal Chem. 42: 476.

Price, J. C. 1968. Pullanase. In "Starch and Its Derivatives," 4th ed., Ed. Radley, J. A. Chapman and Hall, Ltd., London.

Robyt, J. F. and Whelan, W. J. 1968a. The α-amylases. In "Starch and Its Derivatives," 4th ed., Ed. Radley, J. A. Chapman and Hall Ltd., London.

Robyt, J. F. and Whelan, W. J. 1968b. The β-amylases. In "Starch and Its Derivatives," 4th ed., Ed. Radley, J. A. Chapman and Hall, Ltd., London.

Schmidt, A. M. 1973. New flour and bread enrichment standards. Baker's Digest 47(6): 29.

Stenvert, N. L. and Kingswood, K. 1977. The influence of the physical structure of the protein matrix on wheat hardness. J. Sci. Food Agric. 28: 11.

Sterling, C. 1968. The structure of the starch grain. In "Starch and Its Derivatives," 4th ed., Ed. Radley, J. A. Chapman and Hall, Ltd., London.

Tipples, K. H. and Kilborn, R. H. 1968. Effect of pin-milling on the baking quality of flour in various breadmaking methods. Cer. Sci. Today 13: 331.

Ziegler, E. and Greer, E. N. 1971. Principles of milling. In "Wheat Chemistry and Technology," Ed. Pomeranz, Y. American Association of Cereal Chemists, St. Paul, MN.

10. Leavening Agents

Batter and dough products are leavened by water vapor, air, and carbon dioxide. The gases are distributed as small bubbles in batters and doughs and are responsible for the grain of the baked products.

LEAVENING GASES

Vapor is formed from water during baking of all batters and doughs. Air is incorporated as preformed gas cells in beaten egg white and also in other ways, such as by creaming the shortening and by beating a viscous batter. Carbon dioxide is formed in foods by the chemical reaction of sodium bicarbonate with an acid or by the biological action of microorganisms.

The three leavening gases seldom, if ever, act singly. Although water vapor is important in leavening pastry, popovers, and cream puffs, it may be assisted by air, especially in pastry. Air usually is thought of as the principal leavening agent in foods containing beaten egg whites, such as angel food cakes; however, Barmore showed, in a classic study (1936), that water vapor actually is responsible for the larger proportion of the expansion of angel food cakes during baking. In foods leavened by carbon dioxide, whether formed by chemical or biological action, all three gases are present because some air is incorporated into batters and doughs during mixing and vapor is formed from water during baking. However, the important function that carbon dioxide plays in leavening some of these products is demonstrated easily by omission of the baking powder or yeast. A heavy product usually results, with a few exceptions such as certain cakes that normally contain a relatively small proportion of baking powder along with a large proportion of eggs.

The roles of the different leavening gases were demonstrated in two other classic studies, that of Hood and Lowe (1948) and that of Carlin (1944). Hood and Lowe studied cakes shortened with butter, oil, and hydrogenated lard and mixed by a modified conventional method. The fat and part of the sugar were creamed together, the flour and milk were added alternately, and a meringue of whole eggs beaten with the remaining sugar was incorporated. Cakes leavened by carbon dioxide, water vapor, and air were made as described, from a formula that included baking powder. Batter made without baking powder was used for cakes leavened by water vapor and air, whereas cakes leavened by water vapor alone were produced by evacuating the air from this batter with a vacuum pump. Cakes made by these three treatments are shown in Figure 10.1 and the proportion of the total volume increase (relative to batter volume) contributed by each of the three gases is shown in Figure 10.2. The major increase in cake volume was produced by carbon dioxide, followed by water vapor. Air was responsible for only a small proportion of the total increase in volume but was important because the effectiveness of water vapor as a leavening agent depended on the presence of air in the batter. Carlin (1944) demonstrated the importance of air distribution also to carbon dioxide's effectiveness; carbon dioxide did not form new gas cells but diffused into and expanded the existing air cells, now called gas cell nuclei.

Gas cell nuclei are the small gas cells that form during mixing. They consist largely of the air that is incorporated and dispersed, but the rapid evolution of some of the carbon dioxide at room temperature also may make a contribution. Gas cell nuclei are important to the ultimate grain of the product because, as mentioned above, carbon dioxide and steam formed during baking diffuse into the gas cell nuclei and expand them rather than forming new cells.

Figure 10.1. Cakes containing oil and leavened by: left, carbon dioxide, air, and water vapor; middle, air and water vapor; right, water vapor (air-evacuated batter). (Hood and Lowe, 1948)

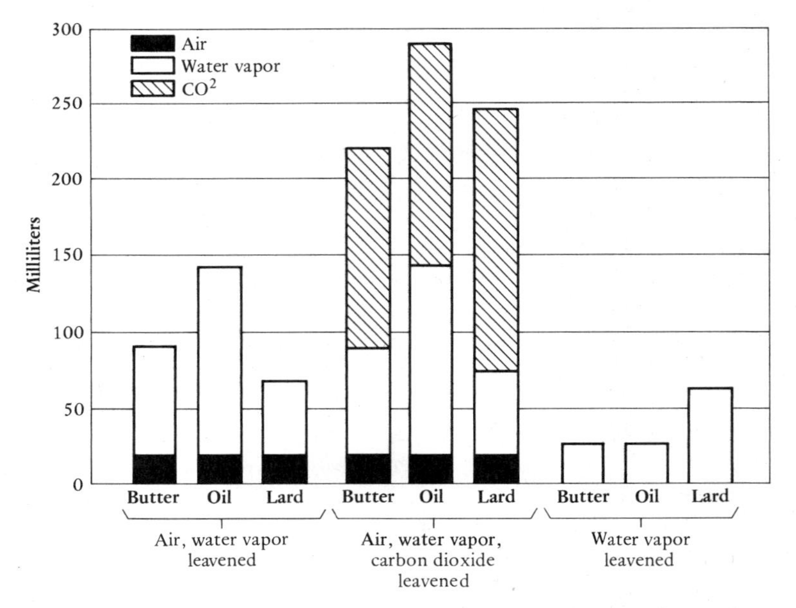

Figure 10.2. Cakes made from butter, oil, and hydrogenated lard; leavened by (1) air and water vapor, (2) carbon dioxide, air, and water vapor, and (3) water vapor. The total volume increase over the initial volume of the batter and the proportion of the increase attributed to carbon dioxide, air, and water vapor are shown. (Hood and Lowe, 1948)

CHEMICAL LEAVENING AGENTS

Chemical leavening agents are used primarily as formulated baking powders in home food preparation. The leavening ingredients are more likely to be added separately in commercial baking and in the production of mixes, but the principles of leavening action are the same in these different situations.

Chemical Nature of Leavening Action

Carbon dioxide can be produced by heat decomposition of certain compounds. For example, the heating of dissolved ammonium bicarbonate results in volatile products.

$$NH_4HCO_3 \rightarrow NH_3 + CO_2 + H_2O$$

Decomposition of ammonium bicarbonate

The absence of a residual salt is advantageous but the ammonia imparts a detectable taste unless it is able to escape completely. Therefore, the one important use of ammonium bicarbonate is in commercially baked cookies that have a large surface relative to their mass and are baked at high temperatures.

Sodium bicarbonate does not require an acid for release of carbon dioxide, as indicated by the reaction

$$2\,NaHCO_3 \rightarrow CO_2 + Na_2CO_3 + H_2O$$

Decomposition of sodium bicarbonate

However, the sodium carbonate produced has an unpleasant taste and its alkalinity has other undesirable effects, for example, on crumb color. Sodium bicarbonate, therefore, is not used alone and ideally is not used in excess.

Chemical leavening nearly always involves carbon dioxide production through the reaction of sodium bicarbonate with an acid. The reaction, where X = any anion, is the following:

$$HX + NaHCO_3 \rightarrow CO_2 + H_2O + NaX$$

Reaction of acid and sodium bicarbonate

The acid may be that of a separate acidic ingredient such as buttermilk. The acidic ingredient is combined with sodium bicarbonate in baking powders.

Baking Powders

All baking powders on the retail market are formulated to yield at least 12%, usually more nearly 14%, carbon dioxide when water is added and heat is applied. Acid salts, being more stable than acids as such, are combined in proper proportions with sodium bicarbonate; cornstarch is added for the purpose of standardization. The cornstarch also has a protective effect against atmospheric moisture.

Baking powders differ from one another as to their specific acidic constituents. They frequently are classified accordingly as SAS-phosphate or phosphate types. Tartrate baking powder formerly was available but disappeared from the market because of its cost. The proportions of sodium bicarbonate and one or more of the various acid constituents to be used in formulation of a baking powder are determined experimentally, by titration, rather than stoichiometrically, because in some cases the exact reactions are not known. The results of such titrations are expressed as neutralizing value (NV), the number of parts by weight of sodium bicarbonate that 100 parts of the leavening acid will neutralize. For example, the NV of sodium aluminum sulfate is 100, whereas that of monocalcium phosphate monohydrate is 80 (Kichline and Conn, 1970); therefore, monocalcium phosphate monohydrate would need to be used in a higher concentration than would sodium aluminum sulfate.

Baking powders differ from one another as to the speed with which carbon dioxide is formed during mixing and baking. This difference is a function of the solubility of the acidic constituents. The dough rate of reaction test, as described by Barackman (1931), is used for predicting the rate of carbon dioxide release during mixing and holding unheated. The results of the test with several leavening acids, as reported by Kichline and Conn (1970), are seen in Figure 10.3. The extremely high rates for cream of tartar and monocalcium phosphate monohydrate

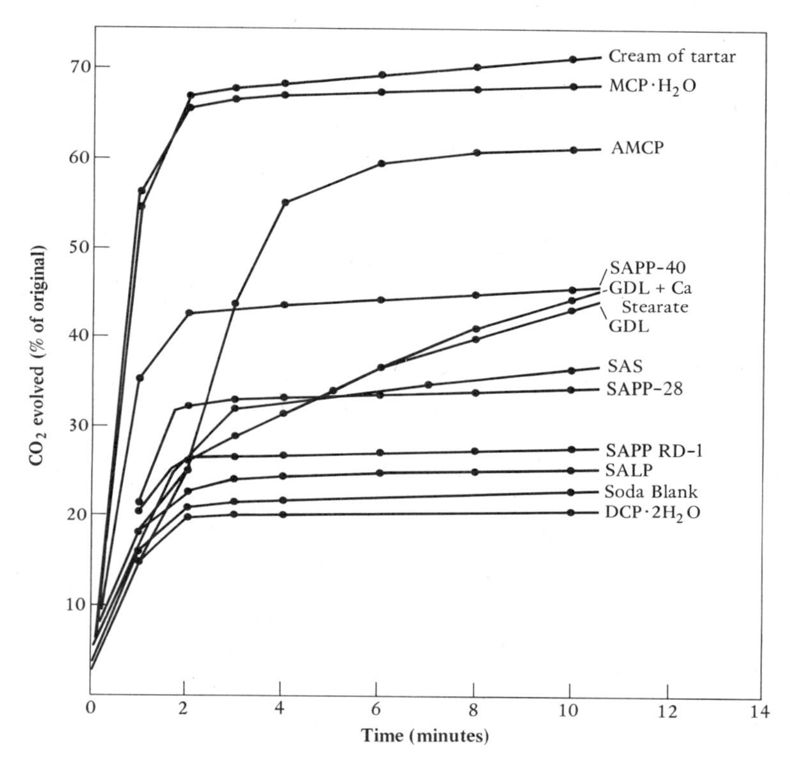

Figure 10.3. Dough rate of reaction curves for various leavening acids at 27°C with 3 min of stirring. (Kichline and Conn, 1970. Courtesy of Baker's Digest.)

(MCP·H_2O) are obvious and show why these acid salts are not used singly. When a tartrate baking powder was made, it contained both tartaric acid and potassium bitartrate.

At the opposite extreme is the reaction rate of dicalcium phosphate dihydrate (DCP·$2H_2O$), which reacts with sodium bicarbonate very little at mixing temperatue and, in fact, reacts very little at all until a late stage of baking. If used with another acid that can carry the major burden of the leavening task, DCP·$2H_2O$ can be useful in the baking industry by helping prevent dipping in the center of cakes.

SAS-phosphate baking powders are most common. It is easy to see from Figure 10.3 why this is so. Monocalcium phosphate monohydrate ($MCP \cdot H_2O$), the usual phosphate in a sulfate-phosphate baking powder, reacts with sodium bicarbonate during mixing, to provide nucleating gas. Sodium aluminum sulfate (SAS) reacts relatively little during mixing. Most of its reaction with sodium bicarbonate occurs during baking, providing carbon dioxide to diffuse into and expand the gas cell nuclei. A phosphate-type baking powder most likely contains a monocalcium phosphate that is coated to delay its dissolving.

The goal for leavening action during baking is release of most of the available carbon dioxide before firming of the product, but not too rapid release. If release of gas is not complete before the structure is firm, volume is relatively low. In a product such as a cake, bulges and cracks may develop in the center portion after the outer portions are firm. If release of gas is too rapid, the fluid or soft batter or dough permits coalescence of bubbles and the result is either a coarse structure or excessive loss of leavening gas (Kichline and Conn, 1970).

The amount of baking powder in a formula affects the properties of the product. The effects on cake quality of varying the level of SAS-phosphate baking powder are shown in Figure 10.4. The volume of the cakes first increases and then decreases as baking powder level is increased. Too little baking powder, because of insufficient carbon dioxide to expand cells properly, results in a fine grain and compact crumb. Too much baking powder causes overextension and breaking of cell walls. The result is an open structure with large cells just below the crust and a compact cell structure near the bottom of the cake when a moderate excess of baking powder is used, and a fallen cake when a great excess is used. The crumb is harsh and crumbly with an excess of baking powder. Maximum palatability does not necessarily coincide with maximum volume.

Types of baking powders were compared extensively by early workers. With tartrate baking powder no longer available, such comparisons are perhaps of less consequence than they once were; however, the reports of some early studies that dealt with types of baking powders, as well as some other factors, constitute interesting and worthwhile reading matter. Those of Briant and Klosterman (1950, Briant et al. (1954), McLean and Lowe (1934), and Moore et al. (1954) are examples.

The successful interchange of sweet and sour milk in flour mixtures is

Figure 10.4. Effects of increasing amounts of baking powder on the volume and cell structure of cakes. The amounts of SAS-phosphate baking powder per cup of flour are 1/2, 1, 1 1/2, 2, and 2 1/2 teaspoons in cakes 1, 2, 3, 4, and 5. Cake 2 is considered ideal. (Courtesy of General Foods Corporation and CALUMET® Baking Powder.)

related to leavening agents. If a product contains no buttermilk or other acidic ingredient, the leavening agent in its formula is baking powder. If buttermilk is substituted for the sweet milk, some sodium bicarbonate is needed to neutralize the lactic acid; however, total substitution of sodium bicarbonate for the baking powder, either on a weight basis or on a volume basis, would result in excess alkalinity and excess leavening. The substitution procedure described here is only a rough guide because of several variables involved, but in practice it works quite well. The acid in one cup of moderately sour milk is neutralized by 1/2 tsp (about 2 g) of sodium bicarbonate and the above combination of sour milk and sodium bicarbonate provides leavening equivalent to about 2 tsp (6–8 g) of baking powder. Thus if a formula calls for 3/4 cup of milk and 2 1/2 tsp of baking powder, the appropriate amount of sodium bicarbonate with the use of 3/4 cup of buttermilk is 3/8 tsp (1.5 g). Leavening equivalent to that of 1 1/2 tsp of baking powder would be taken care of and 1 tsp (3–4 g) of baking powder would need to be used along with the buttermilk and sodium bicarbonate.

BIOLOGICAL LEAVENING SYSTEMS (YEAST FERMENTATION)

Many breads containing yeast not only are leavened by yeast-produced carbon dioxide but also owe their distinctive flavor to the metabolic products of yeast. Other breads are yeast-leavened (and flavored) and also have a sour flavor that results from bacterial production of acid.

Yeast Alone

Baker's yeast is compressed yeast containing 71% moisture. It is the most active form of yeast for breadmaking but is unstable at temperatures above 30°C for even short periods and is stable at room temperature only for about two weeks (Hautera and Lovgren, 1975). Compressed yeast formerly was the major form of yeast also at the retail level, but has been essentially replaced by active dry yeast (ADY). ADY is prepared from special strains of baker's yeasts that are unusually stable to drying. The low moisture level makes room temperature storage possible, though storage life may be extended by refrigeration.

The yeast species that is responsible for carbon dioxide production by either the compressed or the active dry product is *Saccharomyces cerevisiae.* Yeast is capable of either aerobic or anaerobic fermentation, but fermentation in bread dough is largely anaerobic (Pomper, 1969). The overall reaction,

$$C_6H_{12}O_6 \rightarrow 2\,C_2H_5OH + 2\,CO_2$$

Overall yeast fermentation reaction

does not show the complex, step-wise nature of yeast fermentation but is descriptive of the total process.

Yeast activity is affected by available substrate. Sugars are available from several sources. Flour normally contains sugars totaling about 1.5% on the dry weight basis. Sucrose usually is added to yeast doughs. Sucrose is hydrolyzed rapidly to glucose and fructose by yeast invertase at the cell wall or just inside. Some maltose is formed through the action of flour amylases on starch. Maltose is fermentable by *S. cerevisiae* but the rate of utilization depends on the supply of glucose and sucrose; little maltose is fermented in the presence of glucose and sucrose, but if the supply of glucose and sucrose is depleted, yeast becomes adapted to fermentation of maltose, which then can proceed quite rapidly.

Yeast activity is affected by pH and temperature. The optimal pH for fermentation by *Saccharomyces cerevisiae,* 4.0–6.0, is attained readily in a fermenting dough, as some of the carbon dioxide formed becomes dissolved in water. A temperature of approximately 35°C is optimal.

Sucrose, which provides a substrate for yeast, also can have a retarding effect on fermentation if used in excess. The effect probably is osmotic, resulting in dehydration of yeast cells and consequent interference with their metabolic activity. Sweet doughs need levels of yeast sufficient to overcome the fermentation-retarding effect of their relatively high levels of sugar. Sodium chloride also has a retarding effect on yeast fermentation as a result of osmotic pressure. Although used in smaller amounts than sugar, sodium chloride on a weight basis has about 12 times the osmotic effect of sucrose and 6 times the osmotic effect of glucose because of its low molecular weight and its formation of two ions. Sucrose and sodium chloride can be balanced against each other in formulation in order to control fermentation rate so that it is neither too rapid nor too slow; for example, a relatively low level of salt can be used to balance the retarding effect of the high level of sucrose in a sweet dough. An alternative to reducing the sodium chloride level in a sweet dough is increasing the yeast concentration. On the other hand, it might be necessary to reduce the level of yeast in a low-sugar bread made without sodium chloride, as for a low sodium diet.

Reviews of fermentation by *Saccharomyces cerevisiae* are those of Magoffin and Hoseney (1974) and Pomper (1969).

Yeast in the Presence of Bacteria

Several breads are products of both yeast and bacterial fermentation. The yeasts produce carbon dioxide for leavening and the bacteria produce the acids that contribute a distinctive sour flavor.

Rye bread is leavened by *Saccharomyces cerevisiae* and soured by acid-forming bacteria. These bacteria are carried by rye flour but, for the sake of control, are more frequently added in the form of a commercial starter. Some rye breads are simulated in that organic acids are added directly rather than produced in the dough (Matz, 1972).

San Francisco sourdough French bread is leavened by another *Saccharomyces, S. exiguus.* The San Francisco sourdough bread system, described by Sugihara et al. (1970), is particularly acidic as a result of the activity of a bacterium that is present along with *S. exiguus* in the natural San Francisco sourdough cultures. The bacterium, for which Kline and Sugihara (1971) suggested the name *Lactobacillus sanfrancisco,* does not compete with the *Saccharomyces exiguus;* in fact, the organisms show great compatibility. *L. sanfrancisco* produces rather large quantities of both lactic acid and acetic acid and *S. exiguus* thrives at low pH levels.

Salt-rising bread represents yet another biologically leavened product. The starter culture that is used carries both "wild yeast" organisms and acid-forming bacteria. This combination sounds similar to that in the San Francisco sourdough French bread, but the organisms are quite different (Matz, 1972).

EXPERIMENTS

I. Leavening Systems in Chocolate Cakes

A series of chocolate cake formulas includes four that are equivalent as to total leavening. Two of these formulas produce batters of approximately neutral pH. One produces a somewhat acidic batter and the other a somewhat alkaline batter. The fifth formula produces excessive leavening as well as excessive alkalinity. Before measuring batter pH, study the formulas and decide which fits each of the above descriptions.

A. Procedures

1. Prepare the batters according to the following formulas:

	Treatments				
	1-a	1-b	1-c	1-d	1-e
Cake flour, g	100	100	100	100	100
Sodium bicarb., g	—	—	0.7	1.3	5.0
Baking powder, g	5.0	5.0	2.2	—	—
Salt, g	1	1	1	1	1
Cocoa, g	18	18	18	18	18
Sugar, g	150	150	150	150	150
Shortening, g	66	66	66	66	66
Egg, g	100	100	100	100	100
Milk, ml	—	68	—	68	68
Buttermilk, ml	68	—	68	—	—
Vanilla, ml	2	2	2	2	2

   a. For each cake put all ingredients into the bowl of an electric mixer. Blend at low speed; then mix at medium speed 2 1/2 min, stopping twice to clean the sides of the bowl with a rubber scraper.
   b. Transfer 300 g of each batter to a pan approx 19 × 9 × 6 cm. Save the remainder of the batter.
2. Bake the cakes at 185°C (365°F).
3. While the cakes are baking, mix 25 g of each batter with 15 ml of distilled water and measure the pH of each slurry.
4. Display a portion of the remaining batter along with each baked cake.

B. Results

1. If feasible, conduct a blindfold test before other members of the class have seen the baked cakes. Ask them to "think chocolate" and rank the cake samples as to preference.

2. Make a chart listing the five cake codes across the top and the following characteristics down the left margin: batter color, batter pH, crumb color, grain, volume, and flavor. Use scales in the Appendix as guides for the sensory scoring.

C. *Questions*

Why does so much variation in the ingredients produce so little variation in volume? Why doesn't the cake with excessive leavening have a great volume? How do variations in pH affect batter color? Cake color? Describe any differences in grain that are apparent. Describe the flavor differences observed.

II. Yeast-Sugar-Salt Proportions in Yeast Dough

Yeast, sugar, and salt, all of which affect fermentation rate, are varied in individual experiments. The basic formula for yeast bread is designed to show the effects of variations in ingredients in a relatively short laboratory period. It is too high in yeast to be the best formula for normal breadmaking; it also contains no milk.

A. *Basic formula and preparation of dough and bread*

Yeast (active dry), g	4
Water, g or ml	118
Sugar, g	13
Salt, g	4
Flour, hard wheat all purpose, g	200
Shortening, g	8

1. Weigh all ingredients except the water. Adjust the water temperature to 45°C just before weighing or measuring and use immediately.
2. Add the yeast to part of the water.
3. Put sugar, salt, and remaining water into the mixing bowl.
4. Add the yeast dispersion and about 1/4 of the flour; mix well.

5. Cut the shortening into about 1/4 of the flour with a pastry blender. Add to batter.
6. Stir while adding enough of the remaining flour to make a ball of dough that is irregular in shape, rough and dull in appearance, and somewhat sticky to handle. Since it is important to know the total amount of flour in the dough, flour to be used in kneading and shaping the dough is taken from the weighed amount that was not needed for mixing. If the original 200 g of flour are not enough to make a dough of the right consistency, weigh additional flour for use in the dough and on the board.
7. Turn the dough out on a lightly floured board and knead 100 strokes.
8. Roll the ball of dough in an oiled bowl to lightly oil its surface. Record the time and allow to rise in a moist warm place, such as in a cabinet or unlighted oven containing a bowl of boiling water, or on a dish drainer over but not touching hot water in a sink, or on a wire rack over a large bowl of hot water. (Cover with a towel.)
9. When the dough has risen until a light finger impression remains in the dough, record the time and gently punch the dough down, turning the edges toward the center. Weigh 300 g.
10. With the hands, flatten the dough on a lightly floured board and shape into a rectangle. Roll from end to end. Pinch to seal the seam and the ends, and place, seam down, in an oiled pan approx 19 × 9 × 6 cm.
11. Allow to rise as in 8. Record the time required for proofing.
12. Bake at 218°C (425°F) for 20–30 min, until the bread sounds hollow when removed from the pan and knocked on the bottom with the knuckles.
13. Remove bread from pan and cool to room temperature on a wire rack before carefully cutting through the center with a bread knife.
14. Weigh the flour not needed for mixing, kneading, and shaping. Subtract from the amount originally weighed to find the amount in the bread.

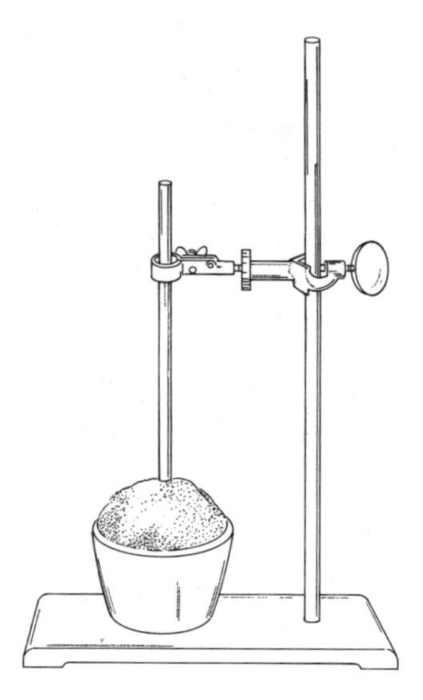

Figure 10.5. Improvised device for checking the height to which dough has risen.

B. Procedures

In order to expedite control of all factors other than the experimental variable in each of the following experiments, the use of an assembly line within a group is suggested. The amount of rising can be standardized by the use of a simple device in which a heavy stirring rod is held vertically by a buret clamp attached to a ringstand. The height of the rod is adjusted so that its tip just touches the top of a properly risen dough, as shown in Figure 10.5. A second ringstand can be arranged for dough that has proofed, or risen in baking pan. Because of the length of time required for making the bread, it might be necessary to freeze it for later examination.

1. Amount of yeast. Make breads using the following amounts of active dry yeast:
   a. 1.8 g
   b. 4.0 g
   c. 5.3 g

2. Amount of sugar. Make breads using the following amounts of sugar:
   a. None
   b. 13 g
   c. 26 g
3. Amount of salt. Make breads using the following amounts of salt:
   a. None
   b. 4 g
   c. 8 g

C. *Results*

Make a table listing the variations down the left margin. Across the top of the table write weight of flour, rising time in bowl, rising time in pan, volume, texture, flavor, and acceptability. Volume can be evaluated visually or by measurement. Score the palatability factors using appropriate scales in the Appendix.

D. *Questions*

How do levels of yeast, sugar, and salt affect rising time? Explain. Do levels of yeast, sugar, and salt affect volume? How do levels of yeast, sugar, and salt affect grain? How do levels of yeast, sugar, and salt affect flavor? Do some other combinations of levels of these three ingredients suggest themselves to you?

SUGGESTED EXERCISES

1. Using an SAS-phosphate baking powder and a high oven temperature, produce muffins that have a crack on the top. See whether the crack can be eliminated by using a lower oven temperature with the same baking power or the high oven temperature and a phosphate baking powder.
2. Compare the reaction rates of different baking powders, including the low sodium baking powder in 4. Quickly mix 4 g of baking

powder with 15 ml of a 1:1 dilution of egg white in a 100-ml mixing cylinder. Record the volume of each dispersion every minute for 5 min. Set the cylinders into a pan of water that has been brought to a boil and removed from the heat. Again record the volume of each dispersion every minute for 5 min.

3. Study the effects of various levels of one or more types of baking powder on the quality of biscuits, muffins, or cakes.
4. The following baking powder formula is recommended for persons on a low sodium diet:

Potassium bicarbonate, g	39.8
Cornstarch, g	28.0
Tartaric acid, g	7.5
Potassium bitartrate, g	56.1

   The formula (KDDA, 1973), which can be prepared by a pharmacist, is accompanied by a statement that 1 1/2 times as much as "regular" baking powder should be used. Test the accuracy of the statement.
5. Plan and conduct an experiment similar to I but with biscuits or muffins.
6. Make yeast doughs with various combinations of the three levels of yeast, sugar, and salt used in II. For example, one combination might be: 4 g yeast, 0 g sugar, 8 g salt.

REFERENCES

Barackman, R. A. 1931. Chemical leavening agents and their characteristic action in doughs. Trans. Amer. Assoc. Cereal Chem. 8: 423.

Barmore, M. A. 1936. The influence of various factors, including altitude, in the production of angel food cake. Tech. Bull. 15, Colorado Agric. Exp. Stn., Fort Collins.

Briant, A. M. and Klosterman, A. M. 1950. Influence of ingredients on thiamine and riboflavin retention and quality of plain muffins. Trans. Amer. Assoc. Cereal Chemists 8: 69.

Briant, A. M., Weaver, L. L. and Skodvin, H. E. 1954. Quality and thiamine retention in plain and chocolate cakes and in gingerbread. Mem. 332, Cornell Univ. Agric. Exp. Stn., Ithaca, NY.

Carlin, G. T. 1944. A microscopic study of the behavior of fats in cake batters. Cereal Chem. 21: 189.

Hautera, P. and Lovgren, T. 1975. The fermentation activity of baker's yeast—its variation during storage. Baker's Digest 49(3): 36.

Hood, M. P. and Lowe, B. 1948. Air, water vapor, and carbon dioxide as leavening gases in cakes made with different types of fats. Cereal Chem. 25: 244.

KDDA. 1973. "The Knoxville Area Diet Manual." Knoxville District Dietetic Assoc., Knoxville, TN.

Kichline, T. P. and Conn, T. F. 1970. Some fundamental aspects of leavening agents. Baker's Digest 44(4): 36.

Kline, L. and Sugihara, T. F. 1971. Microorganisms of the San Francisco sour dough bread process. II. Isolation and characterization of undescribed bacterial species responsible for the souring activity. Appl. Microbiol. 21: 459.

Magoffin, C. D. and Hoseney, R. C. 1974. A review of fermentation. Baker's Digest 48(6): 22.

Matz, S. A. 1972. Formulations and procedures for yeast-leavened bakery foods. In "Bakery Technology and Engineering," 2nd ed., Ed. Matz, S.A. Avi Publ. Co., Westport, CT.

McLean, B. B. and Lowe, B. 1934. Plain cakes. VI. The determination of the optimum quantity of three types of baking powder. J. Home Econ. 26: 523.

Moore, R., Meyer, B. and Buckley, R. 1954. The effect of freezer storage temperatures on cake quality and on the carbon dioxide content of cake batters. Food Res. 19: 590.

Pomper, S. 1969. Biochemistry of yeast fermentation. Baker's Digest 43(2): 32.

Sugihara, T. F., Kline, L. and McCready, L. B. 1970. Nature of the San Francisco sour dough French bread process. II. Microbiological aspects. Baker's Digest 44(2): 51.

11. Yeast Breads and Quick Breads

Flour mixtures often are classified as doughs or batters. Doughs are too thick to be beaten; frequently they are kneaded. Batters, on the other hand, usually are beaten during their preparation and are of a consistency to be dropped or poured. Doughs and batters are complex because of interactions among the ingredients and because of possible variations in the nature and proportions of the ingredients and in the mixing method.

The success of experiments on doughs and batters depends on the control of a number of variables. The nature, proportion, and temperature of ingredients must be controlled. Ingredients often are brought to room temperature before mixing, because the solution of certain ingredients, the distribution of fat, the hydration of gluten proteins, and the beating of eggs proceed more readily at room temperature than at refrigerator temperature. Regardless of the actual ingredient temperature, control is essential. The order in which the ingredients are mixed and the amount of stirring or beating also are important. In most research, the amount and kind of mixing are controlled by use of an electric mixer and a stop watch. When hand mixing is necessary in a class, control involves such procedures as counting strokes. The size of the bowl and the type of mixing utensil are important, as are the size, shape, and material of the pan used for baking. Oven temperature, placement in the oven, and extent of baking also must be controlled.

YEAST BREADS

Yeast breads are particularly dependent on flour. The individual gluten proteins of flour interact during mixing to provide a framework of gluten that will undergo prolonged and extensive stretching during fermentation and that will coagulate in the oven to help form the structure of the loaf.

Ingredients

The ingredients of yeast dough have been discussed in previous chapters but will be considered further in relation to their respective roles in breadmaking. Flour, water, and yeast are the only essential ingredients of yeast dough, though salt, shortening, sugar, and milk solids are commonly added, and sometimes egg as well.

Flour As indicated previously, the proteins of flour are of prime importance to the development of bread dough. Other flour components also have important roles, however.

Role of proteins and lipids in dough development The differences between gliadins and glutenins as to disulfide bonding were discussed in Chapter 9. Disulfide bonds are affected by mechanical stress, and disulfide-sulfhydryl relationships are responsive to the action of reducing and oxidizing agents.

Sulfur groups are involved in more than one type of change during the mixing of flour and water. Some of the existing disulfide bonds (—S—S—) are broken through the applied mechanical energy and through the presence of small amounts of reducing agents such as cysteine. This type of reaction results in interchange between —S—S— and sulfhydryl (—SH) groups. Some oxidizing agents, such as bromates and peroxides, release oxygen to react with hydrogen of —SH groups. Other oxidizing agents, such as azodicarbonamide and dehydroascorbic acid, are hydrogen acceptors; they remove hydrogen atoms from —SH groups. With either type of oxidant, the result is the formation of new —S—S— bonds. Atmospheric oxygen, incorporated during mixing, also contributes to oxidation of —SH groups to —S—S—. The breaking of some —S—S— bonds and the formation of new ones contribute to the development of gluten, the complex between gliadin and glutenin.

Gluten has neither the stickiness and flow properties of hydrated gliadin nor the extreme cohesiveness and elasticity of hydrated glutenin; rather, it has properties intermediate between those of gliadin and glutenin hydrated individually. Properly developed, gluten makes the dough easy to handle, either by hand or by machine, and permits a large amount of expansion during fermentation and the early stages of baking. It should be recognized that although disulfide bonding claims much attention in the literature, hydrophobic interaction, hydrogen bonding, and ionic bonding also make important contributions to dough development during mixing (Pomeranz, 1971).

The actual breadmaking potential of flours varies, and much research has been devoted to the exact nature of the proteins of wheats from which flours of poor and good baking quality are produced. It is recognized that both the glutenin and gliadin fractions of gluten are important in breadmaking. Glutenin's functional role in contributing elasticity and in controlling the time required for dough development was reviewed by Bushuk (1974). Bietz et al. (1973) also reviewed research concerned with glutenin. Gliadin proteins are responsible for loaf-volume potential of a flour. Hoseney et al. (1969) reported that differences in breadmaking potential of flours reside in the gliadin fraction. Flours of poor and good baking quality were fractionated and the gliadin- and glutenin-rich protein fractions were interchanged in reconstitution of flours for baking tests; the gliadin-rich fraction from the flour of good baking quality was required for volume comparable to that obtained with the high-quality unfractionated flour (Figure 11.1).

Differences among flours as to inherent breadmaking quality necessitate variations in oxidizing treatment at the mill or at the bakery. Jackel (1977) summarized the role of oxidants in commercial breadmaking. Flour on the retail market has been treated for its intended purpose; the average consumer's role, therefore, is simply to make a reasonable selection of flour for a given use. The student of food science, however, should have some understanding of the chemical processes that contribute to the functional performance of flour.

Figure 11.1. Cut loaves of bread baked from unfractionated flours of differing baking quality and from reconstituted flours with interchanged gliadin-rich and glutenin fractions. Loaves represent, from left to right: top row, unfractionated flours C.I. 12995 and K501099; bottom row, gliadin-rich fraction of C.I. 12995 with glutenin of K501099, and gliadin-rich fraction of K501099 with glutenin of C.I. 12995. The reconstituted flours contained 12.5% protein and C.I. 12995 starch and water-solubles. (Hoseney et al., 1969)

It has been known for some time that the concentration of lipid in gluten is higher than that in flour; in other words, gluten proteins bind lipid during doughing. Lipid binding occurs even in the absence of added fat. Much study has gone into the nature of the protein-lipid complex. Wheat endosperm is unique in its content of glycolipids, primarily monogalactosylglycerides and digalactosylglycerides. Hoseney et al. (1970) reported that the glycolipids are bound simultaneously to the glutenin proteins by hydrophobic interaction and to gliadin by hydrophilic bonding.

Contribution of starch to structure Starch granules become closely associated with gluten during dough mixing. By its very presence in the gluten structure, starch has a diluting effect and possibly prevents an excessively cohesive structure (Sandstedt, 1961). If the starch has been damaged extensively during milling, it absorbs water at room temperature, thus competing with gluten proteins for available water. The action of flour amylases on damaged starch during dough fermentation provides sugar for yeast. Sugar beyond that utilized by yeast is available for carbonyl-amine browning during baking. Gelatinization of starch during baking contributes to structure of the baked loaf.

Replacement of wheat starch with cornstarch in a flour reconstituted from fractions resulted in inferior baking properties in studies by Sandstedt (1961), d'Appolonia and Gilles (1971), and Hoseney et al. (1971). Hoseney et al. (1971) also used rye, barley, milo, rice, oat, and potato starches, as well as starches from several hard and soft wheat varieties, in reconstituted flours in which gluten and water-solubles were constant. Of the nonwheat starches, only rye and barley starches came close to performing as well as wheat starches in baking tests. Of the wheat starches, only that from durum wheat performed poorly. Performance in breadmaking was not related to gelatinization temperature of the starch or to size of starch granules, and the reason for differences in baking performance of starches was not clear. Medcalf and Gilles (1968) reviewed the role of starch in dough.

Liquid Water obviously is essential for hydrating flour proteins during mixing, and the optimal level of water is important to dough properties. The optimal level of water varies directly with protein quantity and quality. This relationship is particularly important commercially because yield of bread increases with increasing water level.

Water also dissolves sugar and salt and serves as a dispersion medium for yeast cells. Water is responsible for partial gelatinization of starch during baking. Milk sometimes is used as the liquid in noncommercial breadmaking, though bakers are more likely to use water and nonfat dry milk. If fluid milk is used, it must be scalded in order to avoid a deleterious effect on dough consistency and loaf volume. The exact cause of the deleterious effect has been elusive. Swanson et al. (1964) reported that the beneficial effect of heat treatment involves an interaction of the serum protein β-lactoglobulin with κ-casein and a component of α-casein. Volpe and Zabik (1975) reported a study that implicated a dialyzable proteose-peptone component designated as component 5.

Yeast The primary function of yeast, as described in Chapter 10, is to produce carbon dioxide for leavening. Yeast cells contain not only the enzymes required for carbon dioxide production from glucose but also those required to make substrate available. Maltase and sucrase are present and hydrolyze, respectively, maltose, which is formed through amylolytic activity, and sucrose, which frequently is added to yeast doughs. Yeast activity also makes an important contribution to bread flavor. As indicated in Chapter 10, the rate of yeast activity is affected by temperature and by the concentrations of sugar and salt present.

Salt From the standpoint of flavor, the use of sodium chloride at a level of 2% of the flour weight is common. As discussed in Chapter 10, salt has a retarding effect on yeast activity through an osmotic effect on yeast cells. When salt is omitted completely, as in a low sodium bread, reduction of the level of yeast is helpful in keeping the rate of fermentation under control.

Sodium chloride also has a stiffening effect on dough. The effect has not been adequately explained (Bloksma, 1971).

Shortening Up to a point (about 3% of flour weight), added shortening results in increased loaf volume, improved grain, and reduced rate of crumb firming (Pomeranz et al., 1966a). Some studies have involved comparisons of different shortenings and attempts have been made to relate the improving effects to factors such as melting point (Pomeranz et al., 1966b), chain length (Elton and Fisher, 1968), and solid fat index (Cooley, 1965; Koren, 1967). A stable solid fat index over a wide temperature range is characteristic of some effective liquid shortenings that contain high-melting solids dispersed in oil and are used commercially.

Sugar The concentration of added sucrose varies from none to about 8% or a little higher. When no sucrose is added, fermentation is slow initially, because sugar for yeast substrate must be formed by amylolytic activity. If sugar content exceeds 10% of the weight of the flour, fermentation is retarded by the osmotic effect of the dissolved solute on the yeast cells. Excessive sugar also may interfere with gluten development by competing with gluten proteins for water. At its higher levels, sugar contributes to flavor. Also, when more sugar is added than is used in fermentation, the residual reducing sugars formed by sucrase are available for participation in carbonyl-amine browning during baking.

Other ingredients Commercial breads frequently contain mold inhibitors, dough improvers, yeast foods, and emulsifiers. Propionates are the common mold inhibitors. Dough improvers are oxidizing agents. Yeast foods are salts that accelerate yeast activity; these may be combined with oxidizing agents and buffers. Emulsifiers, which actually contribute to fat emulsification in batters, have a crumb-softening effect in bread.

Mixing Methods

The straight-dough and sponge methods are basic in making bread. They will be described briefly, along with some of their modifications.

Straight dough The straight-dough method, in which all the ingredients are mixed together at one time and allowed to rise, is used by most homemakers and by some bakers. The order in which ingredients are added during mixing is of little importance if care is taken that fresh fluid milk, if used, is scalded and that yeast is not destroyed by excessive heat. Shortening can be melted or not as desired. Since the absorption of flour cannot always be predicted, it is usually more convenient to add flour to liquid than liquid to flour. Flour should be added to give a soft, rather sticky dough. A dough temperature of about 28°C is desirable if compressed yeast is used, but should be a few degrees higher if active dry yeast is used (Chapter 10). The temperature of the liquid required to produce dough at the desired temperature depends on the temperatures of the room and the ingredients. Mixing and kneading are done as a single machine operation in bakeries, whereas bread made at home is mixed and kneaded by hand unless a mixer with a sufficiently powerful motor and a dough hook attachment is available. This type of equipment is useful in experimental work. When hand kneading is necessary, it should be done with a light, folding motion in which the fingers do the folding and the heels of the hands do the pushing of the folded dough. After each pushing motion, the dough is given a one-quarter turn. Only a light film of flour is used on the board to avoid making the dough too stiff. As kneading progresses, the tendency for the dough to stick decreases and the dough becomes smooth and elastic (Figure 11.2). When a dough has been kneaded sufficiently, small blisters can be seen on the surface. The time required for kneading varies with the type of motions used, with the size of the dough mass, and with the

Figure 11.2. Stages in dough formation: left, flour moistened and stirred; center, gluten partly developed; right, gluten fully developed. (Courtesy of Wheat Flour Institute.)

flour, being longer for hard than for soft flours. Dough that has not been kneaded enough does not hold the gas well. The bread from such a dough has a coarse, irregular crumb, small volume, and an uneven break along one side of the loaf. Insufficient kneading probably is common in homemade bread. Indeed, it seems difficult to overknead by hand dough made from hard wheat flours by the straight-dough method, but it is possible with soft flours or with the sponge method, in which the gluten of part of the flour has been softened by fermentation. In a bakery where the entire process of mixing and kneading is done by machine, it is easy to overmix. Overmixed dough results in bread with heavy cell walls and poor volume.

The straight-dough method can be modified to omit kneading. This has been called the *no-knead* or *batter method.* The dough can be dropped directly into the pans, before or after rising, or chilled and shaped with the hands. It is always allowed to rise in the pan before baking. Sweet rolls and coffee cakes are made more often without kneading than is bread because the best results are obtained with this method when a soft, rich dough is used. The method results in a more open grain than is obtained with kneading.

It is sometimes convenient to hold dough in a refrigerator and to bake portions of it at intervals. A roll dough made by a standard or a no-knead method is put into a refrigerator immediately after mixing. Care is taken to prevent crust formation, and the dough is punched periodically, as some rising occurs even under refrigeration. Because the dough is cold, the rising time after shaping is longer than with freshly made dough. The storage time for such doughs is less than a week.

Sponge The sponge method of mixing bread dough is used widely by commercial bakers and occasionally at home. The dough requires more handling but is more tolerant of variations, especially of fermentation time and temperature, than is a straight dough. The flavor of bread made by the two procedures is slightly different because of the longer fermentation time of the sponge method. The sponge is made by mixing yeast, liquid, and about half of the flour. The amount of yeast may be less than that for the straight dough. Speed of fermentation can be regulated by the temperature and by the addition of part of the sugar and/or salt. In rising, the sponge becomes light and frothy, and sometimes falls back because viscosity is too low for gas retention. The method derives its name from the spongelike appearance of this mixture. After the sponge has risen, the remaining ingredients are added and the dough is handled as a straight dough. The rising time of the dough is rather short. Because of the long total fermentation period, the sponge method is not desirable with soft wheat flour.

Continuous process A continuous-process mixing method that frequently is used in the baking industry might be considered a modification of the sponge method. The ingredients, one of which is a pre-fermented sponge, are metered into the mixer. The mixed ingredients are passed on to the dough developer and from there to the divider and extruder. The extruded dough is ready for fermentation in the pans, or proofing. Bulk fermentation is bypassed because of the addition of a pre-ferment.

Dough Structure

Most of the literature on gluten proteins deals with glutenin and gliadin separately, while most of the literature on dough structure deals with the gluten complex in relation to the other dough constituents, particularly starch. It is difficult, therefore, to resolve all the pieces of information into a single view of the nature of dough structure. A further complication is the lack of agreement among various researchers as to the significance of some of the findings of research.

The nature of the glutenin macromolecule apparently is basic to the structure of the gluten network and to certain of its properties. Ewart (1972) proposed that the glutenin subunits are linked into long strings, or concantenations (Figure 11.3), and that these concantenations, in which each subunit is joined to the next by two disulfide bonds, are partially coiled and also are entangled with one another. The cross-links at the points of entanglement vary as to type, depending on the chains involved, and are broken at different rates. Greenwood and Ewart (1975) postulated that the cross-links at the points of entanglement, rather than the disulfide bonds, are responsible, along with the coiled structure, for viscoelasticity, which they consider to be inherent in glutenin molecules. During the stress of mixing, the chains might become partially unfolded; then with removal of stress, they could tend to return to their native free state. According to the theory, the effect of

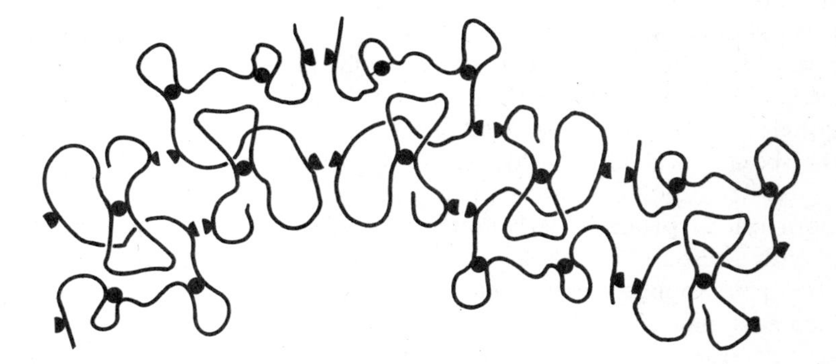

Figure 11.3. Linear linkage of glutenin subunits, with each subunit joined to the next by two disulfide bonds. (Ewart, 1972. Courtesy of Baker's Digest.)

reducing agents in decreasing dough structure is a breaking down of the concantenations. If an excessive amount of breaking of disulfide bonds occurs during mixing, as a result of either excessive reducing agents or overmixing, the glutenin chains will have been reduced in length so that the glutenin behaves more like a viscous liquid than an elastic solid.

Orth et al. (1973) examined isolated glutenins from several varieties of wheat and from one variety of rye by scanning electron microscopy (SEM). The glutenins from bread wheat varieties formed uniform, long fibers (Figure 11.4); glutenins from durum wheat, which is used for semolina products, formed flat, ribbonlike structures (Figure 11.5); and glutenin from rye formed short, relatively thick structures (Figure 11.6). When the disulfide bonds of one of the bread wheat glutenins were chemically reduced by β-mercaptoethanol, the fibrous structure was lost (Figure 11.7). They concluded that the long, thin fibers of bread wheat glutenins probably are important to gluten elasticity.

The above view of the nature of glutenin is not incompatible with the interaction between glutenin and gliadin to form gluten. Nor is it incompatible with a role of galactolipids in binding glutenin and gliadin during doughing, as mentioned previously. The gluten formed through these interactions is viewed as existing in developed dough in the form of thin sheets of parallel fibrils, in which starch granules are trapped. Starch granules that are elliptical in shape tend to become oriented

Figure 11.4. SEM micrograph of purified glutenin of the bread wheat variety Manitou. (Orth et al., 1973)

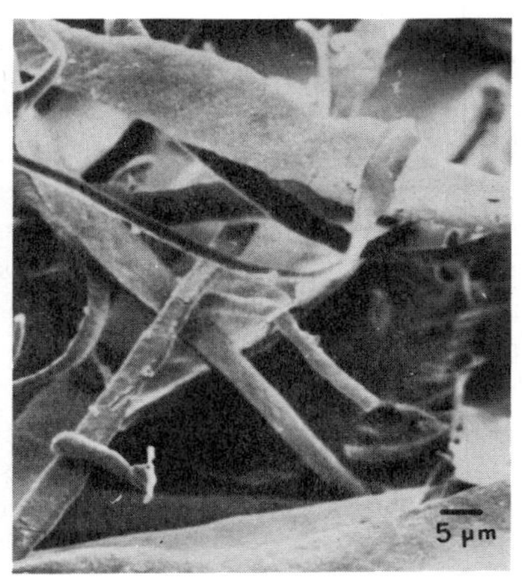

Figure 11.5. SEM micrograph of purified glutenin of the durum wheat variety Stewart 63. (Orth et al., 1973)

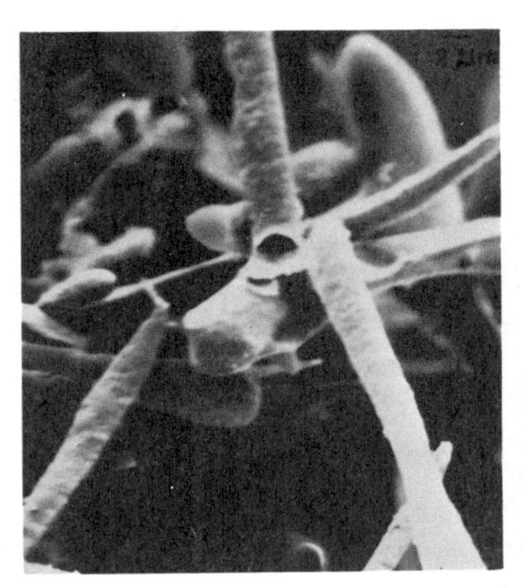

Figure 11.6. SEM micrograph of purified glutenin of the spring rye variety Prolific. (Orth et al., 1973)

Figure 11.7. SEM micrograph of reduced glutenin of the bread wheat variety Manitou. (Orth et al., 1973)

longitudinally in gluten sheets during dough development. Moss (1974), who studied dough microstructure as affected by oxidizing and reducing agents, observed that starch granules were likely to be completely surrounded by gluten under extreme reducing conditions and relatively free of gluten under extreme oxidizing conditions. In optimally developed dough, the situation probably is intermediate, with many starch granules surrounded by interconnected sheets of gluten, and other starch granules trapped between gluten sheets but not associated with the gluten. According to Pomeranz (1971), there apparently is limited interaction between glycolipids and starch granules in the dough. (On the other hand, the glycolipid-starch interaction is greater than the glycolipid-protein interaction in the baked bread.) Also observed in the dough microstructure are discrete masses of fat and yeast cells rather uniformly distributed through the spaces between sheets of gluten. Gas cell nuclei, representing air incorporated during mixing, also are distributed throughout the spaces between sheets of gluten and become enlarged during fermentation as a result of diffusion of carbon dioxide into them.

Fermentation

Doughs mixed by the straight dough or sponge process are subject to bulk fermentation. The dough is allowed to rise in the bowl at a temperature of 27–35°C, depending on the type of yeast used (Chapter 10). Crust formation should be avoided. Putting the dough in a cabinet or unheated oven with a pan of hot water or putting the bowl of dough on a rack over a container of hot water and covering it with a towel can facilitate temperature control and minimize surface drying. Crust formation also can be decreased by oiling the dough, but only the thin film of fat that adheres when dough is turned upside down in a lightly oiled bowl should be used, as a thicker film may cause streaks in the bread. The dough is allowed to rise until it has doubled in bulk and a slight depression remains when the surface is touched gently with a finger, as shown in Figure 11.8. Fermentation time is influenced by the levels of yeast, sugar, and salt, the temperature, and the type of flour. Dough made from strong flour requires a longer rising period than that made from soft flour. The use of a sponge decreases fermentation time. Care is necessary that the fermentation time is neither too short nor too long.

Figure 11.8. Test showing that dough has risen long enough. (Courtesy of Wheat Flour Institute.)

If the dough does not ferment long enough, it retains too much cohesiveness, and if it rises too long, a sour flavor and weakened structure might develop.

Straight doughs made of bread flour are sometimes punched one or more times and allowed to rise again. In this step, gas is removed from the dough by gently plunging the fist into the middle of the dough, folding the edges to the center, and turning it over. This redistributes the yeast cells and their food supply, thus promoting further fermentation. The dough is allowed to rise again until its bulk has almost doubled, which requires a shorter time than the first fermentation. This additional fermentation is not essential for hard wheat all-purpose flours, and is undesirable for soft wheat flours because it would weaken the gluten. If a delay is necessary before shaping the dough for the pan, however, it is better to punch it down than to let it rise too long. Methods of shaping dough are well described in recipe books and elementary textbooks and will not be discussed here. The final fermentation, proofing, occurs in the pan, and the criterion of readiness for the oven is an approximate doubling in bulk.

Baking

Bread is baked in an oven preheated to 204 or 218°C to ensure proper oven spring. The higher temperature may be preferable if many loaves are put into a small oven, or if the loaves are small. The temperature can be held constant or can be reduced to 177°C after the first 15 min of baking if the crusts are becoming too brown. Too high a temperature is to be avoided because it causes a crust to form before rising is complete. Too low a temperature is undesirable because it permits the dough to rise excessively, and possibly to fall, before coagulation takes place. Rich doughs often are baked at a constant temperature of 177 or 190°C to avoid excessive browning. Baking is continued until the bread sounds hollow when tapped. Brown-and-serve rolls are baked at a low temperature just long enough to attain full volume and rigidity, but not long enough to brown the crust. They can be made at home by baking for about 25 min at 135°C, then freezing (Charles and Van Duyne, 1953). Excessive rising and possible falling at such a low oven temperature are avoided by using less yeast and sugar or a shorter proof time than for standard rolls.

The sudden increase in volume of the dough during the first 10–12 min of baking is called oven spring. This increase may amount to about 80% of the volume of the risen dough when a very hard wheat flour is used (Bailey and Munz, 1938). Opening the oven door during this period should be avoided as it prevents some of the increase in volume. Oven spring is caused in part by the expansion of the gases in the heat of the oven and in part by the increased rate of fermentation brought about by the increased temperature of the dough. At first the action by flour enzymes and yeast cell enzymes in the dough is greatly accelerated; then activity ceases as the enzymes are inactivated at their critical temperatures. The oven spring is improved if yeast nutrients adequate to permit rapid fermentation are present when the dough is put into the oven, and if the condition of the gluten permits retention of the rapidly expanding gases. Because good oven spring can be expected only if the gluten is able to expand greatly without breaking, better oven spring can be expected from hard than from soft wheat flour, and from dough that has been fermented enough but not too much. Gluten is capable of especially rapid expansion in the oven because of the initial softening effect of heat on gluten.

As baking continues, gases, including carbon dioxide, alcohol, and water vapor, leave the structure. Starch gelatinizes, receiving water from gluten, which coagulates, forming the rigid structure of bread. The moisture content is reduced from about 45% in bread dough to about 35% in the baked bread. An internal temperature of about 98–99°C is attained (Swortfiguer, 1968). Heating curves for different portions of the loaf are shown in Figure 11.9.

A crust forms as the surface of the bread dries during baking. Crust color is darker than crumb color, largely because of carbonyl-amine browning. Because the browning reaction involves two of the most important nutrients of yeast, sugars and nitrogen compounds, excessive fermentation causes a light crust unless the dough contains liberal

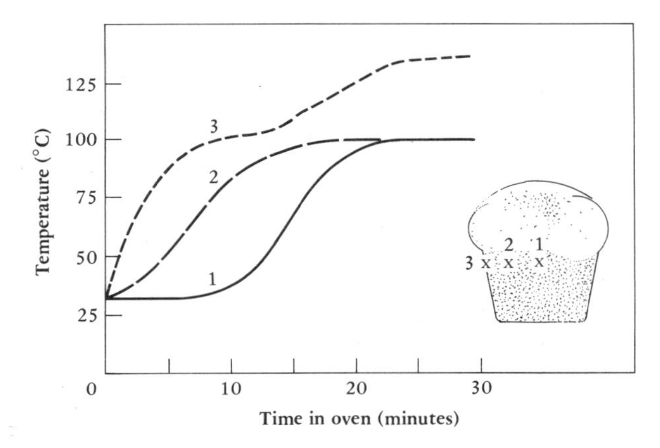

Figure 11.9. Temperature-time profiles from a dough piece baked at 235°C. (Marston and Wannan, 1976. Courtesy of Baker's Digest.)

amounts of these nutrients. Milk helps produce a brown crust primarily because the lactose it contains is not fermented by yeast. Its nitrogen compounds also contribute to the effect. Caramelization of the sugar is not appreciable at the temperature reached in the crust during baking (Baker et al., 1953). The crust can be made relatively soft by brushing it with shortening on removal from the oven. It can be made relatively crusty by brushing the tops of the loaves with water as soon as they are removed from the oven and even during baking after at least 10 min. Breads baked in a pan usually break and shred on one side of the crust just above the pan. This is caused by stretching of the gluten strands during the oven spring. An evenly shredded break is normal, but a wild break with uneven shreds is undesirable. Factors that interfere with the ability of gluten to stretch during the early stages of heating contribute to uneven shred. Loaves should be cooled uncovered on racks.

Bread Quality

Criteria of bread quality include loaf volume, grain, compressibility, and flavor. Rate of staling is a significant factor with regard to deterioration of quality.

Effects of ingredients The effects of variations in the usual ingredients have been mentioned in the discussion of their roles. The effects of substituting other flours for wheat flour have not yet been discussed. A relatively recent development that also should be mentioned is public interest in food fiber from a health standpoint. This interest has stimulated considerable research involving the effects of added fiber on the quality of bread and other baked products.

Breads containing rye flour have been made for many years, even though the proteins of rye have quite different properties from those of wheat. Rye flour has poor breadmaking characteristics as compared with wheat flour, but it imparts a unique flavor to breads containing it. Triticale flour has been available for several years and is of interest from the standpoint of its breadmaking potential. Triticale is a recently developed wheat-rye crossbreed. When it was being developed, the hope was that it would combine enough of the good qualities of both wheat and rye to be useful in breadmaking. Triticale is of interest to the baking industry and to nutritionists because it has a higher protein

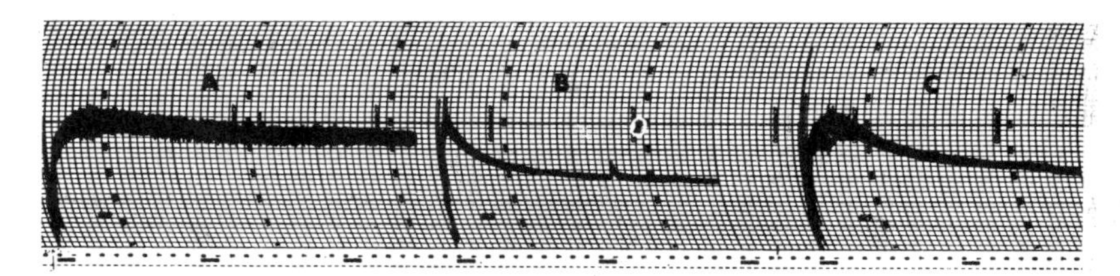

Figure 11.10. Farinograph curves for (left to right) HRW wheat, rye, and triticale flours. (Haber et al., 1976. Courtesy of Baker's Digest.)

content than wheat and because the protein has a higher concentration of lysine, which is a limiting amino acid in wheat.

Haber et al. (1976) reported a comparison of flours from a hard red winter wheat, rye, and triticale having protein contents of 10.7, 7.0, and 11.8% respectively (on a 14% moisture basis). The triticale flour was intermediate between the other flours in water absorption, mixing time, and dough stability. Rye flour was lowest in these properties. The differences in time required for dough development and the differences in dough stability are seen in the farinograph curves in Figure 11.10. The triticale flour produced an even lower loaf volume than the rye flour. The grain of bread made from triticale flour was somewhat better than that of bread made from rye flour but was poor as compared with that of bread made from wheat flour. The one characteristic in which the triticale flour surpassed the wheat flour was in its content of lysine. Haber and coworkers concluded that satisfactory results with triticale in breadmaking will necessitate modification of mixing and fermentation procedures.

Lorenz (1974) pointed out that the apparent advantage of triticale with regard to protein and lysine content is lost during milling because of a relatively low extraction rate. In addition, triticale cultivars vary even more in baking quality than do wheat varieties. Lorenz stated that, in general, triticale flours have particularly high α-amylase activity; they require less mixing than wheat flours and the mixing time is critical; their bread formulas at the 100% level of substitution need dough conditioners such as sodium stearoyl-2-lactylate (SSL); their dough fermentation times need to be shortened; and the baked products stale rapidly as compared with wheat products. While pointing out the above problems, Lorenz (1974) did report consistently satisfactory performance of flour from one triticale variety, and the flavor of triticale flour generally is well accepted. Lorenz (1974) also reported that the substitution of triticale meal, as opposed to flour, at the 50% level in specialty breads shows some promise.

Soy flour is of interest as a protein-rich bread additive with a composition that improves amino acid balance when soy flour is added to wheat in breadmaking. Defatted soy flour provides a means of improving protein nutrition of some segments of the population; full-fat soy flour has potential value in poverty areas where both protein and calories are limiting. Pomeranz (1966) pointed out the inability of soy flour proteins to contribute to the viscoelastic property of wheat gluten. He also discussed an undesirable side effect of heating soy products to inactivate the trypsin inhibitor and to modify the "beany" flavor; the heat treatment has a deleterious effect on breadmaking performance.

Workers who have studied defatted soy flour in bread have found that several modifications in formula and procedure are necessitated by substitution of soy flour for some of the wheat flour. These include increased absorption (water level), increased oxidant, decreased mixing time, and shortened fermentation time.

The addition of dough conditioners improves baking performance. Tsen and Hoover (1973) used full-fat soy flour in bread at levels up to 28%, with and without sodium stearoyl-2-lactylate (SSL). In the absence of SSL, the specific volume of breads was low and grain was poor. In the presence of SSL, acceptable bread could be obtained with full-fat soy flour added at levels up to 24%. SSL also had a positive effect on retention of softness of crumb during storage.

Pomeranz et al. (1969) used another type of additive to improve the characteristics of breads containing soy flour. Sucrose esters are synthetic glycolipids that are available commercially. They are prepared by esterification of either free fatty acids or natural glycerides with sucrose, and they function as nonionic surfactants. They apparently are particularly useful in doughs containing noncereal protein and/or little or no added shortening. For example, Pomeranz and coworkers (1969) reported that the addition of 0.5% sucroglycerides to dough counteracted the deleterious effects of as much as 16% soy flour. The improving effect was at least equal to that of 3% vegetable shortening.

Khan and Rooney (1977) made breads in which wheat flour was replaced by cottonseed, peanut, and sesame flours at levels such that each blend had a protein content of 17.5%. Breads were made with and without SSL and with straight dough versus short-time processes. The breads containing the oilseed flours differed from the all-wheat control considerably when made without SSL by the straight-dough method. For the oilseed flours, the short-time procedure resulted in greater loaf volume than was obtained with the straight-dough method. The short-time method also resulted in increased crumb grain score for the cottonseed and sesame flours. The response to SSL was not the same for the different flours and the two methods.

Hamed et al. (1973) reported the effects of adding sweet potato flour to wheat flour for use in standard dough tests and baking tests and in making *Balady* bread, a bread used in the United Arab Republic. At all levels of addition, the sweet potato flour resulted in a weakening of the dough and a decrease in dough stability. The sweet potato flour significantly decreased acceptability of white bread but increased acceptability of Balady bread when added to wheat flour at the level 10 parts sweet potato flour per 100 parts wheat flour.

Many sources of fiber have been studied in a rather short period of time. Pomeranz et al. (1977) compared the effects of wheat bran, oat hulls, and seven commercial celluloses on bread when used at levels of up to 15% replacement of wheat flour. The fibers differed in their effects on water absorption and mixing time. The level of substitution had a greater effect on loaf volume than did the type of fiber. At the high fiber levels, volume reduction was considerably greater than was expected simply on the basis of gluten dilution. According to Pomeranz and coworkers (1977), the results suggested that gas retention, rather than gas production, was affected. Microscopic study showed that crumb structure of the baked loaves was disrupted by fiber. Although oat hulls were found unsuitable, acceptable breads were obtained with wheat bran and microcrystalline celluloses at a replacement level of about 7%.

Lorenz (1976), recognizing the large amount of bran that is available because of the low extraction rate necessary in the milling of most triticale flours, studied the feasibility of using the bran in high-fiber breads. Triticale bran was used at two degrees of coarseness and at replacement levels of 0, 5, 10, and 15% of the wheat flour in bread made by the straight-dough procedure. The 10 and 15% levels of triticale bran resulted in reduced proof time, increased absorption, and increased mixing time, and also resulted in a color difference in the baked breads. All loaf volumes were comparable. The fine bran produced a much better grain than did the coarse bran, but the fineness of the bran did not affect loaf volume. The total sensory scores for the breads made with fine triticale bran at the 5 and 10% replacement levels were similar to those for the all-wheat control, and the quality of bread at the 15% replacement level was considered good. The samples containing 10 and 15% fine bran retained softness during storage to a greater extent than did the controls.

Further use of fiber in baked flour mixtures will be mentioned in connection with other baked products.

Effects of storage The staling of bread has been studied for many years, and many questions remain unanswered. All of the changes that occur in bread after baking constitute the phenomenon of staling. The gross manifestations of the process include a change in the character of the crust from dry crispness to soft cohesiveness and an increased firmness of the crumb, along with the development of a harsh texture and a stale flavor. Among the changes that can be quantified, along with compressibility, are decreased water absorption capacity, increased starch crystallinity, changes in x-ray diffraction patterns, and decreased enzyme susceptibility of the starch. The effects on sensory properties, of course, are the really important results of staling. The other changes provide means of studying the effects of various treatments.

All of the major components and some minor components of bread have been studied in relation to their potential roles in staling. Starch has been of special interest because it is a major component and because the changes in crystallinity of starch during staling long ago seemed to link starch to the staling process. There is a large amount of evidence of a relationship between starch crystallization and bread-crumb firming. The persistent question has been the mechanism: the relative participation of amylose and amylopectin and the relative roles of starch molecules within and outside the granules. The speed with which amylose retrogrades and the low amount of energy required to reverse the staling changes led Schoch (1965) to theorize that retrogradation of the amylose is complete by the time the freshly baked loaf is cool and that the further changes recognized as staling involve amylopectin chains that associate within the granules. Schoch believed that the firming of the crumb results from the firming of the granules themselves, and he considered the associations within the granules to be the effects that are reversed by heat. Others consider it more likely that both amylose and amylopectin chains are involved (Kim and d'Appolonia, 1977).

The gluten content appears to be inversely related to the staling rate of bread, and gluten once was credited with a direct retarding effect on staling. However, its role now is recognized as probably being indirect. Gluten affects specific volume, which in turn influences the amount of starch present in a given volume of the crumb; in other words, the greater the specific volume of the loaf, the lower is the concentration of starch. It is possible that flour proteins have another indirect effect by influencing the relative roles of amylose and amylopectin in staling.

Lipids, like proteins, have been credited with retarding staling. The binding of lipids to gluten proteins during doughing results in increased volume, however, and again the effect may be indirect. Another possibility, which would suggest a more direct effect, is the finding, mentioned previously, that the galactolipid that becomes associated with gluten proteins during mixing becomes associated with starch during baking (Pomeranz, 1971). Such association might increase the difficulty of association between starch chains, thus interfering with crystallization. This probably is also the mechanism of the protective effect of surfactants, which frequently are added to commercial breads for their crumb-softening effect (Birnbaum, 1977). Examples are monoglycerides, propylene glycol esters, succinyl esters, and sodium-stearoyl-2-lactylate. Some, such as SSL, also are dough conditioners because of their ability to bind flour components.

Moisture levels in bread have been of interest because of several early findings: Increasing absorption (water level) of bread dough retards firming; complete prevention of moisture loss from a loaf of bread does not prevent staling; moisture gradients exist in the loaf, and migration of moisture from the crumb to the crust occurs during storage. The possibility of moisture transfer from starch to gluten has been argued extensively with no definitive answer.

Not only does heating reverse crumb firming, but storage of bread at refrigerator temperature permits staling at a maximum rate. There is evidence that as storage temperature increases, the role of starch in staling decreases and other constituents play a greater role (Kim and d'Appolonia, 1977).

Bread staling is a complex and interesting subject. The recent review by Kim and d'Appolonia (1977), with the bibliography that accompanies it, is a good starting point for further study of this important deteriorative process that occurs in a staple dietary item.

QUICK BREADS

Less development of gluten is desired in quick breads than in yeast breads, and the function of flour in some cases is overshadowed by that of other ingredients. Leavening action is faster than in yeast breads. The volume of research dealing with quick breads is small as compared with that on yeast breads, and only muffins, biscuits, and popovers will be considered in this chapter.

Muffins

Muffins contain flour, liquid, egg, sugar, salt, shortening, and baking powder. The liquid, usually milk, represents about half the amount of flour on a volume basis and approximately equals the amount of flour on a weight basis. The amounts of egg, sugar, and shortening are low in a basic muffin formula as compared with the proportions in a cake; therefore, muffin batter cannot tolerate the amount of mixing to which cake batters are subjected. The mixing method that is referred to as the muffin method was developed for the purpose of permitting minimal mixing. It involves stirring a mixture of milk, egg, and melted shortening or oil into the combined dry ingredients. A satisfactory alternative is cutting the fat rather finely into the dry ingredients and then adding the milk and egg mixture. In either case, mixing should be just sufficient after the liquid is added to dampen the dry ingredients; the batter should not be smooth.

Muffins from properly mixed batter are large and symmetrical, without peaks or tunnels. Their crust is pebbled rather than rough or smooth, they have a fairly even, open grain, and they are tender.

Figure 11.11. Properly mixed muffins (top) and overmixed muffins (bottom). In each case, both sides of the same cut muffin are shown. (Courtesy of Foods and Nutrition Dept., Oregon State University.)

Properly mixed muffins rise evenly and well during the early part of the baking period because the batter contains many gas cells in readily extensible gluten. Overmixing decreases tenderness and produces a muffin with a crust that is light in color, smooth, and peaked, with tunnels that go toward the peak, and with a fine grain between the tunnels. Properly mixed and overmixed muffins are shown in Figure 11.11.

The volume of muffins becomes larger with moderate overbeating and smaller with excessive overbeating. The peaks and tunnels characteristic of overmixed muffins are probably the result of overdevelopment of gluten and loss of carbon dioxide during mixing. Some carbon dioxide is lost during overmixing, and the batter changes from a lumpy mixture that does not hold together to one that is smooth and falls in strands from the spoon, indicating gluten development. Both the strong gluten and the decreased amount of carbon dioxide prevent the muffin from rising normally during the early part of the baking period. After a crust has formed, however, sufficient steam pressure develops internally to push through the batter, forming tunnels and a peak at the soft center of the crust. If some of the dough is pushed out of the crust, a knob is formed. The peak usually is at the point where the spoon left the batter as it was dropped into the pan because the gluten strands follow the spoon. More tunnels are formed if an excessively high oven temperature causes a crust to form early in the baking period. Anything that interferes with gluten overdevelopment, such as extra sugar and shortening or the substitution of cornmeal or a nonwheat flour for part of the flour, decreases tunnel formation. Commercial muffin mixes usually have relatively high proportions of sugar to make them more nearly foolproof. They tend to have a more cakelike crumb than do basic muffins. Richer mixtures frequently are used also in large-quantity food production because of the relative difficulty of avoiding overmixing.

Agitation should be minimized as muffin batter is being put into the pans. The use of an ice cream dipper has been suggested for this purpose and, in any case, the batter should be placed rather than dropped into the pans. In experimental work, the portions are weighed; an ice cream dipper is helpful in achieving the desired weight with a minimum amount of agitation.

Bowman et al. (1973), who were interested in developing some hypoallergenic foods, reported the successful substitution of rice flour totally for wheat flour in muffins, as well as in waffles, banana bread, and date-nut bread. The latter two products also were free of milk, but all contained egg. Of the four quick breads, the muffins compared least well with wheat flour controls, but they were acceptable.

Biscuits

There are at least two types of biscuits in the United States: one, a crusty biscuit having a soft, tender crumb that is not flaky; and the other, a biscuit of larger volume with a crumb that will peel off in flakes and of a more breadlike texture. The first, which is especially popular in the southeastern states, is attained by using a soft flour and handling the dough as little as possible; the second is attained by using a harder flour, stirring the dough well, kneading lightly with folding, and rolling thicker than for the first type. The larger volume of the second type of biscuit can be accounted for by the thicker dough and by gluten development, which enables the dough to hold carbon dioxide well. Gluten development through folding, in combination with the cutting in of fat, also produces a layered crumb when layers of dough are separated by the expansion of steam during baking. Another regional difference in biscuit-making is the more frequent use of buttermilk with baking soda for the first type of biscuit.

Zaehringer et al. (1956) fractionated flour and recombined the fractions in various ways. When the fractions were combined in proportions typical of hard wheat bread flour, biscuits were darker in crust color, larger in volume, and less tender than when the proportions of soft wheat pastry flour were used.

Biscuit characteristics are affected by the type of milk used. In an experiment on rolled biscuits, those made with diluted evaporated milk, reconstituted dry whole milk, and water alone were similar in quality, whereas biscuits made with fresh whole milk were more tender in crust and crumb, more pleasing in flavor, more compressible, and of better volume (Briant and Hutchins, 1946). Differences among milks were less pronounced in experiments on drop biscuits (Kirkpatrick et al., 1961) than in the earlier experiment with rolled biscuits. Longer mixing for hydration of dry ingredients and more milk than the standard amount were needed with some of the processed milks than with fresh whole or skim milk.

As with yeast bread, it is impossible to specify in a formula an amount of liquid that is appropriate for all flours. It is best to find an amount that gives a soft dough with the flour to be used, and then to add it all at once to the blended mixture of dry ingredients and shortening in subsequent batches.

Too little mixing and kneading are more frequent faults in making the flaky type of biscuit than in overmixing. Undermixed biscuits are small in volume with a rough, sometimes spotted crust. Their interior is coarse in texture and not flaky. Overmixing also is to be avoided, because it overdevelops the gluten and makes biscuits hump on top because of the pressure of gases held by the cohesive structure near the end of the baking period. Such biscuits are dry and tough.

The mixed dough can be rolled and folded rather than kneaded, if desired. The dough is rolled into a sheet about 1 cm thick, folded double, rolled, and folded again. It then is rolled to the desired thickness (with rolling guides in experimental work) and cut. Cut biscuit dough can be covered and allowed to stand on a baking sheet for about 30 min at room temperature, or longer in a refrigerator, before baking.

Popovers

Popovers are shells that are almost hollow, moist but not soggy inside, and crisp and brown outside. They are made of a very thin batter that usually contains equal volume of milk and flour and a high proportion of egg. The leavening agent, steam, is produced from the liquid and is held by the shell if the temperature is right for almost simultaneous rapid expansion and coagulation. Within limits, the size of popovers increases with the amount of egg. Although fat is not an essential ingredient, a small amount usually is included for flavor. The amount of liquid in the batter far exceeds the proportion for development of a cohesive gluten structure. For this reason, the amount of beating has little effect on the quality of the product, though it should be sufficient to make the batter smooth.

The pans in which popovers are baked can be made of various materials, such as glass, tin, aluminum, or iron, but they should be deep and have rather straight sides. Custard cups made of glass or ovenware are satisfactory if they are deep. The pans should be oiled but need not be heated before the batter is added. During baking, it is essential for some crusting to occur early to hold in steam. As baking continues, the steam expands the batter and the volume of the popovers increases. After maximum volume is reached, the popovers must become firm enough that they do not collapse when removed from the oven. Although a moderate oven temperature can be used if the heat is maintained evenly or increased during the expansion of the popovers, a hot oven usually is used, and the temperature may be decreased toward the end of the baking period to permit the crusts to dry out without becoming too brown.

EXPERIMENTS

I. Yeast Dough Ingredients and Dough Treatment

The basic formula for yeast bread that was used previously for studying the effects of yeast, sugar, and salt is used here for observing the effects of varying flour and nonfat milk and the effects of varying kneading, bulk fermentation, and proof time.

A. *Formula and method*

Use the formula and general procedures described in Experiment II, Chapter 10. Measure loaf volume by one of the methods described in Chapter 16.

B. *Variations*

1. Amount of flour
   a. Use 20% less flour than used in the basic formula. It might be necessary to beat this dough in the bowl instead of kneading it. If it is too soft to shape after bulk fermentation, stir to remove the gas and spoon or pour it into the pan. This dough may not attain the desired volume in rising; therefore, allow it to rise the same length of time as required for 1b.
   b. Basic formula.
   c. Use 20% more flour than used in the basic formula.
2. Kind of flour. Make doughs substituting rye flour for the following amounts of wheat flour:
   a. None
   b. 20 g
   c. 40 g
3. Nonfat dry milk. Add the milk to the first portion of flour in the basic formula.
   a. None
   b. 12 g
   c. 20 g

   Toast a slice of each loaf and compare the toasted slices as to color.

4. Bulk fermentation. Make triple the basic formula. Weigh three 275-g portions for the following treatments:
   a. Shape and proof after *no* bulk fermentation.
   b. Basic procedure.
   c. Carry out two bulk fermentations.
5. Proofing. Triple the basic formula and weigh three 275-g portions.
   a. Bake with no proofing.
   b. Basic procedure.
   c. Allow to rise 1 1/2 times as long as in 5b.

C. *Results*

Make a table as in Experiment IIC in Chapter 10.

D. *Questions*

What effect does too soft or too stiff a dough have on the quality of bread? How does rye flour affect the volume of bread? What effect do milk solids have on the quality of bread? Of toast? What is the effect of underfermentation in the bulk state? Overfermentation? What is the effect of underproofing? Overproofing?

II. Baking Powder Biscuits

The experiments on baking powder biscuits show the effects of the amount of mixing, of the types of flour and leavening, and of standing before baking on the quality of biscuits. The basic formula includes enough mixing to produce the large, flaky type of biscuit. The optimum amount of milk can be determined in preliminary trials. It is recommended that each experiment be done by a team.

A. *Formula and method*

All-purpose flour, g	110
Baking powder (SAS-phosphate), g	5.4
Salt, g	3
Fat, g	25
Milk, ml	81 (approx)

1. Sift the flour, baking powder, and salt into a mixing bowl. Set oven at 218°C (425°F).
2. Cut the fat into the dry ingredients using a pastry blender. Continue cutting until no fat particles are larger than peas.
3. Pour all the milk into the flour-fat mixture at one time. Mix vigorously with a fork until the dough is stiff, cutting through the center of the dough with the fork several times. Total number of mixing strokes should be about 30.
4. Lightly flour a rolling pin and bread board or pastry cloth. Turn the dough out onto the board and knead, with or without folding (be consistent), for 15 strokes.
5. Roll the dough until it is 1–1.5 cm thick. Use rolling guides. Cut with a biscuit cutter about 5 cm in diameter, flouring the cutter each time it is used.
6. Place biscuits on an unoiled baking sheet, using a spatula. Leave 1–1.5 cm of space between biscuits.
7. Bake at 218°C until the biscuits are golden brown (about 12 min).

B. Variations

1. Amount of mixing. Triple the basic recipe, but mix only until the dry ingredients are moistened. Count the strokes used.
   a. Minimum mixing. Turn 1/3 of the dough out on a board, roll without kneading, cut, and place on a baking sheet.
   b. Moderate mixing. Mix the remaining dough enough to make the total number of strokes equal to 30. Turn half of this dough out on a board, knead 15 strokes, roll, cut, and place on a baking sheet with biscuits from 1a.
   c. Excessive mixing. Knead the remaining dough 85 additional strokes, roll, cut, and place on the baking sheet with the other biscuits. Bake at 218°C until biscuits from 1b are golden brown.
2. Type of flour
   a. Make biscuits according to the basic formula, using the all-purpose flour ordinarily used in the laboratory.
   b. Make biscuits according to the basic formula, using 110 g of all-purpose flour that is harder or softer than that

used in 2a, and altering the amount of milk if necessary. Record the amount of milk used.

c. Make biscuits according to the basic formula, using 128 g of cake flour. If it is necessary to change the milk from the amount used in the basic formula, record the amount used.

3. Type of leavening and effect of standing. For each type of leavening, make biscuits according to the basic formula with the exception noted. Place half of the biscuits from each lot on each of two baking sheets. Bake those on one baking sheet immediately, but allow those on the other baking sheet to stand at room temperature for 30 min before baking.
   a. Basic formula.
   b. Use 6.5 g phosphate baking powder.

C. *Results*

Score shape, volume, crust appearance, crumb characteristics, tenderness, and flavor, using appropriate scales. (See the Appendix.) Tabulate the scores.

D. *Questions*

What effect does the amount of mixing have on the quality of biscuits? Were all of the flours suitable for biscuits? Were any differences noted in the flavor of biscuits made with the two baking powders? Were spots noticed on the crusts of any of the biscuits? If so, how could they be avoided? Can biscuits be allowed to stand at room temperature before baking without serious loss of quality?

III. Muffins

The effects of different amounts of mixing on the quality of muffins made with two types of baking powder, the effects of the types of flour and fat, and the effects of the method of adding shortening are studied. It is recommended that each experiment be done by a team.

A. *Formula and method*

Flour, all purpose, g	110
Sugar, g	12
Baking powder, SAS-phosphate, g	5.4
Salt, g	3
Egg, g	24
Milk, ml	125
Oil, g	21

1. Oil six muffin cups. Set oven at 218°C (425°F).
2. Sift the flour, sugar, baking powder, and salt into a mixing bowl.
3. Beat the egg slightly; add milk and oil and mix well.
4. Make a well in the dry ingredients. Add the liquid ingredients and stir immediately with a rubber scraper for 16 strokes, or until the dry ingredients are just dampened. The batter should look lumpy at the end of the mixing period.
5. Transfer the batter to the oiled muffin cups with as little agitation as possible. Each cup should be a little more than half full.
6. Bake the muffins at 218°C until they are golden brown, about 20 min. Remove from the pans immediately.

B. *Variations*

1. Amount of mixing with two types of baking powder. Muffin batter made with each of two types of baking powder is mixed to different extents. Because it will be necessary to weigh batter into the individual muffin pans, counterbalance the pans before mixing the batter. Label the sections of the pans.
   a. Prepare muffins according to the basic formula, stirring only until the ingredients are dampened. Weigh two 40-g portions into muffin cups. Mix the remaining batter for 30 sec and weigh two more 40-g portions. Mix the remaining batter 30 sec more and weigh two more 40-g portions. Bake at 218°C until some of the muffins are

golden brown. Muffins beaten the most probably will have a pale color.
   b. Repeat 1a, using 6.5 g of phosphate baking powder.
2. Type of flour
   a. Basic formula.
   b. Basic formula with 110 g of cake flour substituted for all-purpose flour.
   c. Basic formula with 110 g of self-rising flour substituted for all-purpose flour. Omit baking powder and salt.
3. Method of adding the fat
   a. Use solid shortening in the basic formula and melt it before adding it to the mixture of egg and milk.
   b. Use solid shortening in the basic formula and cut it into the sifted dry ingredients with a pastry blender until the mixture looks like cornmeal.

C. *Results*

Score shape, volume, crust appearance, crumb characteristics, tenderness, and flavor, using appropriate scales. (See the Appendix.) Tabulate the scores.

D. *Questions*

Did the effects of overbeating muffins vary with the type of baking powder? How did the different flours affect the properties of the muffins? Did the method of adding the shortening affect the properties of the muffins?

SUGGESTED EXERCISES

1. Compare different types of wheat flour in basic yeast doughs.
2. Compare different shortening levels in yeast doughs, biscuits, and/or muffins.
3. Repeat IB2 with soy flour, triticale flour, peanut flour, whole wheat flour, or any other flour of interest.

4. Repeat IB2 with a series of fibers such as wheat bran or microcrystalline cellulose.
5. Vary the level of sugar in muffins.
6. Batter and dough proportions are expressed as weight percentages based on the weight of flour. Find the appropriate information and complete the following chart:

	Doughs			Batters		
	Pastry	Bread	Biscuits	Muffins	Cake	Popovers
	%	%	%	%	%	%
Flour	100	100	100	100	100	100
Liquid	____	____	____	____	____	____
Eggs	____	____	____	____	____	____
Sugar	____	____	____	____	____	____
Salt	____	____	____	____	____	____
Shortening	____	____	____	____	____	____
Baking powder	____	____	____	____	____	____
Active dry yeast	____	____	____	____	____	____

7. Substitute nonwheat flours and/or fibers in biscuits and/or muffins.
8. Substitute soy and/or whole wheat flour for wheat flour in pizza dough. For a second experiment, add bran to the dough containing soy flour.
9. Develop a formula and procedure for making starch bread or other products that normally contain flour.

REFERENCES

Bailey, C. H. and Munz, E. 1938. The march of expansion and temperature in baking bread. Cereal Chem. 15: 413.

Baker, J. C., Parker, H. K. and Fortmann, K. L. 1953, Flavor of bread. Cereal Chem. 30: 22.

Bietz, J. A., Huebner, F. R. and Wall, J. S. 1973. Glutenin. The strength protein of wheat flour. Baker's Digest 47(1): 26.

Birnbaum, H. 1977. Interactions of surfactants in breadmaking. Baker's Digest 51(3): 16.

Bloksma, A. H. 1971. Rheology and chemistry of dough. In "Wheat Chemistry and Technology," Ed. Pomeranz, Y. American Association of Cereal Chemists, St. Paul, MN.

Bowman, F., Dilsaver, W. and Lorenz, K. 1973. Rationale for baking wheat-, gluten-, egg- and milk-free products. Baker's Digest 47(2): 15.

Briant, A. M. and Hutchins, M. R. 1946. Influence of ingredients on thiamine retention and quality in baking powder biscuits. Cereal Chem. 23: 512.

Bushuk, W . 1974. Glutenin—functions, proberties and genetics. Baker's Digest 48(4): 14.

Charles, V. R. and Van Duyne, F. O. 1953. Effect of freezing and freezer storage upon quality of baked rolls, brown-and-serve rolls, and shaped roll dough. Food Technol. 7: 208.

Cooley, J. A. 1965. Role of shortening in continuous dough processing. Baker's Digest 39(3): 37.

d'Appolonia, B. L. and Gilles, K. A. 1971. Effect of various starches in baking. Cereal Chem. 48: 625.

Elton, G. A. H. and Fisher, N. 1968. Effect of solid hydrocarbons as additives in breadmaking. J. Sci. Food Agric. 19: 178.

Ewart, J. A. D. 1972. Recent research and dough visco-elasticity. Baker's Digest 46(4): 22.

Greenwood, C. T. and Ewart, J. A. D. 1975. Hypothesis for the structure of glutenin in relation to rheological properties of gluten and dough. Cereal Chem. 52(3II): 146r.

Haber, T., Seyam, A. A. and Banasik, O. J. 1976. Rheological properties, amino acid composition and bread quality of hard red winter wheat, rye and triticale. Baker's Digest 50(3): 24.

Hamed, M. G. E., Refai, F. Y., Hussein, M. F. and El-Samahy, S. K. 1973. Effect of adding sweet potato flour to wheat flour on physical dough properties and baking. Cereal Chem. 50: 140.

Hoseney, R. C., Finney, K. F. and Pomeranz, Y. 1970. Functional (breadmaking) and biochemical properties of wheat flour components. VI. Gliadin-lipid-glutenin interaction in wheat gluten. Cereal Chem. 47: 135.

Hoseney, R. C., Finney, K. F., Pomeranz, Y. and Shogren, M. D. 1971. Functional (breadmaking) and biochemical properties of wheat flour components. VIII. Starch. Cereal Chem. 48: 191.

Hoseney, R. C., Finney, K. F., Shogren, M. D. and Pomeranz, Y. 1969. Functional (breadmaking) and biochemical properties of wheat flour components, III. Characterization of gluten protein fractions obtained by ultracentrifugation. Cereal Chem. 46: 126.

Jackel, S. S. 1977. The importance of oxidation in breadmaking. Baker's Digest 51(2): 39.
Khan, M. N. and Rooney, L. W. 1977, Baking properties of oilseed flours—evaluation with a short-time dough system. Baker's Digest 51(3): 43.
Kim, S. K. and d'Appolonia, B. L. 1977. The role of wheat flour constituents in bread staling. Baker's Digest 51(1): 38.
Kirkpatrick, M. E., Matthews, R. H. and Collie, J. C. 1961. Use of different market forms of milk in biscuits. J. Home Econ. 53: 201.
Koren, P. M. 1967. The shortening requirements of the continuous mixing process. Baker's Digest 41(5): 104.
Lorenz, K. 1974. Triticale—a promising new cereal grain for the baking industry? Baker's Digest 48(3): 24.
Lorenz, K. 1976. Triticale bran in fiber breads. Baker's Digest 50(6): 27.
Marston, P. E. and Wannan, T. L. 1976. Bread baking—the transformation from dough to bread. Baker's Digest 50(4): 24.
Medcalf, D. G. and Gilles, K. A. 1968. The function of starch in dough. Cereal Sci. Today 13: 382.
Moss, R. 1974. Dough microstructure as affected by the addition of cysteine, potassium bromate, and ascorbic acid. Cereal Sci. Today 19: 557.
Orth, R. A., Dronzek, B. L. and Bushuk, W. 1973. Studies of glutenin. IV. Microscopic structure and its relations to breadmaking quality. Cereal Chem. 50: 688.
Pomeranz, Y. 1966. Soy flour in breadmaking—a review of its chemical composition, nutritional value and functional properties. Baker's Digest 40(3): 44.
Pomeranz, Y. 1971. Composition and functionality of wheat-flour components. In "Wheat Chemistry and Technology," Ed. Pomeranz, Y. American Association of Cereal Chemists, St. Paul, MN.
Pomeranz, Y., Shogren, M. D. and Finney, K. F. 1969. Improving breadmaking properties with glycolipids. I. Improving soy products with sucroesters. Cereal Chem. 46: 503.
Pomeranz, Y., Rubenthaler, G. L., Daftary, R. D. and Finney, K. F. 1966a. Effects of lipids on bread baked from flours varying widely in bread-making potentialities. Food Technol. 20: 1225.
Pomeranz, Y., Rubenthaler, G. L. and Finney, K. F. 1966b. Studies on the mechanism of the bread-improving effect of lipids. Food Technol. 20: 1485.
Pomeranz, Y., Shogren, M. D., Finney, K. F. and Bechtel, D. B. 1977. Fiber in breadmaking—effects on functional properties. Cereal Chem. 54: 25.
Sandstedt, R. M. 1961. The function of starch in the baking of bread. Baker's Digest 35(3): 36.
Schoch, T. J. 1965. Starch in bakery products. Baker's Digest 39(2): 48.

Swanson, A. M., Sanderson, W. B. and Grindrod, J. 1964. The effects of heat treatments given to skimmilk and skimmilk concentrate before drying. Cereal Sci. Today 9: 292.

Swortfiguer, M. J. 1968. Dough absorption and moisture retention in bread. Baker's Digest 42(4): 42.

Tsen, C. C. and Hoover, W. J. 1973. High-protein bread from wheat flour fortified with full-fat soy flour. Cereal Chem. 50: 7.

Volpe, T. and Zabik, M. E. 1975. A whey protein contributing to loaf volume depression. Cereal Chem. 52: 188.

Zaehringer, M. V., Briant, A. M. and Personius, C. J. 1956. Effects on baking powder biscuits of four flour components used in two proportions. Cereal Chem. 33: 170.

12. Cakes and Pastry

Cakes and pastry, baked flour mixtures that are used primarily for dessert, are discussed in this chapter. Much of the research literature deals with commercial products but has general application.

CAKES

Angel food and sponge cakes, which do not contain added fat or leavening agents, were included in Chapter 3. Shortened cakes will be discussed in this chapter.

Ingredients

The major ingredients of plain shortened cakes are similar to those of muffins, though the proportions are quite different. Egg, shortening, and sugar levels all are higher in cakes than in muffins. Vanilla and other flavoring ingredients that are not used in muffins are common cake ingredients.

Balance of cake ingredients is of long-recognized importance, but what constitutes balance undergoes change as new ingredients are developed and old ones are modified. This is particularly true in the baking industry, but also affects home baking. For example, the addition of surfactants to shortenings is an ingredient modification that affects cake formulation in the baking industry; the most likely effect on home baking is improved performance of the available shortenings and of cake mixes.

The rules for cake formulation are based largely on balance between tenderizing ingredients, sugar and fat, and structural ingredients, flour and egg. Today's consumer apparently prefers "high-ratio" (high sugar) cakes, in which the weight of sugar exceeds that of flour. If a ratio of 120–140 parts, by weight, of sugar to 100 parts of flour is used, the proportions of other tenderizing and structural ingredients should be established so that the weight of shortening does not exceed the weight of eggs (Lawson, 1970). In white cakes, the weight of egg white might well be higher than that of shortening because the protein concentration in egg white is lower than that in whole egg or in yolk.

Balance between liquids and the batter constituents that have an affinity for water is the basis for the rule that the weight of the liquids, including fluid milk, water, and eggs, should equal or exceed the weight of the sugar; otherwise the high sugar level interferes with hydration of proteins and gelatinization of starch.

The ingredients of cakes are discussed individually in other chapters. They will be considered here only in relation to their respective roles in cakes.

Flour Cake flour, a chlorinated short patent soft wheat flour, is the usual flour chosen for cakes. As mentioned in Chapter 9, the chlorine treatment of cake flour improves performance in cakes. Tsen et al. (1971) related effects of chlorine on flour protein to cake quality, and Kulp et al. (1972) studied the effects of chlorine on the starch of the same flour as that studied by Tsen et al. Kulp (1972) summarized the results of those and other studies. Chlorine was found to increase the dispersibility of flour proteins. The effect on starch appeared to involve only surfaces of granules. The authors theorized that the improving effect of chlorine might be facilitation of stabilizing interactions between starch and protein and between starch and lipid.

Soft wheat all-purpose flour, which is available in some areas of the United States, performs quite similarly to cake flour. Hard wheat all-purpose flour can be used in cakes if substituted on a weight basis or if used in smaller volumes than soft wheat flours (1 cup minus 2 tbsp of hard wheat all-purpose flour replacing 1 cup of a soft wheat flour). Even with appropriate substitution, cakes made from hard wheat all-purpose flour have less even grain, less velvety crumb, and less compressibility than cakes made from soft wheat flours. Although flour is an important structural ingredient of cakes, the flour proteins are dispersed as discrete particles in cake batter, rather than forming a continuous structure as in yeast dough. Thus starch plays a larger role in cake structure than do flour proteins.

The complete role of starch in cakemaking would be better understood if it were clear why some starches perform about as well as wheat starch and others perform less well. Reconstitution and baking studies, with wheat starch of cake flour replaced by starches from other sources, have been conducted. In one such study (Sollars and Rubenthaler, 1971), rye starch performed as well and barley starch nearly as well as wheat starch; potato and corn starches performed fairly well; and rice starch gave poor results in lean-formula cakes made from the formula developed by Kissell (1959) for accentuating treatment effects. Howard et al. (1968) related differences among starches to differences in their thermal setting behavior during baking of cakes; however, this single property does not appear to account totally for the differences among starches as to performance in cakes.

The fractionation, reconstitution, and baking tests conducted by Baldi et al. (1965) showed all of the flour fractions to contribute to the normal structure of white cakes.

Liquid The liquid in cakes serves as a solvent for sugar, salt, and chemical leavening. It hydrates flour proteins and gelatinizes starch. The extent of starch gelatinization is greater in cake than in bread. Milk, the usual liquid in cakes, contains carbonyl-amine reactants, thus contributing to crust browning. As with bread, relatively high levels of liquid contribute to keeping quality.

Leavening As in bread doughs, the carbon dioxide that is formed in cake batters diffuses into gas cell nuclei that consist of small air bubbles incorporated during mixing. Expansion of these gas cells during baking results in the volume increase. Leavening systems in cakes are similar to those in biscuits and muffins except that lower levels of leavening agent can be used in cakes. Cake batters are less cohesive and thus offer less resistance to expansion than quick bread batters and doughs. If an acidic ingredient such as buttermilk replaces fresh milk in the formula, sodium bicarbonate is substituted for a portion of the baking powder, as described in Chapter 10.

Eggs The readily coagulable proteins of egg contribute to structure of the baked cake, though the rather high lipid concentration in egg yolk must be considered a tenderizing factor. Egg also emulsifies added fat. The role of egg in emulsification of shortening is attributable to the lipoprotein of the yolk.

Sugar Sugar obviously affects flavor. It also contributes to moisture retention of the baked cake because of its hygroscopicity. In addition, sugar has a tenderizing effect through its competition with flour proteins for water during mixing and through its effect in raising the coagulation temperature of flour and egg proteins during baking. The sugar that ordinarily is used in cakes is sucrose, which is nonreducing; therefore, the contribution to browning is minimal.

Shortening In addition to its tenderizing effect, shortening enhances apparent moistness, fineness and uniformity of grain, and keeping quality. If either margarine or butter is used, flavor also is imparted; hydrogenated shortening ordinarily is used in cake, however. The effectiveness of shortenings has been increased in recent years by the addition of surfactants, which facilitate emulsification in batters. Shortenings on the retail market usually contain added monoglycerides. Commercial shortenings are likely to contain other surface-active lipid derivatives, such as propylene glycol-fatty acid esters and glycerol-lactic acid esters. With these emulsifiers, liquid shortenings can be used in the baking industry, where their fluidity is advantageous from the standpoint of bulk handling operations such as pumping and metering (Lawson, 1970). Liquid shortenings differ as to type. In a common type, the emulsifier, such as a mixture of propylene glycol monoester and glyceryl lactopalmitate, is dissolved in an oil that is hydrogenated sufficiently to increase stability without eliminating fluidity at room temperature. The partial hydrogenation results in some suspended solids at room temperature. These solids make a contribution to crumb strength in the baked product (Petricca, 1976), and the technology of liquid shortening includes means of assuring small, stable crystals (Chapter 8). Further developments, affecting the consumer at least indirectly, undoubtedly will occur.

Other ingredients Many of the additional ingredients that might be in cakes are flavoring materials such as vanilla, spices, and synthetic flavors, which are used in small amounts. These ingredients have little effect other than that for which they are added. Other ingredients that also are added primarily for flavor affect other properties. For example, cocoa, chocolate, fruit juices, apple sauce, other forms of fruit, and molasses affect batter pH. Unless adjustments are made, texture and other properties in turn are affected. Ingredients such as cocoa and chocolate also influence the amount of liquid needed.

Mixing

The conventional method or one of its modifications is a time-honored and successful method of mixing cakes. When mixing is done by hand, plastic fat is creamed and sugar is added gradually while creaming is continued. For whole-egg cakes, either eggs or yolks are beaten into the creamed mixture. Gradual addition of ingredients is important in hand mixing to permit proper incorporation of air without excessive exertion, but it is not essential with an electric mixer, with which the fat and sugar, with or without eggs, can be beaten at one time. Flour sifted with baking powder and salt is added alternately with milk. If the eggs have been separated or if white cake is being made, beaten egg whites are folded into the batter after the other ingredients have been added. Detailed directions for the conventional method are given by Halliday and Noble (1946). In a modification called the *conventional-meringue method,* about half of the sugar is beaten into the egg whites before they are added to the batter. In another modification, called the *conventional-sponge method,* about half of the sugar is beaten into whole eggs until the mixture is foamy and stiff; then the mixture is folded into the batter after all other ingredients have been added. Both of these modified methods have proved successful, especially with soft fats. They make creaming easier, especially if the amount of fat is small, because less sugar is added to the fat, and they incorporate additional air at the end of the mixing process in the form of a meringue or sponge.

In the *muffin method,* egg, milk, and melted fat are combined, then beaten with the dry ingredients until smooth. This rapid method was probably more popular before cake mixes were available than it is now. Although muffin cakes are acceptable, especially when warm, they are usually inferior in eating and keeping qualities to cakes made by other methods. A modification of the muffin method, in which egg whites beaten with or without part of the sugar are added after the other ingredients are combined, is especially successful if the fat is oil. When sugar is beaten into the egg whites, this modification is called the *muffin-meringue method.*

In the *pastry-blend method,* the fat and flour are beaten together first. The remaining steps include blending a mixture of sugar, baking powder, and 1/2 the milk into the flour-fat mixture and finally adding the egg and remaining milk. The baking powder can be sifted with the flour if desired.

The *one-bowl method* is especially adapted to electric mixers but also is satisfactory with hand mixing. All the ingredients, including plastic fat at room temperature, are combined in one or two steps. If a second step is used, part of the liquid and usually the eggs are added after some of the beating has been done. Leavening may be added at this time instead of with the dry ingredients. This method also has been called the *quick-mix, single-stage,* or *dump method.* It is particularly successful with high-ratio cakes and has received general acceptance, undoubtedly because of a combination of its convenience and the quality of the products produced.

The choice of a method for mixing a cake depends on convenience, on whether an electric mixer is available, and on the type of fat to be used. In many instances, any one of several methods used correctly would give excellent results. Hunter et al. (1950) compared the one-bowl, conventional, and pastry-blend methods, using formulas containing three levels of sugar (100, 125, and 140% of the weight of the flour). The formulas were balanced by the use of increased amounts of fat, egg, and milk with the higher levels of sugar. The appearance of the cakes is shown in Figure 12.1. The authors reported that, although these three methods of mixing produced no marked differences in palatability, the pastry-blend method appeared to be adapted to a wider range of conditions, such as formula, kind of fat, and temperature, than the other methods investigated. The appropriateness of the one-bowl method, particularly for high-ratio cakes, was demonstrated.

Comparisons of mixing by hand and by machine are difficult and not

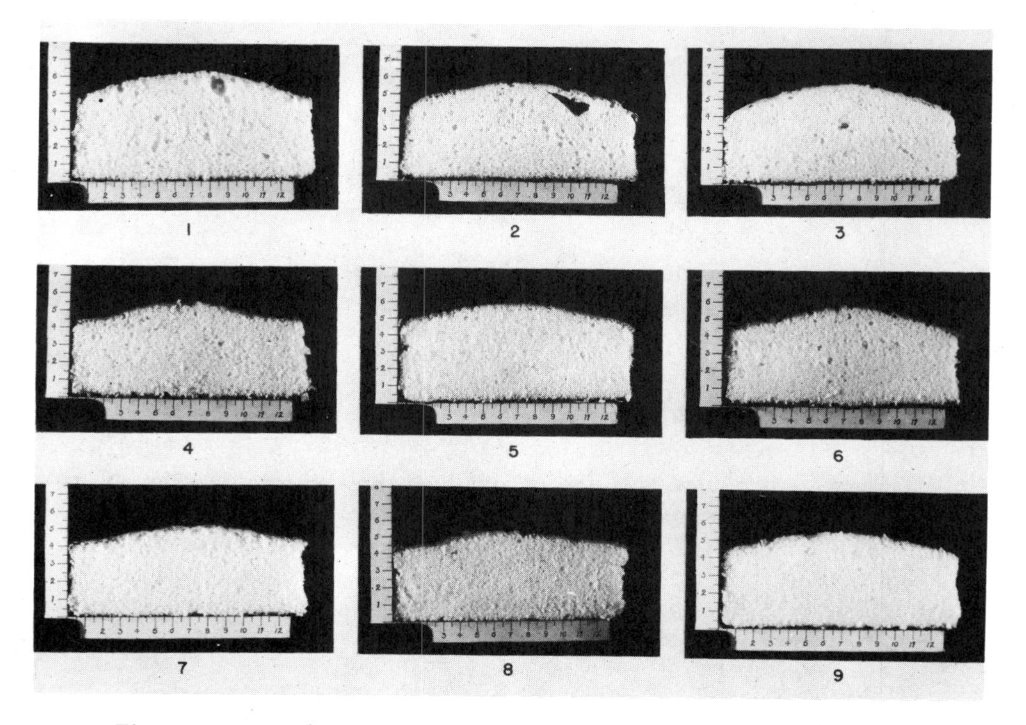

Figure 12.1. Cakes prepared by three methods of mixing from three formulas having different levels of sugar. Cakes are arranged by number as follows:

	Method of mixing		
Sugar level, %	Conventional	Pastry-blend	One-bowl
100	1	2	3
125	4	5	6
140	7	8	9

(Hunter et al., 1950. Reprinted with permission of the New York State College of Human Ecology, a statutory college of the State University, Cornell University, Ithaca, NY.)

applicable to all conditions because the results depend on the thoroughness of hand mixing and the type of electric mixer used. Cakes made by the one-bowl method were satisfactory when mixed by hand but better when mixed by machine (Redfield, 1947), probably because vigorous beating is especially necessary for this method. The wire whip available on certain electric mixers is a better tool than the flat paddle for mixing cake batters, except for creaming fat and sugar, for which the paddle is recommended. For hand mixing, a shallow wooden spoon and a heavy mixing bowl with sloping rather than straight sides are desirable (Halliday and Noble, 1946). The size of the mixing bowl should be adequate but not excessive.

It usually is recommended that ingredients be at room temperature when cake is mixed unless the day is very warm. Dunn and White (1937) suggested that the temperature of shortening be 21–24°C. However, the unfavorable effects on cake quality commonly attributed to high and low ingredient temperatures were not evident in the study by Hunter et al. (1950). Ingredients at 8, 22, and 30°C were used for cakes made of hydrogenated vegetable shortening, margarine, and lard by the conventional and pastry-blend methods. Under the conditions of the experiment, the temperature of the ingredients produced no important differences in the quality of the cakes. The authors pointed out that with hand mixing or another formula, the effect of temperature might have been greater. In another study (Ohlrogge and Sunderlin, 1948), temperatures high enough to melt the fat had a harmful effect on cake quality.

Structure

Two aspects of cake structure are the structure of batter and the structure of crumb. Most studies of structure have dealt with the batter because of its particular usefulness in gaining insights into relationships among cakemaking variables, certain aspects of batter structure, and cake quality.

Batter The continuous phase in a cake batter is aqueous. In this aqueous phase, some batter constituents, such as sugar, salt, and leavening salts, are dissolved. Other constituents are colloidally dispersed, such as proteins, or suspended, such as starch granules, fat globules, and gas cells. Howard (1972) emphasized the large number of interfaces resulting from the large number of different phases, and the consequently great opportunity for interactions to occur during mixing and baking.

Among the most important occurrences during mixing are the dispersal of shortening and the incorporation and dispersal of air. Emulsifiers, or surfactants, have a direct role in the dispersal of fat. According to Howard (1972), the added emulsifiers do not affect aeration directly but, by forming films around fat globules, they keep the fat from interfering with the foam-stabilizing properties of the protein. The increased viscosity resulting from increased dispersal of fat undoubtedly contributes to retention of incorporated air.

Crumb The cake crumb consists of an open, fibrous network of coagulated protein in which gelatinized starch granules are embedded. Ideally, the network forms thin walls surrounding small, uniform spaces in which the gases expanded during baking. Some irregularities usually are present, though the extent of their development is variable. These irregularities occur mostly in the form of relatively large holes that result from coalescence of some of the gas bubbles during the early stage of baking. In some cakes, tunnels running diagonally from the bottom toward the center top of the cake are seen. Possibly more tunnels form during baking than are apparent in the baked crumb; the tunnels might form and then collapse if the development of rigidity is delayed (Trimbo and Miller, 1973).

The fat, which melts during baking, probably forms films on the structure surfaces. Water vapor that does not leave during baking condenses on internal surfaces upon cooling. Sugar apparently is dissolved in this layer of water. Several workers have studied the dynamics of the development of cake structure during baking (Handleman et al., 1961; Howard et al., 1968; Wilson and Donelson, 1963).

Cake Quality

Effects of ingredients Variation as to fineness of granulation of cake flour affects cake quality (Yamazaki and Donelson, 1972). Fine granulation is generally considered desirable. Pin-milling (Chapter 9), therefore, has been applied to cake flour, though the extent of starch damage and the effect of starch damage on flour performance frequently have not been considered. Miller et al. (1967) studied both starch damage and particle size in relation to baking quality of pin-milled flour used in cakes. Their data suggested that the beneficial effects of pin-milling are attributable to the reduction in particle size and not to the accompanying starch damage. In fact, starch damage was negatively correlated with cake quality in the study. Potential beneficial effects of pin-milling thus seemingly depend on there not being excessive starch damage.

Nonwheat flours were used by Bowman et al. (1973) in the development of hypoallergenic cakes. Chocolate cake was made successfully from a formula that included a mixture of rice and rye flours. The formula did contain milk and egg. Orange cup cakes containing rice flour and no milk or eggs also were acceptable as to appearance and texture and had a pleasing flavor. They did not have sufficient structural strength to be baked as whole cakes. Plain rice flour pound cakes were similar to conventional pound cake except in texture. All three products were considered acceptable.

The addition of fiber to bread was discussed in Chapter 11. Various forms of fiber also have been substituted for some of the flour in cakes. Bran and middlings, singly and in combination, were substituted for wheat flour in white layer cakes at replacement levels up to 16% bran or 12% middlings or 28% bran and middlings combined. Although color and some other physical properties were affected, the other physical properties and the sensory scores were not affected adversely by the substitutions (Brockmole and Zabik, 1976). In another study (Rajchel et al., 1975), the same substitutions of bran and middlings for wheat flour were made in banana, chocolate, nut, and spice cakes. The substitutions affected color and volume in particular, but sensory evaluation showed most cakes to be acceptable. The same group (Springsteen et al., 1977) then tried higher levels (up to 70% substitution) of wheat bran in white layer cakes. The 50% and 70% levels of substitution resulted in excessive effects on sensory scores and on volume, but cakes of the 30% replacement level were acceptable. When the effect of fineness of grind was observed at the 30% level, the finer grind gave the better results. Flour in white cakes was replaced by microcrystalline cellulose at levels of 0, 20, 40, and 60% of the flour volume (Brys and Zabik, 1976). Cakes of the 60% replacement level were gummy and excessively moist, but cakes of the 20 and 40% levels were comparable to control cakes.

Both the amount and the kind of liquid affect cake properties.

Trimbo and Miller (1973) reported that low liquid levels are conducive to tunnel formation during the baking of cakes from an institutional mix. The type of liquid used in a cake can have effects that depend on the formula. For example, the effect of using buttermilk depends not only on total leavening equivalency but also on the resulting pH. With balance between buttermilk and sodium bicarbonate along with equivalent total leavening, the effects of the substitution of buttermilk may be slight. With excessive acidity or excessive alkalinity, flavor, crumb color, crust browning, cake grain, and volume may be affected. Flavor ranges from somewhat sour with excessive acidity to bitter or soapy with excessive alkalinity. Crumb color is darker in excessively alkaline plain cake and redder and darker in excessively alkaline chocolate cake than in other cakes. Crust browning occurs increasingly as alkalinity increases; this represents the effect of pH on carbonyl-amine browning. Cake grain, within limits, tends to increase in fineness with decreased pH and to become coarse, with thick cell walls, with increased pH. Volume tends to be relatively low with excessive acidity and to increase with alkalinity up to a point beyond which volume decreases. Flavor and color effects are noted before changes in grain and volume become drastic. Assuming total leavening equivalency, insufficient sodium bicarbonate to neutralize the acid is preferable to excessive sodium bicarbonate when buttermilk is used. A study in which differences in cake characteristics were observed when various combinations of fresh whole milk, buttermilk, baking powder, and sodium bicarbonate were used was that of Briant et al. (1954). Crumb pH ranged from 5.45 to 7.60.

Miller et al. (1957) varied pH in white cakes from 4.8 to 7.7 (crumb pH) through the use of leavening systems with varying proportions of potassium bitartrate and sodium bicarbonate. Their formula included honey, which contains reducing sugars; the cakes, therefore, were susceptible to carbonyl-amine browning of the crumb. Miller and coworkers (1957) observed that the extent of browning of the crumb could be controlled with a pH of about 6.3 without an appreciable loss in volume. The effect was achieved by the use of the leavening acid in a coated form so that its reaction with sodium bicarbonate was slowed sufficiently to reduce carbon dioxide loss during mixing. The coated acid became available during the later stages of baking, when crumb browning would have occurred. Ash and Colmey (1973) discussed pH in cakes in more detail.

The level of baking powder (SAS-phosphate) was one of the variables in the previously mentioned extensive study conducted by Hunter et al. (1950). Cakes were prepared from low-, medium-, and high-sugar formulas by three mixing methods, at three ingredient temperatures, with three different shortenings, and with three levels of baking powder (4, 6, and 8% of flour weight). As the level of baking powder was increased in the medium-sugar formula with hydrogenated shortening, specific volume and compressibility of the crumb increased. The 8% level of baking powder was excessive in cakes shortened with margarine and with lard under some mixing conditions. Evidence of a structure-weakening effect was observed in these instances.

Briant et al. (1954) varied the type of baking powder in plain and chocolate cakes. Although some effects on batters were observed, cake characteristics were affected little.

The effects of eggs on cake quality can be appreciated by comparing shortened cakes that are high and low in egg content. In a series of cake experiments (Pyke and Johnson, 1940), the level of whole fresh egg was increased without any other change in the formula. The cakes became tougher and the volume first increased, then decreased. However, if the liquid content of the batter was kept constant by reducing it according to the moisture content of the added eggs, the volume of the cakes continued to increase with the amount of egg over the range studied. In actual cake formulas, the sugar and fat levels are such as to compensate for the toughening effect of egg.

Funk et al. (1970) observed the effects of different market forms of eggs in yellow layer cakes. The use of frozen, foam–spray-dried, freeze-dried, and spray-dried whole eggs resulted in cakes that did not differ significantly according to sensory evaluation, though some differences in batter properties occurred.

The level of sugar in a formula is subject to considerable variation. Formulas for shortened cakes often are divided into two groups: those in which the weight of sugar is less and those in which it is greater than the weight of the flour. In traditional recipes for cakes containing a low to medium level of eggs and mixed by the conventional method, the weight of sugar seldom exceeds 100% of the weight of the flour; in other words 1 cup or less of sugar is used for 2 cups of cake flour. More recently, the higher levels of sugar have come into common use, along with mixing methods other than the conventional method. The popularity of cake formulas having high levels of sugar is based on the moist, tender quality of the cakes.

Formulas that were identical except for the level of sugar were used in one study (de Goumois and Hanning, 1953). The addition of 15% sugar to the basic formula, which contained 100% sugar, did not improve the overall quality of the cake, but the addition of 30% sugar made the batter more viscous and the cakes larger, softer, and more tender. Flavor scores were similar for the three levels of sugar. In the study of Hunter et al. (1950), the levels of liquid, egg, and fat were increased as the level of sugar was increased from 100 to 140%. As in the first study, cakes made with the higher levels of sugar were more tender, moist, and velvety than those made with 100% sugar. Although the judges seemed to prefer the flavor of cakes low in sugar, their total scores favored the cakes high in sugar. The superior scores of cakes high in sugar resulted from factors other than flavor.

A moderate increase in sugar without other changes is possible with some recipes. More often, however, higher levels of sugar are accompanied by increased levels of total liquid, egg, and fat. The liquid and egg are increased for reasons mentioned in the discussion of formulation earlier in the chapter. The reason for increasing the level of shortening is related to batter aeration. Without increased fat, batter viscosity is likely to be relatively low as a result of the other changes, and retention of air is correspondingly reduced. Emulsification of added fat, on the other hand, results in increased viscosity and thus greater batter aeration as compared with that obtained when the sugar and liquid levels are increased but the shortening level is not.

Honey, when substituted for part of the sugar in a cake, contributes to moisture retention during storage. This is because honey contains fructose (levulose), which is extremely hygroscopic. The effect of the reducing sugars of honey on crumb browning were discussed earlier in this chapter, along with the problem's solution achieved by Miller et al. (1957). That solution is applicable to a commercial baking situation. At the consumer level, the use of honey might need to be restricted to products in which the crumb is brown anyway. If the use of honey in a white cake batter is preferred, carbonyl-amine browning might be controlled by slight acidification of the batter, such as by addition of cream of tartar, if one is willing to accept the flavor effect and the possibly reduced volume.

High-ratio cakes made from institutional cake mix developed less tunneling during baking than did cakes with a low sugar level (Trimbo and Miller, 1973). Trimbo and Miller postulated that the higher sugar level, by delaying the development of batter rigidity, permitted tunnels to collapse upon forming.

The quality of cake is greatly affected by the amount and type of fat used in making it. Increasing the level of fat from 20 to 40% in a formula containing 100% sugar with no other change improved the quality of cakes made by the one-bowl method (Hanning and deGoumois, 1952). The higher levels of fat increased tenderness, improved flavor, and made the grain finer and more uniform and the texture more moist and silky.

Matthews and Dawson (1966) studied the effects of both the kind and level of shortening in white cakes. Each of five shortenings was used at five levels ranging from 12.5 to 100% of the weight of the flour. Specific gravity of the batters decreased as the level of shortening increased, and the batters containing the hydrogenated shortenings or butter had lower specific gravities than those containing margarines. Batter viscosity increased with increased fat level up to 50%. Beyond 50%, the shortenings differed as to their effect on batter viscosity. Shear force values of the cakes decreased with increased fat and were highly correlated negatively with sensory tenderness scores.

The properties of lard have been changed through technological developments in recent years, so that results of early studies would not necessarily obtain today. Lard has not been considered to contribute to batter aeration to the extent that other shortenings do, though the mixing method influences the effect of the kind of shortening (Figure 12.2).

Batters containing oil also do not ordinarily hold air well. The addition of emulsifiers might be helpful. The incorporation of air through the addition of beaten whole egg or egg white as the last mixing step improves results. Sugar beaten with the egg stabilizes the foams (Ohlrogge and Sunderlin, 1948).

Jooste and Mackey (1952) reported increased overall acceptability of cakes to which monoglycerides were added. Guy and Vettel (1973) studied cakes made from butter, with and without emulsifier. As compared with the control, which contained hydrogenated shortening, the butter cake had a low volume. Addition of emulsifier to the butter cake resulted in a volume equal to that of the control (Figure 12.3). Addition

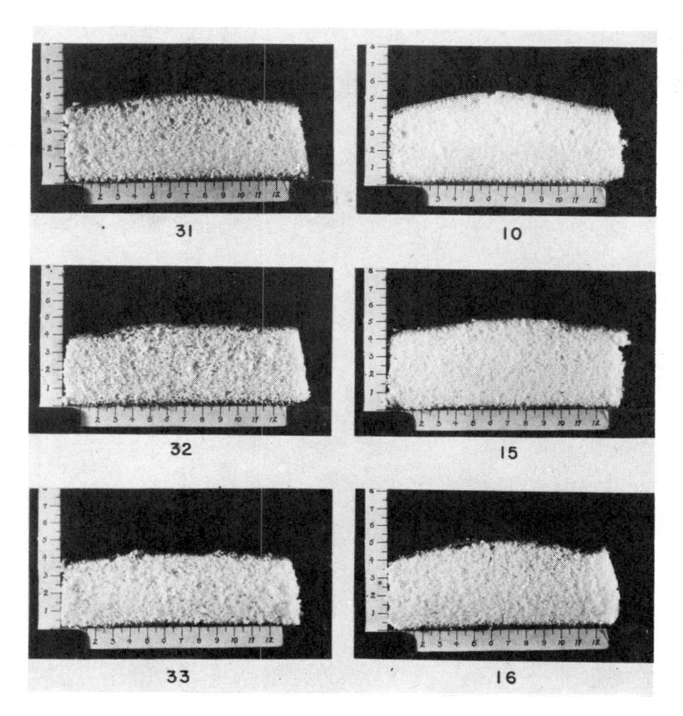

Figure 12.2. Cakes prepared by two methods of mixing from high-ratio formulas. Cakes are arranged by number as follows:

	Method of mixing	
	Conventional	Pastry-blend
Hydrogenated shortening	31	10
Margarine	32	15
Lard	33	16

(Hunter et al., 1950. Reprinted with permission of the New York State College of Human Ecology, a statutory college of the State University, Cornell University, Ithaca, NY.)

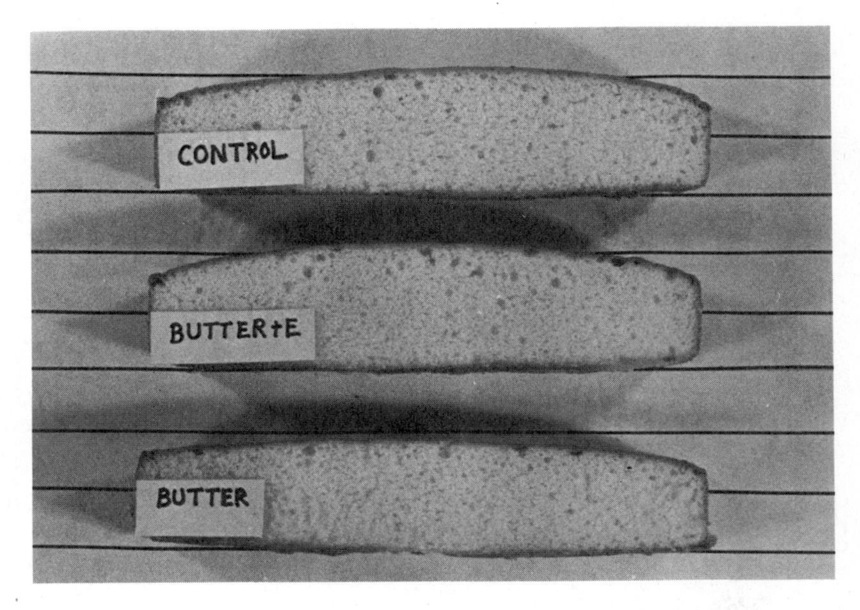

Figure 12.3. Effect of emulsifier on butter cakes. (Guy and Vettel, 1973. Courtesy of Baker's Digest.)

of emulsifier to the control cake made from shortening that already contained emulsifier did not affect volume.

Interest in the possible benefits of medium-chain triglycerides (MCT) to patients who do not absorb fats normally stimulated interest in the effects of MCT on the characteristics of foods containing them. MCT oils, prepared from coconut oil, contain mostly octanoic (8:0) and decanoic (10:0) acids. Bowman (1973) reported the effect of substituting MCT oil at levels of 0–100% for other shortenings in several products, including white and yellow cakes. Special cake formulas having corn oil as shortening were used. The white cake in which MCT oil was substituted for 50% of the corn oil were not acceptable, but the yellow cakes at the 50% level of substitution were reasonably good.

Dried whey is a nutritious and readily available byproduct of cheesemaking. As indicated in Chapter 4, it is available in such large quantities that it threatens to become a serious environmental pollutant. Considerable effort has been devoted to the use of dried whey in foods. It was shown to be a desirable ingredient of yellow cakes, in which it tended to improve flavor, tenderness, browning, and keeping quality and to produce a soft, moist crumb (Hanning and deGoumois, 1952).

Effects of baking conditions Success in making cake depends not only on the ingredients and mixing method but also on the way in which the cake is baked. The way in which the pan is filled, the size and shape of the pan, the material of which the pan is made, and the oven temperature all affect the speed of heat penetration and hence the quality of the cake. On the whole, factors that make heat penetration faster improve cake quality.

It is important that neither too little nor too much batter be used in a pan. If the correct amount is used, the cake will fill the pan completely at the end of the baking period. If desired, air that may have been entrapped in the batter during filling of the pan can be released by tapping the pan sharply on the work surface.

The effect of the size and shape of the baking pan on cakes was studied by Charley (1952). Two series of pans were used: one varying in shape but not in capacity, the other varying in depth. The amount of batter used in all pans was proportional to their capacity. Probably because of more rapid and more even heat penetration, cakes baked in shallow pans tended to be larger, more tender, less prone to crust browning, and flatter topped than those baked in deeper pans. In deep pans, the batter tended to become firm around the edges while staying soft in the center. Expansion of the soft center then produced a rounded or humped crust with a crack. Crusts were browner in the deep pans because the baking time was longer than in shallow pans. Cakes baked in shallow pans scored significantly higher than those baked in deeper pans. Those baked in round and square pans of the same depth were similar.

The material of which the baking pan is made also affects the speed of baking. Baking is rapid in pans with a dark color, a dull finish, or both because they absorb and transmit heat readily. Baking is comparatively slow in metals with bright surfaces because they tend to reflect heat. Comparing cakes baked in pans of different materials, Charley (1950) found that the faster-baking pans produced cakes having larger volumes and better overall crumb quality than did the less efficient pans. These advantages were gained with some sacrifice of appearance, because cakes from the faster-baking pans tended to be peaked and browner on the sides than on top.

Oven temperature is an important factor affecting speed of heat penetration. If the temperature is too low, coagulation of proteins and gelatinization of starch are slow and gas is lost from the batter. As gas is lost from the cells, the remaining cells enlarge, their walls thicken, volume is reduced, and the cake may settle in the middle. When the oven temperature is too high, a crust forms on the cake before it has risen fully. The soft batter in the middle of the cake then rises and forms a hump or even a crack. These effects were noted when oven temperatures of 149, 163, 190, and 218°C were used with a formula containing a high ratio of sugar to flour (Jooste and Mackey, 1952). Each increase of temperature within this range resulted in improved palatability and, if the fat contained an emulsifier, in increased volume as well. The cakes are shown in Figure 12.4. Cakes baked at the higher temperatures had small air cells with thin cell walls and thin, tender crusts. They showed the most browning and least adherence of the crust to the pan, but at 218°C they had some tendency to peak. Cakes baked at the lower temperatures showed white dots on the crusts that appeared to be sections of crust raised by gas and then dehydrated. Under the conditions of the experiment, oven temperatures of 190 to 218°C were better for shortened layer cakes than 149 or 163°C.

PASTRY

The term *pastry* includes a wide variety of products made from doughs containing medium to large amounts of fat. Some pastries, such as certain sweet rolls, are made with a medium amount of fat and leavened with yeast. Another type, called puff pastry, contains so much fat that a portion of it is folded into the dough. In this chapter, only the plain pastry ordinarily used for pies will be considered.

Ingredients

The shortening value of different fats was discussed in Chapter 8. Fats that are the softest while being incorporated into a dough seem to make the most tender pastry. Butter and margarine, largely because of their

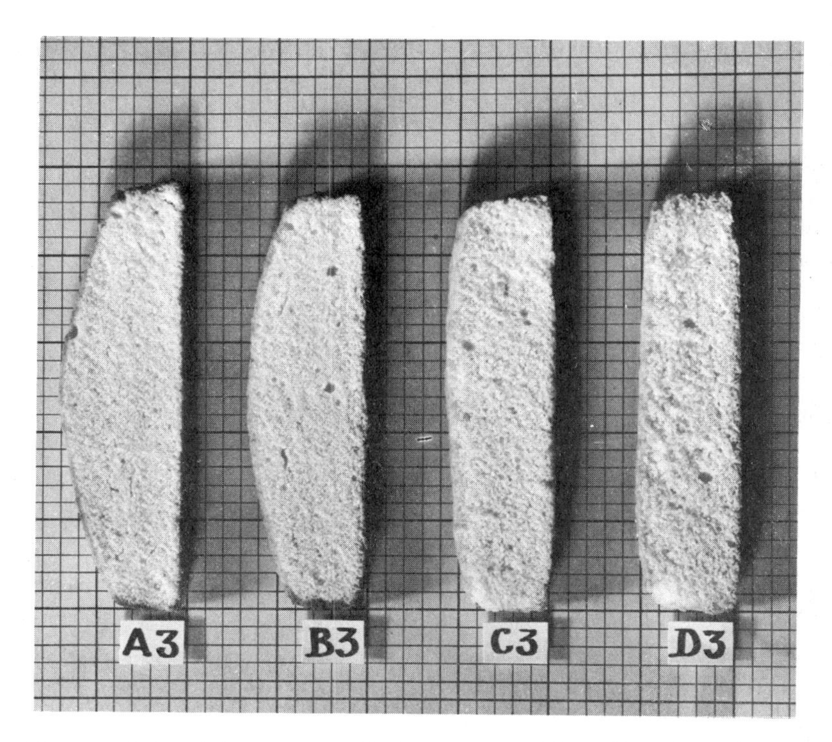

Figure 12.4. Cakes made with butter containing 6% glyceryl monostearate and baked at four temperatures. Cake A3 was baked at 218°C (425°F); B3 at 191°C (375°F); C3 at 163°C (325°F); and D3 at 149°C (300°F). (Reprinted from *Food Research,* Vol. 17, No. 3, pp. 185–196, 1952. Copyright © by Institute of Food Technologists.)

water content, produce less tender pastry than hydrogenated vegetable shortening when the fats are used on an equal-weight basis. Oil usually produces more tender pastry than plastic fats when the conventional mixing method is used, but the pastry may be crumbly and greasy.

As would be expected, the tenderness of pastry increases with the amount of fat in the formula, and an excessive amount makes pastry too tender to handle easily. In many of the research formulas, about 40 g of fat are used for 100 g of all-purpose flour. This proportion is equivalent to about 1/4 cup of hydrogenated fat per cup of flour. With this relatively lean formula, differences in tenderness attributable to the type of shortening and to other variables are shown more clearly than with the higher proportion of fat in formulas used in home baking.

All-purpose flours are widely and successfully used to produce pastries that are both tender and flaky, but flour of lower protein content will make more tender pastry. In one study (Denton et al., 1933), the breaking strengths of pastries made from different flours were as follows: cake or pastry flour, 346–400 g; all-purpose flour, 564–751 g; bread flour, above 950 g. To produce pastries of equal tenderness, less fat is required with pastry than with all-purpose flour.

The amount of water in pastry is important because pastry dough that is too dry is crumbly and difficult to roll, whereas an excessive amount of water results in tough pastry. Striking results were obtained by Swartz (1943), who made pastries with 100 g of pastry flour, 44 g of fat, and 25, 32.5, or 40 g of water. When the ingredients were at 21–24° C and blended at low speed in a mixer for 30 sec, breaking strengths for pastries with these three amounts of water were 394, 683, and 1035 g respectively.

Methods

The conventional method of mixing pastry involves cutting the solid fat into the flour and salt and gradually adding just enough water to make the dough manageable for rolling. Several variations in the method of combining the ingredients of pastry are possible. In one procedure, half of the shortening is cut into the flour until the mixture is as fine as corn meal and the remainder is cut in until the large particles are the size of peas. In another, the water is mixed with part of the flour, the fat is cut into the remaining flour until it is the size of peas, and the two mixtures are combined. A modified puff pastry method is possible in which about 2 tablespoons of the flour-fat mixture are removed before the addition of water and later are sprinkled over the pastry sheet. The sheet is rolled up like a jelly roll, then cut and rolled out for the pie pan. For the hot-water method, boiling water and fat are beaten together and the dry ingredients are added last. Pastry made in this way by machine (Rose et al., 1952) or by hand (Rose, 1953) was found to be crumbly but more tender than pastry made by the conventional method. Breaking strengths were more uniform throughout the sample of hot-water pastry, perhaps because of gluten formation and uneven distribution of fat in pastry made by the conventional method.

The properties of tenderness and flakiness can occur together or separately. Thorough mixing of fat and flour usually results in more tender pastry than does minimal mixing. Pastry usually is more tender when the ingredients and dough are warm than when they are cold. During mixing, warm fat coats the flour particles, whereas cold fat is broken into discrete particles. Increased tenderness at relatively high temperatures has been observed with seasonal fluctuations in room temperature (Denton et al., 1933) and also with the use of fat and water at room temperature, 21–24°C, rather than at refrigerator temperature, 4–7°C (Swartz, 1943). In one study (Hornstein et al., 1943), temperature was found to affect the shortening value of some fats more than others. Although most of the fats produced more tender pastry when mixed at 27°C than at 18°C, this was especially pronounced with butterfat. Cutting fat into the flour as rather large particles usually is considered to promote flakiness of the baked pastry. Thin horizontal layers are evident in flaky pastry. The thorough blending of flour and fat obtained when pastry is made by the hot-water method or with oil in the conventional method usually produces pastry that is mealy rather than flaky, though it is possible to produce flaky pastry with intimate mixing of fat and flour.

Lengthening the mixing time after water is added to the dough usually toughens pastry because of excessive gluten development. In a study of pastry made with lard, the toughening effect of added mixing with water was much more pronounced when the fat and water were at refrigerator temperature than when they were at room temperature (Swartz, 1943). This might reflect the greater plasticity of fat at the higher temperature.

EXPERIMENTS

I. Shortened Cakes

In these cake experiments, several mixing methods are compared, and the effects of variations in the level and type of individual ingredients are studied. Each experiment has only one variable. The experiments that are included can be expanded or altered to meet the interests of the group.

Control measures should include machine mixing, if possible, and a team approach, with one student performing the same operation on all cakes that are to be compared directly. It often is convenient to weigh the flour, salt, sugar, and fat in advance and store in tightly covered containers. Baking powder is best weighed and sifted with the other dry ingredients just before the cake is made. Mixers differ as to bowls, beaters, and speeds; therefore, the mixing needs to be standardized for the specific mixers used. During mixing, the beater and the sides and bottom of the bowl should be scraped often enough to ensure thorough mixing. Mixing should be planned so that after baking powder and liquid have both been added, work will be continuous until the batter is in the pan. Use loaf pans 19 × 9 × 5 cm, if available. These pans hold about 800 ml. If the available pans are of a different size, use about 0.375 g of batter per ml capacity. When the batter is in the pan, it can stand a few minutes before it is baked. The oven shelf should be placed slightly below the center of the oven and all cakes to be baked in an oven should be put in at once without crowding. Baked cakes should be cooled on a rack at least 10 min before removal from the pan. If it is impossible to score the cakes on the day they are baked, they can be frozen for future examination. The use of some objective measures of batter and cake properties is desirable (Chapter 16).

A. *Formula and methods*

Cake flour, g	78
Baking powder (SAS-phosphate), g	2.9
Salt, g	1.5
Vanilla, ml	1.2
Milk, ml	79
Shortening (hydrogenated), g	41
Sugar, g	100
Egg, g	36

1. Conventional method (by machine or by hand)
    a. Prepare pan by oiling and fitting the bottom with waxed paper; set oven at 190°C (375°F). Sift together the flour, baking powder, and salt. Add vanilla to the milk.
    b. Mix the cake by machine.

1) Cream shortening 1 min; add sugar gradually for 1 min; continue creaming for 90 sec, using medium high speed.
2) Add egg gradually for 1 min; continue creaming for 1 min. Stop the mixer and scrape sides of bowl thoroughly to the bottom.
3) Add 1/3 of the flour mixture and 1/3 of the milk and beat for 45 sec, using low speed. Repeat twice. Using medium high speed, blend batter for 45 sec.

Or mix the cake by hand.

1) Cream fat in a mixing bowl with a wooden spoon or a rubber scraper until it is plastic and very light. Add sugar, about a tablespoon at a time, beating after each addition until mixture looks fluffy.
2) Add egg in three portions, beating well after each addition. Beat for 1 min after the last addition of egg.
3) Add a heaping tablespoon of the sifted flour mixture; stir until flour is dampened, and beat 40 strokes.
4) Add 1/3 of the remaining flour mixture and 1/3 of the milk. Stir until flour is dampened and beat 40 strokes. Repeat twice. Beat an additional 50 strokes after all ingredients have been added.

c. Weigh 300 g of batter into a loaf pan and bake at 190°C until the cake springs back under the pressure of a finger, about 30 min.

2. Conventional-sponge method (by machine or hand)
   a. Prepare pan by oiling and fitting with waxed paper; set oven at 190°C (375°F). Sift together flour, baking powder, and salt. Add vanilla to milk.
   b. Mix the cake by machine.
      1) Cream fat for 1 min, using medium-high speed; add half the sugar gradually for 1 min; continue creaming for 90 sec. Stop the mixer and scrape sides of bowl thoroughly to the bottom.
      2) With a rotary beater, beat egg until it forms soft peaks. Add remaining half of the sugar gradually, beating well after each addition. Continue beating

until mixture is thick and looks like meringue. If a hand-operated beater is used, the total number of turns should be at least 250.

3) Add 1/3 of the flour mixture and 1/3 of the milk to the fat and sugar and beat for 45 sec, using low speed. Repeat twice. Stop the mixer and scrape.
4) Add the egg-sugar mixture and fold with a wire whisk until thoroughly combined.

Or mix the cake by hand.

1) Cream the fat in a mixing bowl with a wooden spoon or a rubber scraper until it is very plastic and light. Add half the sugar, about a tablespoon at a time, beating after each addition until the mixture looks fluffy.
2) With a rotary beater, beat egg until it forms soft peaks. Add remaining half of the sugar gradually, beating well after each addition. Continue beating until mixture is thick and looks like meringue. If a hand-operated beater is used, the total number of turns should be at least 250.
3) To the fat-sugar mixture, add a heaping tablespoon of the sifted flour mixture, stir until the flour is dampened, and beat 40 strokes.
4) Add 1/3 of the remaining flour mixture and 1/3 of the milk. Stir until flour is dampened and beat 40 strokes. Repeat twice.
5) Add the egg-sugar mixture and fold until well combined.

c. Weigh 300 g of batter into a loaf pan and bake at 190°C until the cake springs back under the pressure of a finger, about 30 min.

3. One-bowl method (by machine or by hand)
   a. Prepare pan by oiling and fitting with waxed paper; set oven at 190°C (375°F). Sift together flour, sugar, baking powder, and salt. Add fat, which must be at room temperature but not melted, vanilla, and about 50 ml of the milk.
   b. Mix the cake by machine.
      1) Beat at low speed for 1 min and at medium speed

for 1 min. Stop the mixer and scrape the bowl from the bottom.

2) Add egg and remaining milk. Beat at medium speed 2 min.

Or mix the cake by hand.

1) Stir with a wooden spoon or rubber scraper until flour is dampened, then beat hard for 2 min. Do not count time taken out to scrape the bowl and spoon or to rest during the 2 min of mixing.

2) Add egg and remaining milk; beat hard for 2 min.

c. Weigh 300 g of batter into a loaf pan and bake at 190°C until the cake springs back under the pressure of a finger, about 30 min.

4. Pastry-blend method (by machine)

a. Prepare pan by oiling and fitting with waxed paper; set oven at 190°C (375°F). Sift together flour, baking powder, and salt. Combine egg, half the milk, and vanilla. Stir remaining half of the milk with the sugar.

b. Mix the cake by machine.

1) Into the bowl of the mixer, put flour mixture and fat, which must be at room temperature but not melted. Beat at medium speed for 1 min and at medium high speed for 2 min. Stop the mixer and scrape.

2) Using medium speed, gradually add the milk-sugar mixture in 2 min; continue beating for 1 min. Stop the mixer and scrape.

3) Add the egg-milk mixture gradually while beating at low speed for 1 min. Stop the mixer and scrape. Beat at medium speed for 2 min.

c. Weigh 300 g of batter into a loaf pan and bake at 190°C until the cake springs back under the pressure of a finger, about 30 min.

5. Solution method (by machine or by hand)

a. Prepare pan by oiling and fitting the bottom with waxed paper; set oven at 190°C (375°F). Sift together flour, baking powder, and salt. Combine egg, half the milk, and the vanilla. Put sugar into mixing bowl; add remaining milk, and stir until most of the sugar dissolves. Add flour mixture and fat, which must be at room temperature but not melted.

b. Mix the cake by machine.
   1) Beat at low speed for 1 min and at medium speed for 1 min. Stop the mixer and scrape.
   2) Add the egg mixture. Beat at medium speed for 1 min and at medium high speed for 1 min.

   Or mix the cake by hand.
   1) Stir with a wooden spoon or rubber scraper until the flour is dampened, then beat hard for 2 min. Do not count time taken out to scrape the bowl and spoon or to rest during the 2 min of mixing.
   2) Add the egg mixture; beat hard for 2 min.
c. Weigh 300 g of batter into a loaf pan and bake at 190°C until the cake springs back under the pressure of a finger, about 30 min.

6. Muffin method (by hand)
   a. Prepare pan by oiling and fitting with waxed paper; set oven at 190°C (375°F). Sift together into a mixing bowl flour, sugar, baking powder, and salt.
   b. Mix the cake by hand.
      1) Beat egg until foamy; add milk, vanilla, and fat that has been heated just enough to melt it; mix well. Use the mixture immediately for step (2).
      2) Make a well in the center of the dry ingredients. Add the liquid ingredients; stir vigorously with a rubber scraper until flour has been dampened, and then beat vigorously for 1 min.
   c. Weigh 300 g of batter into a loaf pan and bake at 190°C until the cake springs back under the pressure of a finger, about 30 min.

B. *Variations in ingredients*

1. Amount of sugar. Prepare three cakes using the one-bowl method and the basic formula with the following amounts of sugar:
   a. 78 g (100% of the weight of the flour)
   b. 100 g (128% of the weight of the flour) (basic formula)
   c. 117 g (150% of the weight of the flour)

2. Amount of fat. Prepare four cakes using the one-bowl method and the basic formula with the following amounts of fat:
   a. 20 g (26% of the weight of the flour)
   b. 30 g (39% of the weight of the flour)
   c. 41 g (53% of the weight of the flour) (basic formula)
   d. 52 g (67% of the weight of the flour)
3. Type of fat. Prepare four cakes by the conventional sponge method using the basic formula with the following fats:
   a. Hydrogenated shortening
   b. Vegetable oil
   c. Butter (reduce salt to 0.8 g)
   d. Lard
4. Amount of egg. Prepare three cakes using the one-bowl method and the basic formula with the following amounts of egg:
   a. 18 g (23% of the weight of the flour)
   b. 36 g (basic formula) (46% of the weight of the flour)
   c. 54 g (69% of the weight of the flour)
5. Amount of milk. Prepare three cakes using the one-bowl method and the basic formula with the following amounts of milk:
   a. 61 g (59 ml) (78% of the weight of the flour)
   b. 81 g (79 ml) (basic formula) (104% of the weight of the flour)
   c. 101 g (99 ml) (130% of the weight of the flour)
6. Type of flour. Prepare two cakes using the one-bowl method and the basic formula with the following flours:
   a. Cake flour
   b. All-purpose flour

C. *Variations in method*

1. Mixing methods. Prepare five cakes using the basic formula and the following mixing methods:
   a. Conventional
   b. One-bowl
   c. Pastry-blend
   d. Solution
   e. Muffin

2. Variations of the conventional method. Prepare five cakes using the basic formula and the following mixing methods:
   a. Conventional.
   b. Conventional with egg added at the end. Follow the directions for the conventional method except do not add egg to the creamed mixture. Beat egg with rotary beater until it forms soft peaks. If a hand-operated beater is used, the total number of turns should be at least 150.
   c. Conventional-sponge.
   d. Conventional with eggs separated. Weigh 12 g of yolk and 24 g of egg white instead of whole egg. Follow the directions for the conventional method except add yolk instead of whole egg to the fat and sugar. Beat egg white until it is stiff enough to hold a peak but is still shiny. After all other ingredients have been added to the batter, fold in beaten egg white until thoroughly combined.
   e. Conventional-meringue. Weigh 12 g of yolk and 24 g of egg white instead of whole egg. Follow the directions for the conventional method except add only half of the sugar to the fat, then add yolk instead of whole egg. Beat egg white to a soft peak, then beat the remaining sugar in gradually. Continue beating until the meringue is stiff enough to hold a peak. After all other ingredients have been added to the batter, fold in the meringue until thoroughly combined.
3. Baking temperature. Prepare four times the basic formula, using the one-bowl method. Bake at 149°C (300°F), 177°C (350°F), 190°C (375°F), and 218°C (425°F), adjusting the baking times as necessary.

D. *Results*

Evaluate volume, shape, grain, crust color, crumb color, tenderness, moistness, and flavor of cakes, using scales of appropriate adjectives. (See the Appendix.) Tabulate the scores along with results of any objective measurements made on batters or cakes.

E. Questions

Discuss the effects of variations in the level of sugar, fat, eggs, and milk. What other adjustments in the recipe are sometimes made when the proportions of these ingredients are changed? Discuss the relative merits of each method of mixing cakes. What proved to be the best oven temperature for the cakes? How might this temperature vary with other recipes and baking pans?

II. Pastry

Factors affecting the quality of pastry are illustrated by variations in a basic pastry formula that is less rich than those usually used in home baking so that it will be more sensitive to factors affecting tenderness. The effects of variations in the type and amount of fat, in the amount of water, and in the mixing method are studied. As usual, a team approach is suggested as an aid to control of mixing. Guides should be used for rolling to uniform thickness. Breaking strength of the pastries should be measured if a shortometer is available. Index to flakiness can be measured as the thickness of a stack of a specified number of wafers. A vernier caliper is a convenient tool for making this measurement.

A. Formula and methods

All-purpose flour, g	41
Salt, g	1
Shortening, hydrogenated, g	18
Water, ml	13

1. Have all ingredients at room temperature. Set oven at 218°C (425°F).
2. Put flour, salt, and fat in a mixing bowl having a diameter of about 15 cm. Cut in the fat with a pastry blender until no particles are larger than peas.
3. Sprinkle water on the flour-fat mixture. Then cut in with a pastry blender. After every five strokes, remove dough from blender with a rubber scraper. Count strokes, using only enough to make a dough that will form a ball when

pressed together. A total of 25 strokes usually is enough. If necessary, add more water. Record the total amount used.

4. Put ball of dough on a piece of waxed paper large enough to roll it on and flatten to a thickness of about 2.5 cm with the hands. Lay a guide strip, approx 3 mm thick, on each side of the dough. Cover with another piece of waxed paper and roll the dough to the thickness of the guides. Keep the dough in a rectangular shape while rolling.
5. Peel off the upper paper and invert the pastry on a baking sheet. Remove the other piece of paper.
6. Cut the wafers 4.5 × 9 cm but do not separate them from each other or remove the extra dough around the edges. Prick the wafers uniformly. (See Experiment III, Chapter 8.) Bake at 218°C until the edges of the dough are light brown. Cool before making measurements.

B. *Variations in ingredients*

1. Type of fat. Make pastry according to the basic formula, keeping the water level constant and using each of the following fats, all at room temperature:
   a. Hydrogenated shortening
   b. Lard
   c. Butter (reduce salt to 0.5 g)
   d. Butterfat
   e. Margarine (reduce salt to 0.5 g)
2. Amount of fat. Make pastry according to the basic formula, keeping the water level constant and using the following amounts of hydrogenated shortening:
   a. 12 g
   b. 18 g (basic formula)
   c. 24 g
   d. 36 g
3. Amount of water. Make pastry according to the basic formula with the following amounts of water:
   a. Same as basic formula
   b. 50% more than in basic formula
   c. Twice as much as in basic formula

C. Variations in method

1. Method of adding water. Make pastry according to the basic formula but add water by each of the following methods, mixing only enough to make a dough that will hold together.
   a. Basic formula.
   b. Pour water in slowly while stirring with a fork.
   c. Pour water in slowly while stirring with a rubber spatula.
   d. Sprinkle water on the flour-fat mixture about 3 ml at a time, tossing with a fork under the drops of water. Try always to sprinkle on a dry portion and remove dampened particles to the side of the bowl.
2. Added manipulation. Triple the basic formula, mix through step 3, and use for the following:
   a. Basic formula. Remove 70 g of dough. Roll and bake as for the basic formula.
   b. Additional mixing. Blend remaining dough 20 additional strokes. Remove 70 g of dough. Roll and bake as for the basic formula.
   c. Additional mixing and kneading. Knead the remaining dough vigorously for 1 min. Then roll and bake as for the basic formula.
3. Mixing method using hydrogenated shortening
   a. Basic formula.
   b. Extra mixing of fat. Make pastry according to the basic formula but cut in fat until mixture has the consistency of fine corn meal.
   c. Flour-suspension method. Mix 34 g of the flour with the salt. Cut in the fat until no particles are larger than peas. Using a fork, mix the water with the remaining flour. Add to flour-fat mixture all at once and blend with a fork until the mixture just holds together. Roll and bake.
   d. Hot-water method. Bring water to a boil in a small container and add immediately to fat in a mixing bowl. Stir with a fork. Then beat until the mixture is smooth,

thick, and peaked when the fork is removed. Add flour and salt and stir rapidly with a fork until all the flour disappears. Roll and bake.

4. Mixing method using vegetable oil. For each variation, have the oil at room temperature and the liquid at refrigerator temperature. Because milk contains 87% water, use 15 g of milk as the equivalent of 13 g of water.
   a. Basic formula but with oil and ice water.
   b. Put oil and water into a small container without stirring. Pour into flour and salt. Stir with a fork until flour is dampened. Roll and bake.
   c. Put oil and 15 g or 15 ml of milk into a small container without stirring. Pour into flour and salt. Stir with a fork until flour is dampened. Roll and bake.
   d. Put oil and 15 g or 15 ml of milk into a small container. Beat with a fork until thickened and creamy. Pour immediately into the flour mixture and stir with a fork until flour is dampened. Roll and bake.

D. *Results*

Score appearance, flakiness, tenderness, flavor, and acceptability, using scales of appropriate adjectives. (See the Appendix.) Tabulate the scores along with breaking strength and index to flakiness.

E. *Questions*

List the fats used in order of the tenderness of the pastries made from them. Compare the color and flavor of pastries made from the different fats. What is the effect of added fat on tenderness? Of added water? Of added manipulation? What method of adding water produced the most tender pastry? How did pastries made by the different mixing methods compare? Did the mixing method have more or less effect on properties of pastry made with oil than of pastry made with hydrogenated shortening? Of all of the factors studied, did any have more effect than the others on flakiness? Did any method of adding vegetable oil result in pastry that was both flaky and tender?

SUGGESTED EXERCISES

1. Note that the muffin method and the one-bowl method, when done in a single step, are the same except that plastic fat is melted for the former. Compare cakes made with the same fat in plastic form at room temperature and melted.
2. Study the effect of under- and over-mixing on cake made by the one-bowl method.
3. Compare cakes made from flours of varying protein content. Also substitute wheat starch for all of the cake flour in one batter. If the cake made with starch is unsatisfactory, try increasing the proportion of egg in a cake containing starch rather than flour.
4. Compare cakes made from different types of flour to which starch has been added as needed to equalize protein concentration. Wheat starch is preferable for this purpose. If it is not available, use cornstarch.
5. Compare plain cakes mixed by the one-bowl method and containing butter, with and without added emulsifier. Distilled monoglycerides, added to the butter at a level of about 2% of the weight of the butter, perform well.
6. Compare cakes to which dried whey has been added at different levels.
7. Develop a satisfactory formula and procedure for increasing the level of polyunsaturated fatty acids in cake through the use of a vegetable oil that is highly polyunsaturated.
8. Use nonfat dried milk to increase the concentration of protein in cake. Determine the maximum level of addition that is feasible for a given set of conditions as to formula and mixing method.
9. Compare cakes containing bran substituted at different levels for flour; for example, 10, 20, and 30% replacement by weight.
10. Carry out Experiment I in Chapter 10, if it was not done during the study of leavening agents.

11. Compare gingerbreads made with combinations of sweet milk, buttermilk, baking powder, and sodium bicarbonate such as to produce a range of pH values.
12. Compare cakes and pastries made from the individual ingredients and from retail commercial mixes. Compare the products as to quality, cost, and time required to produce them.
13. Conduct Experiments III and IV in Chapter 8 if they were not done earlier.

REFERENCES

Ash, D.J. and Colmey, J.C. 1973. The role of pH in cake baking. Baker's Digest 47(1): 36.

Baldi, V., Little, L. and Hester, E.E. 1965. Effect of the kind and proportion of flour components and of sucrose level on cake structure. Cereal Chem. 42: 462.

Bowman, F. 1973. MCT cookies, cakes, and quick breads: quality and acceptability. J. Amer. Dietet. Assoc. 62: 180.

Bowman, F., Dilsaver, W. and Lorenz, K. 1973. Rationale for baking wheat-, gluten-, egg-, and milk-free products. Baker's Digest 47(2): 15.

Briant, A.M., Weaver, L.L. and Skodvin, H.E. 1954. Quality and thiamine retention in plain and chocolate cakes and in gingerbread. Mem. 332, Cornell Univ. Agric. Exp. Stn., Ithaca, NY.

Brockmole, C.L. and Zabik, M.E. 1976. Wheat bran and middlings in white layer cakes. J. Food Sci. 41: 357.

Brys, K.D. and Zabik, M.E. 1976. Microcrystalline cellulose replacement in cakes and biscuits. J. Amer. Dietet. Assoc. 69: 50.

Charley, H. 1950. Effect of baking pan material on heat penetration during baking and on quality of cakes made with fat. Food Res. 15: 155.

Charley, H. 1952. Effects of the size and shape of the baking pan on the quality of shortened cakes. J. Home Econ. 44: 115.

deGoumois, J. and Hanning, F. 1953. Effects of dried whey and various sugars on the quality of yellow cakes containing 100% sucrose. Cereal Chem. 30: 258.

Denton, M.C., Gordon, B. and Sperry, R. 1933. Study of tenderness in pastries made from flours of varying strengths. Cereal Chem. 10: 156.

Dunn, J.A. and White, J.R. 1937. Factor control in cake baking. Cereal Chem. 14: 783.

Funk, K., Conklin, M.T. and Zabik, M.E. 1970. Use of frozen, foam–spray-dried, freeze-dried, and spray-dried whole eggs in yellow layer cakes. Cereal Chem. 47: 732.
Guy, E.J. and Vettel, H.E. 1973. Effects of mixing time and emulsifier on yellow cakes containing butter. Baker's Digest 47(1): 43.
Halliday, E.G. and Noble, I.T. 1946. "Hows and Whys of Cooking," 3rd ed. The University of Chicago Press, Chicago.
Handleman, A.R., Conn, J.F. and Lyons, J.W. 1961. Bubble mechanics in thick foams and their effects on cake quality. Cereal Chem. 38: 294.
Hanning, F. and de Goumois, J. 1952. The influence of dried whey on cake quality. Cereal Chem. 29: 176.
Hornstein, L.R., King, F.B. and Benedict, F. 1943. Comparative shortening value of some commercial fats. Food Res. 8: 1.
Howard, N.B. 1972. The role of some essential ingredients in the formation of layer cake structures. Baker's Digest 46(5): 28.
Howard, N.B., Hughes, D.H. and Strobel, R.G.K. 1968. Function of the starch granule in the formation of layer cake structure. Cereal Chem. 45: 329.
Hunter, M.B., Briant, A.M. and Personius, C.J. 1950. Cake quality and batter structure. Bull. 860, Cornell Univ. Agric. Exp. Stn., Ithaca, NY.
Jooste, M.E. and Mackey, A.O. 1952. Cake structure and palatability as affected by emulsifying agents and baking temperatures. Food Res. 17: 185.
Kissell, L.T. 1959. A lean-formula cake method for varietal evaluation and research. Cereal Chem. 36: 168.
Kulp, K. 1972. Some effects of chlorine treatment of soft wheat flour. Baker's Digest 46(3): 26.
Kulp, K., Tsen, C.C. and Daly, C.J. 1972. Effect of chlorine on the starch component of soft wheat flour. Cereal Chem. 49: 194.
Lawson, H.W. 1970. Functions and applications of ingredients for cake. Baker's Digest 44(6): 36.
Matthews, R.H. and Dawson, E.H. 1966. Performance of fats in white cake. Cereal Chem. 43: 538.
Miller, D., Nordin, P. and Johnson, J.A. 1957. Effect of pH on cake volume and crumb browning. Cereal Chem. 34: 179.
Miller, B.S., Trimbo, H.B. and Powell, K.R. 1967. Effects of flour granulation and starch damage on the cake making quality of soft wheat flour. Cereal Sci. Today 12: 245.
Ohlrogge, H.B. and Sunderlin, G. 1948. Factors affecting the quality of cakes made with oil. J. Amer. Dietet. Assoc. 24: 213.
Petricca, T. 1976. Fluid bakery shortenings. Baker's Digest 50(5): 39.
Pyke, W.E. and Johnson, G. 1940. Relation of mixing methods and a balanced formula to quality and economy in high-sugar-ratio cakes. Food Res. 5: 335.
Rajchel, C.L., Zabik, M.E. and Everson, E. 1975. Wheat bran and middlings. A

source of dietary fiber in banana, chocolate, nut and spice cakes. Baker's Digest 49(3): 27.
Redfield, G.M. 1947. Mixing a plain cake. J. Home Econ. 39: 518.
Rose, T.S. 1953. Supplementary study on pastry methods. J. Home Econ. 45: 337.
Rose, T.S., Dresslar, M.E. and Johnston, K.A. 1952. The effect of the method of fat and water incorporation on the average shortness and the uniformity of tenderness of pastry. J. Home Econ. 44: 707.
Sollars, W.F. and Rubenthaler, G.L. 1971. Performance of wheat and other starches in reconstituted flours. Cereal Chem. 48: 397.
Springsteen, E., Zabik, M.E. and Shafer, M.A.M. 1977. Note on layer cakes containing 30 to 70% wheat bran. Cereal Chem. 54: 193.
Swartz, V. 1943. Effect of certain variables in technique on the breaking strength of lard pastry wafers. Cereal Chem. 20: 121.
Trimbo, H.B. and Miller, B.S. 1973. The development of tunnels in cakes. Baker's Digest 47(4): 24.
Tsen, C.C., Kulp, K. and Daly, C.J. 1971. Effects of chlorine on flour proteins, dough properties, and cake quality. Cereal Chem. 48: 247.
Wilson, J.T. and Donelson, D.H. 1963. Studies on the dynamics of cake-baking. I. The role of water in formation of layer cake structure. Cereal Chem. 40: 466.
Yamazaki, W.T. and Donelson, D.H. 1972. The relationship between flour particle size and cake-volume potential among Eastern soft wheats. Cereal Chem. 49: 649.

13. Sugars and Crystallization

Sugars have been mentioned many times in previous chapters because of their contributions to the properties of many types of food. The chemical and physical nature and the functions of sugars in foods will be summarized in this chapter. Discussion of one of the functions, responsibility for the structure of crystalline candies, will be expanded. Finally, the other type of crystalline structure in foods, that which is developed during freezing, will be considered.

SUGARS

Chemical Nature

The monosaccharides of particular interest in food science are the *hexoses* (glucose, fructose, and galactose), though the pentoses (arabinose, xylose, and ribose) also are of interest. Glucose and galactose are aldohexoses and fructose is a ketohexose. Their formulas are represented in the Haworth cyclic structures, shown on page 406.

α-D-glucopyranose α-D-fructofuranose

A discussion of the significance of the α (versus β) and D (versus L) notations can be found in any biochemistry textbook. The Haworth structures for the sugars are not realistic in that all of the atoms in the ring are represented as being in a single plane; actually, different arrangements in space are possible among sugars, and structures such as the "chair" and the "boat," each with its own variations,

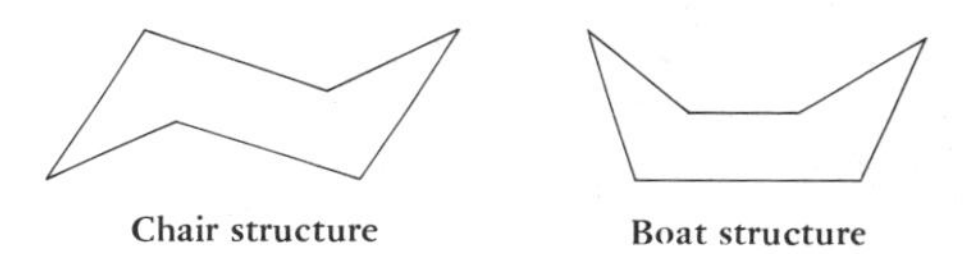

Chair structure Boat structure

are becoming common in the literature as sugar chemists continue to learn of the possible arrangements and their significance. A specific example is the following representation of α-D-glucopyranose:

Chair structure, α-D-glucopyranose

Birch and Shallenberger (1973) believe that configurations, or arrangements of atoms and groups about individual carbon atoms, dictate the possible conformations, or molecular shapes, and that configuration and conformation determine the properties of sugars. Although recognition of these aspects of the molecular structure of sugars is important, further detail is beyond the scope of this book. The Haworth structures will be used herein as a matter of convenience.

The carbonyl groups (or potential carbonyl groups) in sugar molecules are important with regard to the chemical reactions in which sugars participate. Note in the formulas presented above that the C1 of glucose is the aldehydic carbon and the C2 of fructose is the keto carbon. These are reducing groups, both of which are "tied up" when glucose and fructose combine (in glycosidic linkage) to form the disaccharide sucrose:

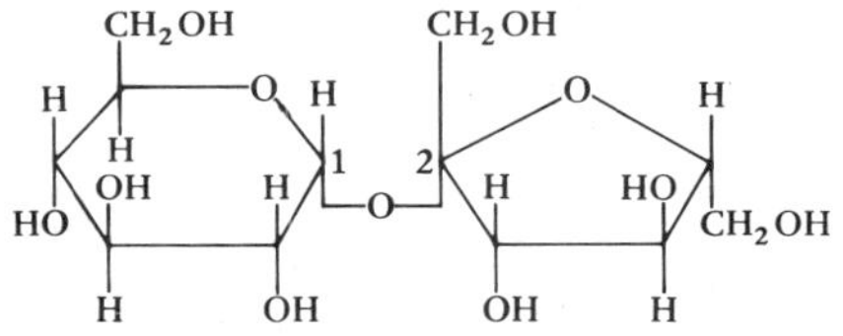

Sucrose (α-D-glucose, β-D-fructose)

Thus the sucrose molecule does not have any reducing ability. A disaccharide is defined as a carbohydrate that yields two monosaccharide units when hydrolyzed. When the disaccharide sucrose is hydrolyzed to glucose and fructose, two reducing groups are freed for each glycosidic bond broken. In maltose, two glucose molecules are joined through the C1 of one and the C4 of the other; in other words, one reducing group is tied up and the other remains free. This disaccharide, therefore, is a reducing sugar, but hydrolysis of maltose results in a mixture with yet greater reducing power than that of the intact molecule. The lactose molecule consists of glucose and galactose linked through the C4 of glucose and the C1 of galactose; thus the situation with regard to reducing ability is the same as that for maltose.

Hydrolysis of disaccharides is an important reaction in foods. From a chemical standpoint, the significance lies not in the increased potential for participation in oxidation-reduction reactions (though this is useful analytically), but in the freeing of carbonyl groups that are potential reactants in browning (Chapter 2). In addition, because sugars differ in properties such as sweetness and solubility, the properties of a reaction mixture are different from those of the disaccharide that is hydrolyzed.

One such property change accompanying hydrolysis is responsible for some common terminology. The hydrolysis specifically of sucrose is called *inversion* because the mixture of products has a different rotatory power from that of sucrose. Sucrose in solution rotates a plane of polarized light to the right—it is dextrorotatory. Fructose, one of the products of sucrose hydrolysis, is so strongly levorotatory that the mixture of products is levorotatory. Because the direction of rotation by the solution is inverted, the enzyme that catalyzes the reaction is called *invertase,* and the mixture of glucose and fructose resulting from the reaction is called *invert sugar.*

Several examples of enzymatic catalysis of sugar hydrolysis were mentioned in earlier chapters. These examples include the hydrolysis of maltose and sucrose by yeast enzymes (Chapter 10) in breadmaking and the treatment of milk with lactase to reduce development of grittiness in ice cream, resulting from low solubility of lactose, and to permit the use of milk by individuals who are intolerant of lactose. Another example, the hydrolysis of sucrose in the candy industry, will be mentioned during the discussion of crystallization of sugars.

Hydrolysis also can be catalyzed by a combination of heat and dilute acid. The significance of this method of hydrolysis in relation to sugars also will be discussed in connection with crystallization.

Caramelization of sugars was discussed briefly in Chapter 2. Like carbonyl-amine browning, caramelization is a series of reactions rather than a single reaction. It should be noted that the browning that occurs when caramels are made is primarily carbonyl-amine browning rather than caramelization.

The hydroxyl groups on sugar molecules permit their reacting with other compounds through either glycosidic or ester linkages. Anthocyanins, discussed in Chapter 6, are examples of compounds in which sugars are in glycosidic linkage with nonsugar moieties. Sinigrin, a flavor component also mentioned in Chapter 6, is another example of a glycoside. Sucrose esters, mentioned in Chapter 11 as surfactants that are useful in breadmaking, are examples of sugar in ester linkage with another compound.

Sugars polymerize in nature to form polysaccharides that have functional importance in foods. These include cellulose (Chapter 6) and starch (Chapter 9). Pectic substances (Chapter 6) are polymers of sugar derivatives.

Further information concerning the chemical reactions of sugars is presented by Aurand and Woods (1973).

Physical Properties

Solubility in water is a property of sugars that is derived from the large number of hydroxyl groups in sugar molecules. It has been recognized for many years that sugars differ as to solubility. Fructose is most soluble at room temperature, followed by sucrose, glucose, maltose, and lactose. Although fructose is the most soluble sugar and lactose the least soluble over a broad temperature range, the solubility rankings of sucrose, glucose, and maltose vary somewhat with the temperature at which solubility is measured. Because sugars are similar with respect to their content of hydroxyl groups, it would seem that their differences in solubility must be related to configuration and/or conformation.

The significance of differences in solubility lies in the appropriateness of different sugars in food applications involving high sugar concentrations. Lactose, for example, has too low a solubility to be used successfully as a food ingredient other than as a component of milk, in which it is in dilute solution.

Hygroscopicity is related to solubility. Apparently the same structural features that cause sugars to differ in solubility also cause them to differ in hygroscopicity. Thus fructose is particularly hygroscopic. Donnelly et al. (1973) studied the hygroscopicity of several sugars. At 24°C the order of hygroscopicity was as follows: fructose>sucrose/glucose (50/50)>sucrose>sucrose/maltose (50/50)>glucose>lactose. Honey, which contains a rather high proportion of fructose, frequently serves as a humectant in home-baked foods because of the ability of its fructose component to absorb moisture from the atmosphere.

Crystallizability is a physical property that varies inversely with solubility. Crystallization will be discussed later in the chapter.

Sweetness obviously is an important property of sugars. Ranking sugars as to sweetness is quite a different matter from ranking them as to solubility because sweetness is a sensory quality. Its perception depends on many conditions of the test, including the concentration at which sugars are compared, the temperature of the solutions, and the viscosity of the carrier medium. Regardless of the conditions, fructose is quite consistently ranked as the sweetest sugar and lactose as the least sweet sugar, though the actual values assigned to designate sweetness relative to that of sucrose do depend on the conditions.

Shallenberger (1963) hypothesized and later (Shallenberger et al., 1965) presented further evidence that sweetness of sugars varies with hydrogen bonding. According to the theory, adjacent intramolecular hydroxyl groups that have the potential of eliciting the sweet taste response are involved in hydrogen bonding with one another. Thus their ability to elicit the response is restricted. Differences in configuration and conformation would result in differences in the distances between groups that have hydrogen bonding potential. Birch and Shallenberger (1973) later suggested that configuration and conformation probably are responsible for all of the properties of the sugars.

Doty (1976) stated that recent technological advances in Europe brought fructose to the forefront as a practical and effective sweetening agent in the food industry. The high degree of sweetness of fructose also is the basis for the recent extensive use of high fructose corn sirups in the food industry. High prices of sucrose resulted in increased dependence on corn sirups. Corn sirups, even those representing the highest levels of conversion of starch to sugar, formerly were less sweet than sucrose; however, the development of a commercially feasible procedure for increasing the sweetness of corn sirups has made available corn sirups with fructose representing more than 40% of the total solids. This is achieved by treatment of a high D.E. corn sirup with immobilized glucose isomerase. The enzyme converts glucose in the sirup to fructose. The principle of enzyme immobilization is described in Chapter 2.

Obviously a high fructose corn sirup also has humectant properties. The use of high fructose sirups in bakery products was reported by Saussele et al. (1976) and Volpe and Meres (1976).

Hodge et al. (1972) reviewed the properties of maltose, which they considered to have some potential advantages over other sweetening agents.

Functions in Foods

The functions of sugars in food have been discussed in various chapters and will only be listed here. Sugars participate in nonenzymatic browning reactions. They modify the properties of egg, starch, and gelatin gels. They dehydrate pectin micelles to permit pectin gels to form.

They stabilize egg white foams. They serve as substrate for yeast fermentation in bread dough and lactic acid fermentation in cultured buttermilk and some other dairy products. They have a large role in the aeration of batters and a tenderizing function in batter and dough products. Sugars contribute sweetness to many foods and are responsible for the structure of candies. In noncrystalline candies, crystals fail to form because of a high degree of sugar concentration, a high concentration of interfering substances, or a combination of these factors. In hard candies the moisture level is so low that the mixture becomes rigid quickly and the sugar molecules do not have an opportunity to become oriented into crystals. Interfering substances function through mechanisms that will be discussed in connection with crystallization.

CRYSTALLIZATION

Both similarities and differences exist between crystallization in candy and crystallization in a frozen dessert. In each case, crystallization occurs from a solution and results in structure formation. In a crystalline candy, the crystals are composed of sugar and their formation depends on supersaturation. As sugar crystallizes out of solution, the solution becomes decreasingly concentrated because of removal of solute. Eventually the solution is no longer supersaturated and crystallization cannot continue. In a frozen dessert, crystallization again occurs from a sugar solution but the crystals are composed of water (ice). As freezing occurs with formation of ice crystals, the sugar solution becomes increasingly concentrated because solvent is removed from the solution. Eventually the solution is so concentrated that its freezing point is too low for further freezing under the existing conditions.

Candy

Different kinds of crystalline candy are discussed in numerous elementary textbooks. The discussion herein will be confined to the underlying principles.

Structure In a crystalline candy, many small crystals of sugar are suspended in a small amount of a concentrated sugar solution. Other ingredients, such as acid, are in the solution. Still others, such as fat, may be adsorbed on crystal surfaces.

Effects of procedures The relatively large crystals of free-flowing sugar are dissolved. The solution is concentrated by boiling. The concentrated solution is supersaturated by cooling and crystallization occurs during beating of the supersaturated solution. The resulting product is sufficiently cohesive and firm to hold its shape, and contains crystals so small as not to be apparent.

Concentration Two points must be remembered: (1) As water is boiled out of an aqueous solution, the solution becomes increasingly concentrated. (2) For a given solute, a given boiling point represents a specific concentration (Chapter 2). The sugar concentration achieved during boiling determines the amount of sugar that ultimately will crystallize and, therefore, the firmness of the product. Underconcentration results in a soft product, whereas overconcentration results in a hard product. If the concentration step is not done properly, succeeding steps cannot compensate. It is important, therefore, that the thermometer be calibrated with boiling water and that it be read carefully at the endpoint. The bulb of the thermometer should be covered with the boiling sirup. If the boiling point exceeds the desired temperature, it is better to add some water and bring the solution back to the proper concentration than to gamble. Unless the mixture contains milk solids, which are subject to scorching, there is no need to stir once the sugar is dissolved.

The rate of heating is not critical unless an acidic ingredient is present, but it should be similar for all samples being compared in an experiment. Heating periods of about 15–20 min are satisfactory for solutions containing a cup of sugar. If an acidic ingredient, such as cream of tartar, is present the rate of heating is extremely important because inversion of sucrose occurs. Inversion, which occurs during heating, affects the ability of the sucrose to crystallize in a later step. The extent of inversion depends on the concentration of the acidic ingredient and on the length of time that it has an opportunity to act. A small amount of cream of tartar can have as great an effect in a solution heated slowly to the final boiling point as a large amount of cream of tartar in a solution heated rapidly to the same boiling point.

Cooling The solution should be undisturbed during cooling; otherwise, premature crystallization is likely to occur, resulting in a grainy product. Although the solution becomes supersaturated when cooling is begun, the *extent* of supersaturation, which is important to the crystallization process, increases as cooling proceeds. However, a factor in opposition to the increasing supersaturation is the increasing viscosity that also occurs during cooling, as will be discussed below.

Beating Crystallization is initiated when beating is begun. If beating is begun when the solution is quite hot, crystallization is rapid, and the product is grainy (Figure 13.1). If beating is delayed until the solution has cooled to room temperature, beating is difficult, crystallization is slow, and the product is smooth. The preceding two statements at first seem to contradict theory, because if the extent of supersaturation increases with cooling, one might expect crystallization to occur more readily in the cooler, more highly supersaturated solution. Viscosity

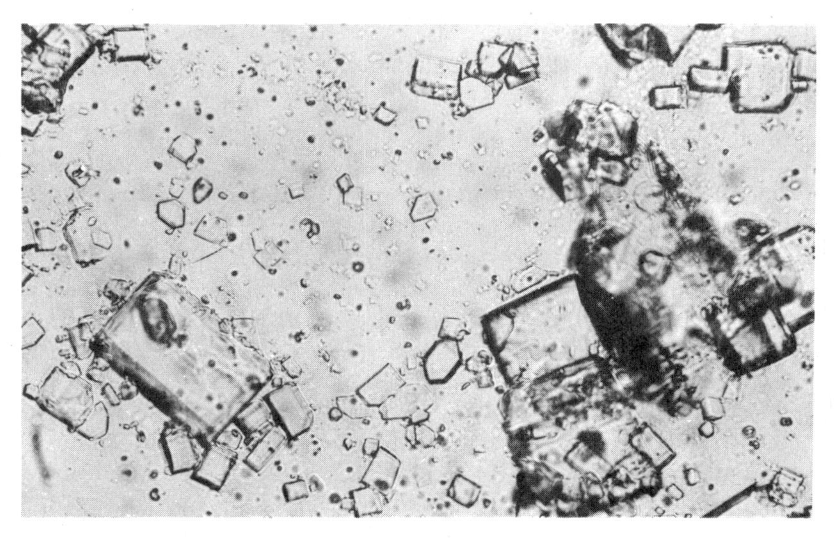

Figure 13.1. Sugar crystals in fondant made from sucrose and water and beaten while hot (103°C). (Courtesy of Vida Wentz.)

also is involved, however. In crystallization, the molecules align themselves into definite patterns to form crystals, each containing many oriented molecules. The molecules can move into their proper positions more readily in a sirup of relatively low viscosity than in one of high viscosity. A moderate extent of cooling provides for adequate development of supersaturation, in a properly concentrated sirup, without too great viscosity.

Once initiated by beating, crystallization continues spontaneously. However, beating is continued for the purpose of keeping the crystals small and the product smooth. If crystallization occurs very rapidly, as in a hot sirup, even vigorous beating may not result in a smooth product. On the other hand, cooling to room temperature would be rather wasteful of time and energy. A temperature of 40–50°C for the beginning of beating favors both reasonably rapid crystallization and the formation of small crystals.

Effects of ingredients In practice one would be likely to have additional constituents present for the sake of palatability. Other constituents may have either chemical or physical effects on crystallization.

Chemical effects As stated above, the chemical effect of cream of tartar is inversion of sucrose, or hydrolysis to a mixture of glucose and fructose. The reaction varies as to extent, depending on the heating conditions. The effect is retardation of crystallization. Crystallization is less rapid from a solution of mixed sugars than from a solution of a single sugar. The different kinds of molecules must sort themselves out during crystal formation. Crystals tend to be small because of the slower crystallization (Figure 13.2 as compared with Figure 13.3). It is possible for hydrolysis to interfere with crystallization sufficiently to produce a sticky or runny product. (Note: Interference does not mean prevention.)

A factor that has not been taken into account in the above discussion is the effect of hydrolysis of sucrose on the boiling point of the solution. After hydrolysis, the same weight of sugar consists of more molecules than originally; therefore, the effect on boiling point is greater and a specified boiling point is reached at a lower sugar concentration than if no hydrolysis occurs. Some of the retardation of crystallization probably results from underconcentration. If the extent of hydrolysis were known, a correction in the final boiling point could be made; however,

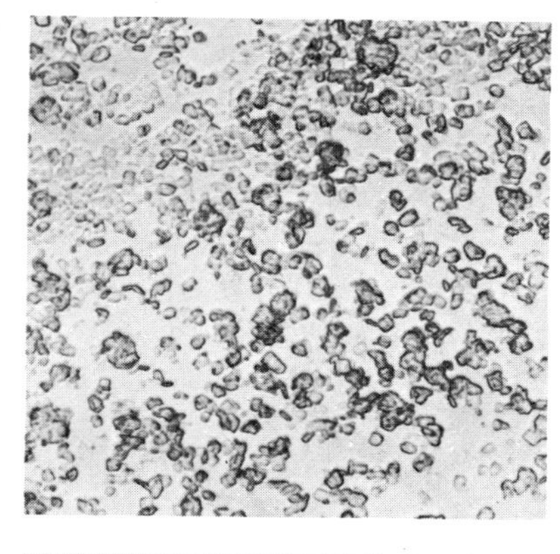

Figure 13.2. Sugar crystals in fondant made from sucrose, water, and cream of tartar and cooled to 40°C before beating. (Courtesy of Vida Wentz)

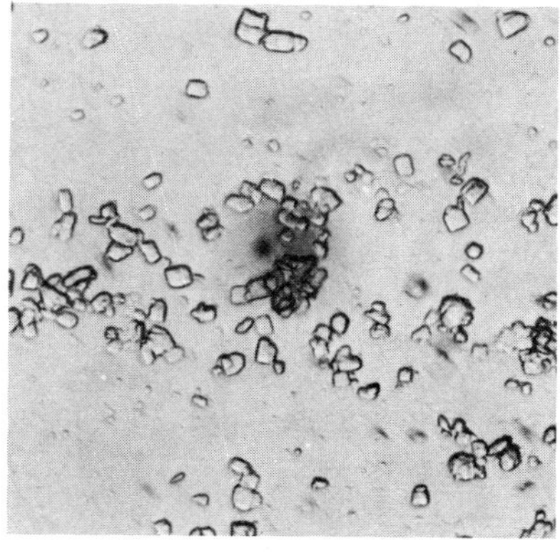

Figure 13.3. Sugar crystals in fondant made from sucrose and water and cooled to 40°C before beating. (Courtesy of Vida Wentz)

that is not the case. Although compensation would involve guesswork and is not practical, this factor at least should be recognized.

Physical effects Fat is an example of a substance that has a physical effect on crystallization. It tends to be adsorbed on surfaces of crystals as they form and thus interfere with growth of large crystals. Protein, such as that in milk solids, acts similarly to fat. These substances exert their effect during the beating stage. Addition of corn sirup or honey also retards crystallization and thus promotes smoothness. With either of these substances, the effect is similar to that of the cream of tartar except that *added* sugars, rather than sugars produced chemically, have the retarding effect.

Frozen Desserts

Frozen desserts include ice cream, for which federal and state regulations require specified minimum levels of fat and milk solids; French ice cream, which contains some egg; ice milk, which contains too little fat to qualify as ice cream; milk sherbet; and water ice. The discussion below pertains primarily to commercially frozen ice cream but the information has rather general application.

Structure The unfrozen ice cream mix is an oil-in-water emulsion. The aqueous phase in which the fat globules are dispersed also serves as a dispersion medium for proteins in colloidal dispersion and sugars and salts in true solution. Commercial ice cream mixes are homogenized and the fat globules, therefore, are relatively small. Their surfaces are coated with a stabilizing layer of coagulated protein.

During freezing, the characteristic structure of ice cream results from the freezing of water, with formation of crystals. Air is incorporated so that the frozen mixture is a foam. Incorporation of air results in an increase in volume, called *overrun.* The air cells not only make the ice cream light and soft but also contribute to smoothness by mechanically hindering the formation of ice crystals. Excessive overrun results in too little body.

The fat globules become solidified and to a large extent they form clusters. The ice crystals form in the unfrozen aqueous phase. The aqueous phase never completely freezes because as ice crystals form, making the solution increasingly concentrated, the freezing point of the remaining solution decreases until the limit of the freezing capability is reached.

Effects of procedures The temperature at which an unfrozen mix is stored affects its stability. Coalescence of fat globules accelerates as the storage temperature is increased. The temperature to which the mix is frozen affects firmness. Decreasing temperature results in increased firmness, partly as a result of further freezing of water and partly as a result of increased solidification of fat.

Agitation of the ice cream mix during freezing contributes to aeration. Some coalescence of fat globules also is brought about by agitation and is desirable to a certain extent because it contributes to body and stiffness. Excessive agitation results in churning, so that relatively large particles of fat are dispersed through a frozen mixture that has a coarse texture and poor body because of poor distribution of fat.

Freshly made ice cream is rather soft; firmness increases with further freezing during the hardening phase, in which the mixture is not agitated. Hardening involves growth of existing crystals rather than formation of new ones; therefore, the development of a fine foam and fine crystal structure during initial freezing is important to the smoothness of the ultimate product.

Temperature fluctuation during continued storage of the frozen and hardened ice cream results in migration of water vapor from smaller ice crystals to larger ones. The result is decreased smoothness. Minimal temperature fluctuation, therefore, is desirable (Figure 13.4).

Effects of ingredients Small ice crystals, not too close together, result in a smooth texture in ice cream; therefore, factors that control crystal size should contribute to smoothness. Such factors include concentrations of nonfat milk solids and of fat that are sufficient to interfere with crystal growth. Sugar lowers the freezing point of the water, resulting in greater distances between crystals because of a

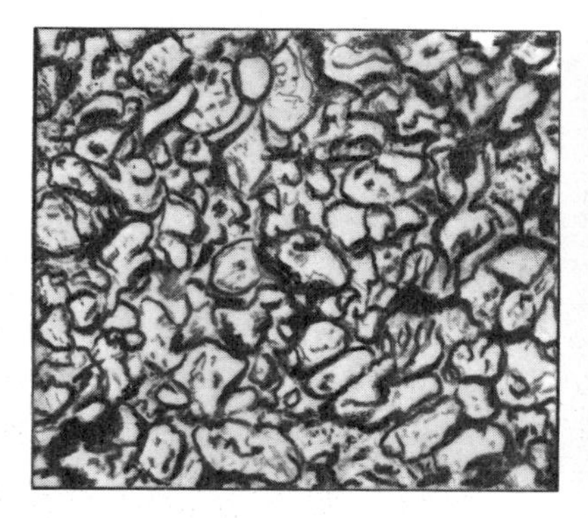

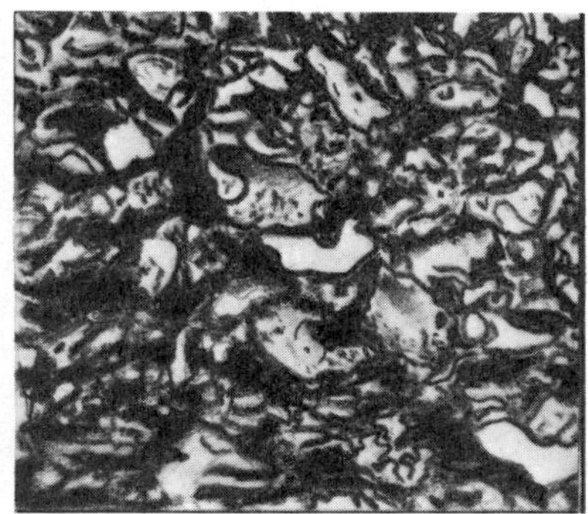

Figure 13.4. Micrographs of the same ice cream stored under different conditions: left, stored for 1 da at a constant low temperature; right, stored for 5 da at widely fluctuating temperatures. (Cole, 1932)

smaller extent of freezing at a given temperature. Other solids, both fat and nonfat, interfere mechanically with crystallization. Hydrocolloids (plant gums) are common ingredients of commercial frozen desserts. Gelatin is added to ice cream occasionally. Corn sirups may be added as sweetening agents. In addition to their sweetening effect, they contribute to smoothness. If used in excess, they impart a chewy quality.

Total solids content may have a profound effect on body. Low concentrations of solids result in excessive overrun and fluffiness, hence, poor body. Very high concentrations of solids, on the other hand, can cause gumminess.

High lactose concentrations promote the development of sandiness resulting from lactose crystallization. It is interesting that lactose crystallization is more likely to occur at temperatures above −18°C than at lower temperatures, possibly because of greater mobility of the lactose molecules at the higher temperatures.

Added emulsifiers contribute to the undesirable churning effect mentioned previously. This negative effect of emulsifiers seems rather strange, but it might reflect an initially fine dispersion of fat in the unfrozen mix so that the protein surrounding the globules is spread too thin. Addition of nonfat milk solids, on the other hand, contributes resistance to churning, as do certain salts.

Keeney and Kroger (1974) discuss ice cream further.

EXPERIMENTS

I. Fondant

Several principles that apply to fondant and to other crystalline candies are illustrated. The effects of cream of tartar, corn sirup, and fat on crystal size are studied, as well as the effects of beating candy at different temperatures.

A. *Basic formula and procedure*

Sugar, g	200
Distilled water, ml	118

1. Determine the boiling point reading in water for each

thermometer that is to be used. The final boiling point of the solutions in the experiment is 114°C; thus the endpoint is indicated by the thermometer reading that is 14° higher than that when the thermometer bulb is immersed in actively boiling water.

2. Put the ingredients in a 600-ml beaker or a pan having a diameter at the base of not more than 10 cm. Stir slightly but not enough to distribute sugar on the sides of the beaker or pan.
3. Place the beaker or pan on a small burner. Suspend the thermometer from a ring stand in such a way that its bulb is completely immersed in the center of the sugar solution. Covering the pan with a piece of aluminum foil pressed against the thermometer and sides of the pan for about 3 min permits steam to dissolve sugar crystals on the sides of the pan.
4. Try to regulate the heat so that the total cooking time is about 15 min. Remove the aluminum foil after the first 3 min of boiling. Cook to a boiling point of 114°C. Remember to correct the reading if necessary.
5. Transfer quickly to a 15-cm casserole in which a calibrated thermometer has been placed. Do not scrape the sirup into the casserole or permit it to drip. Do not move the thermometer in the solution or otherwise agitate the sirup during cooling. Allow to cool to 50°C. Note whether crystals form on the surface while the sirup is cooling.
6. Record the time and beat rapidly and vigorously with a wooden spoon until the mixture loses its gloss and becomes stiff. This might happen suddenly. Record time.
7. Immediately work the fondant vigorously with hands for 60 sec.
8. Store in a tightly covered container until microscopic examination is carried out.
9. Prepare a slide for microscopic observation by putting a drop of turpentine and a tiny grain of candy from center of the ball on a slide. Mix with a toothpick. Add a cover slip, and slide it back and forth to further disperse crystals. Observe under a magnification of about 200×. Move the slide about until a good, uncrowded field is found. Sketch

crystals from samples representing the different treatments.

B. Variations

1. Added substances. Make six fondants according to the basic formula and with the following additions:
   a. None (basic formula)
   b. 0.2 g cream of tartar
   c. 0.8 g cream of tartar
   d. 10 g white corn sirup
   e. 40 g white corn sirup
   f. 12 g butter or margarine
2. Added substances in fondants made with tap water. Make three fondants according to the basic formula but with tap water and the following additions:
   a. None (basic formula, tap water)
   b. 0.2 g cream of tartar
   c. 0.4 g cream of tartar
3. Beating temperature. Prepare three fondants according to the basic formula. Beat the sirups as follows:
   a. After cooling to 70°C
   b. After cooling to 50°C
   c. After cooling to 40°C

C. Results

Score color, consistency, smoothness, and taste, using appropriate scales. (See the Appendix.) Tabulate the scores, along with beating times and findings with regard to crystal size.

D. Questions

What would be the result of not removing the aluminum foil in step 4 of the basic procedure? Why is turpentine used rather than water or glycerol in preparation of the slides of sugar crystals? Compare the quality of fondants made without an interfering agent, with an optimum amount, and with an excessive amount. What are the comparative merits of the interfering agents used? Explain the mechanisms by which the different interfering agents exert their effect. What influence did the alkalinity of the water have on the effect of cream of tartar? Compare the size of the crystals in the fondants beaten at different temperatures and explain the differences. What would be wrong with comparing the crystallization of sucrose and glucose (dextrose) from solutions concentrated to a boiling point of 114°C? How might a more reasonable comparison be made?

II. Ice Cream

Liquids, stabilizers, and freezing rates are varied in the ice cream experiments. The experiments were written with recognition of the widespread availability and use of electric ice cream freezers. If they are not available, the experiments can be modified for use in hand-cranked freezers or even in freezer trays of refrigerators.

A. *Basic formula and procedure*

Whipping cream, ml or g	236
Milk, ml or g	236
Sugar, g	62
Vanilla, ml	5

1. Scald freezer cans and dashers before use.
2. Combine the ingredients. Stir to dissolve the sugar.
3. Pour mixture into freezer can of premeasured inside depth; measure the distance from the top of the can down to the surface of the unfrozen mixture. Calculate the depth of the mix by difference.
4. Pack the space between the freezer can and the outer container with an 8:1 ice:salt mixture. Freeze according to the freezer instructions.
5. Remove the dasher, smooth the surface of the frozen mixture, and measure from the top of the can down to the surface of the frozen mixture. Obtain the new depth by difference.
6. Calculate % overrun: $100\left(\frac{\text{final depth} - \text{initial depth}}{\text{initial depth}}\right)$.

Although the percentage increase in depth is calculated, overrun represents percentage increase in volume because of the relationship between depth and volume of a cylinder.

B. Variations

1. Fat content. Make three ice creams, varying the proportions of whipping cream and milk as follows:
   a. 118 ml whipping cream and 354 ml milk
   b. 236 ml whipping cream and 236 ml milk (basic formula)
   c. 354 ml whipping cream and 118 ml milk
2. Stabilizing ingredients. Make five ice creams, making the following additions:
   a. None (basic formula).
   b. 2 g gelatin, hydrated in about 60 ml of the milk and then dispersed by heating. Cooling can be speeded by dissolving the sugar in the hot gelatin dispersion.
   c. 48 g whole egg, incorporated into a stirred custard made from the milk, sugar, and egg. Cool before adding cream and vanilla.
   d. The same as b except use the cream-milk proportions of B1a.
   e. The same as c except use the cream-milk proportions of B1a.
3. Market forms of milk in ice milks. Make six ice milks, using the following as the total liquid in the formula:
   a. 472 ml fluid whole milk
   b. 472 ml evaporated milk, diluted 1:1
   c. 472 ml evaporated milk, undiluted
   d. 472 ml nonfat dry milk (NFDM), reconstituted according to the package directions
   e. 472 ml NFDM, reconstituted to double strength
   f. 472 ml NFDM, reconstituted to triple strength
4. Freezing rate. Make three ice creams, using the ice-salt mixtures listed below. Replenish as necessary. Record freezing times.
   a. 12:1
   b. 8:1 (basic procedure)
   c. 4:1

C. Results

Score smoothness, body, and flavor, using appropriate scales. (See the Appendix). Tabulate the scores, along with % overrun and, for B4 only, freezing time.

D. Questions

Explain any differences in overrun observed. Did overrun affect palatability? Explain any differences in smoothness. What factor(s) most affected body? What was the effect of the ice:salt ratio on freezing rate? What factor not varied in these experiments also would have affected the freezing rate?

SUGGESTED EXERCISES

1. Make the Benedict's test on a sucrose solution (200 g sucrose and 118 ml distilled water), removing a drop of solution when the sucrose is just dissolved, another at 100°C, and another after boiling the solution to 114°C. Repeat with a solution to which 0.8 g cream of tartar is added along with the sugar. Test an unheated solution of glucose. Explain the results. Why is only the unheated solution of glucose tested?
2. Make a series of fondants, using 200 g sugar in each and varying the amount of water from 59 to 236 ml. Weigh each finished fondant and calculate its total water content (fondant weight − weight of sugar used). Does the initial amount of water determine the final amount of water? If not, why not? (Note: the weight of finished fondant also may be estimated by weighing the concentrated sirup in the pan on a top-loading balance and subtracting the weight of the empty pan.)
3. Compare the properties of two fondants containing 0.8 g cream of tartar/200 g sucrose, one heated as rapidly as possible to 114°C and the other heated very slowly to 114°C. Explain the results.
4. Make caramels from a modified formula containing no corn sirup. (Formula: 200 g sucrose; 14 g margarine; 127 ml milk; 110 ml half-and-half. Stir constantly while heating to a boiling point of 120°C.) Repeat with glucose (dextrose). Note the color of the glucose

mixture at a boiling point of 120°C but continue heating to a boiling point of 138°C. What would happen if an attempt were made to make caramels or any other candy of lactose?

5. Repeat IIB3, adding 2 g gelatin.
6. Make ice creams to which varying levels of dried whey are added.
7. Develop ice cream or ice milk formulas for special dietary needs, for example, reinforced with extra milk solids; made with lactase-treated milk; or made with a highly polyunsaturated oil, beaten into NFDM that has been reconstituted to triple strength.

REFERENCES

Aurand, L.W. and Woods, A.E. 1973. "Food Chemistry." Avi Publ. Co., Westport, CT.

Birch, G.G. and Shallenberger, R.S. 1973. Configuration, conformation and the properties of food sugars. In "Molecular Structure and Function of Food Carbohydrate," Eds. Birch, G.G. and Green, L.F. Applied Science Publishers, Ltd., London

Cole, W.C. 1932. A microscopic study of ice cream texture. J. Dairy Sci. 15: 421.

Donnelly, B.J., Fruin, J.C. and Scallet, B.L. 1973. Reactions of oligosaccharides. III. Hygroscopic properties. Cereal Chem. 50: 512.

Doty, T.E. 1976. Fructose sweetness: a new dimension. Cereal Foods World 21: 62.

Hodge, J.E., Rendleman, J.A. and Nelson, E.C. 1972. Useful properties of maltose. Cereal Sci. Today 17: 180.

Keeney, P.G. and Kroger, M. 1974. Frozen dairy products. In "Fundamentals of Dairy Chemistry," 2nd ed., Eds. Webb, B.H., Johnson, A.H. and Alford, J.A. Avi Publ. Co., Westport, Ct.

Saussele, H. Jr., Ziegler, H.F. and Weideman, J.H. 1976. High fructose corn syrups for bakery applications. Baker's Digest 50(1): 32.

Shallenberger, R.S. 1963. Hydrogen bonding and the varying sweetness of the sugars. J. Food Sci. 28: 584.

Shallenberger, R.S., Acree, T.E. and Guild, W.E. 1965. Configuration, conformation, and sweetness of hexose anomers. J. Food Sci. 30: 560.

Volpe, T. and Meres, C. 1976. Use of high fructose syrups in white layer cake. Baker's Digest 50(2): 38.

II. FOOD EXPERIMENTATION

14. Approaching the Experiment

An independent problem is an integral portion of many advanced courses in food science. The student, having become accustomed to following instructions in conducting experiments, ultimately is expected to independently select and define a problem, plan and conduct the study, and write a report. This chapter deals with the preparations for conducting an individual study. The principles involved apply to the gamut of studies, ranging from a problem constituting only a small portion of a course to a really extensive project.

SELECTING AND DEFINING THE PROBLEM

Ideally a student arrives at the stage of selecting a problem for independent study having been well exposed to the experimental approach and to the food science literature. These previous academic experiences, plus everyday encounters with practical food-related questions, may have resulted in a store of ideas awaiting just such an opportunity. More often the unanswered questions are submerged and the ideas for first problems are elusive.

The search for ideas should not be left until the last minute. A leisurely browsing through the shelves of a food market may produce ideas involving use of newly available ingredients or new uses and/or treatments for established products. Information concerning current developments, obtained through the news media and popular magazines, frequently leads to ideas. Conversation with friends and family members may expose questions concerning ingredients and methodology that could be answered through experimentation. Product development for modified diets provides opportunities that are of special interest to future dietitians. Reading research reports and textbooks should make one aware of gaps in the available information and of apparent discrepancies or inconsistencies in accepted theory.

Not every independent problem makes an original contribution to knowledge, but every independent problem should be worth doing. Careful selection of a problem enhances its value as a learning experience and contributes to the personal satisfaction derived from conducting the study.

Successful selection of a problem involves consideration of the available resources, including time, and of the likelihood that the question that is posed really can be answered, or the problem solved, under the prevailing conditions. In a class situation it is not unusual for each student to be asked for more than one problem suggestion because the instructor must take a broader view than the students, considering the total space, equipment, and financial resources available, as well as the total learning experience. Each student learns from observing others as well as from conducting an individual study; therefore, the instructor is concerned with the range of experiences provided through all of the problems undertaken within a group.

Further definition of a problem usually is necessary after the initial selection. Suppose the functional properties of different market forms of egg are of interest. This obviously is an inclusive topic. Further definition might result in a decision to study angel food cake volume as affected by the use of fresh, frozen, or dried egg white. Not only are the market forms now specified, but so also is the functional property (foaming) that is to be observed, as well as the major criterion of foaming behavior that is to be used. If an extensive project involving a large array of functional properties and many criteria of functional performance were being undertaken, careful definition of each phase of the project would be important.

REVIEWING THE LITERATURE

A review of pertinent literature frequently aids in problem definition and in development of the specific plan. Whether or not a need for such aid is felt, a literature review *prior to* undertaking the work is important and frequently prevents regrets later. If the problem has been selected and defined in advance of the literature search, the review should provide information as to methodology used in previous related studies. Such information should be helpful in the development of a specific plan.

Early establishment of an orderly procedure for conducting a literature review is of lasting usefulness. Textbooks, with their usual lists of references, provide a convenient starting point for reading on a given subject. Review articles such as those found in *Advances in Food Research* have rather inclusive bibliographies. Research articles, however, constitute the primary source of information and should be used as extensively as possible. Indexes and abstract journals are used for locating pertinent references. Each such publication provides a list of the periodicals indexed or abstracted.

Some of the most pertinent indexes and abstract journals will be described briefly. The *Biological and Agricultural Index* is a cumulative subject index to periodicals in the fields of biology and agriculture and related sciences. Entries are alphabetical by subject. The *Applied Science and Technology Index,* as its name suggests, has a technological emphasis in its listings. For example, journals such as *Food Technology and Food Engineering* are included in this index, but the *Journal of Food Science* is not. *Current Contents* is not exactly an index but neither is it an abstract journal. Published weekly, it consists entirely of tables of contents. Since its initiation in 1958, it has expanded into five separate titles. The one that covers food science periodicals currently is titled *Current Contents: Agriculture, Biology and Environmental Sciences.*

Food Science and Technology Abstracts is a particularly useful abstract journal. Published monthly since 1969, it comprises 19 sections of abstracts, some devoted to specific food groups, such as cereals and bakery products, and some to other food-related categories, such as food microbiology and basic food science. Each issue has an author index and a detailed subject index. In addition, there is an annual cumulative index. Depending on the subject of interest, other journals that may be helpful are *Biological Abstracts, Chemical Abstracts,* and *CRC Critical Reviews in Food Science and Nutrition.* One of the many features of the *Journal of the American Dietetic Association* each month is a group of abstracts of pertinent articles from a variety of journals. This collection is particularly useful in a situation where library resources are limited.

The most recent indexing and abstracting issues cannot cover the most recent research literature. Therefore, it is necessary also to go through recent issues of the research journals.

Effective use of indexing and abstracting journals, as well as of the index of a single issue of a research journal, involves some skill in selecting subject entries to look under. In studying tables of contents, recognition of pertinence in rather obscure titles also is a skill that may be developed with experience and increasing knowledge of the subject matter.

The actual mechanics of taking notes from the research articles can be handled with varying degrees of efficiency. Separate index cards for individual references are convenient for note taking because they can be arranged as needed when the review actually is being written. The format for a single article should begin with the complete reference. Consistently writing the reference according to the exact bibliographical form to be used in the final report helps one become accustomed to that form and also helps prevent a necessary return to the library to fill in an accidental gap. (See the bibliographical form for the references used in this book.)

Effective use of the time spent in the library involves exercise of judgment as to when to read for detail and when to scan. Scanning ability is developed with practice and saves considerable time. If the reader, having scanned an article, is uncertain as to its pertinence to the study at hand, a notation on the reference card as to the type of information included will prevent loss of the reference, while requiring little time. An obviously pertinent article will be read carefully, and adequately detailed notes will be taken. It is advantageous to make notes in such a way as to expedite avoidance of plagiarism in the writing of the review. Direct quotations should be either avoided during note taking or identified clearly with quotation marks.

The total learning experience in reviewing the literature is enhanced if the habit of reading critically is developed early. Raising questions concerning experimental plans, specific procedures used, presentation and interpretation of data, and even writing style is mental exercise that can benefit one's own work.

Further discussion of the review of literature is presented in *Foundations of Home Economics Research* (Compton and Hall, 1972).

PLANNING THE EXPERIMENT

Before proceeding with a plan, the investigator should write the specific question that is to be answered if this has not been done previously. With the literature review completed, a final judgment now is made as to whether and how the question can be answered. A written plan is prepared.

The written plan includes first a meaningful title, and then a justification for the proposed study. The justification frequently is based on the related literature or on the lack thereof and leads into a statement of the objective(s). It is important that the objective(s) be attainable. The planned procedure for meeting the objective(s) follows and includes the experimental design, details of the procedures to be used, and copies of any evaluation forms to be utilized in collection of data.

The experimental design includes the plan for collecting data—the amount of work to be done each day, extent of replication planned, and randomization to be involved. The procedures include the formulas to be used, the sources of materials, the controls to be applied to the procurement and use of materials, and the standardization and controls to be applied to any food preparation and evaluative processes.

Some preliminary work usually is necessary before all decisions pertaining to the plan can be made. The preliminary work not only provides information as to whether the plan is realistic from the standpoint of time, but it also makes the worker aware of decisions that must be made in advance of the actual study. Preweighing of staple supplies contributes to efficiency as well as to control of experimental conditions but requires a plan for storage of the preweighed supplies.

Conscious choices sometimes must be made between alternative control measures. For example, suppose that beaten egg white is an ingredient in an experimental product and that the only experimental variable involving the egg white is the market form. Obviously other things about the egg white, including the extent of beating, must be controlled. Control of the extent of beating conceivably could be achieved through the use of an electric beater for a predetermined constant time at a constant rate. However, the egg white of different market forms might not respond similarly to that controlled beating procedure and perhaps it would be preferable to beat the egg whites to a specific stage. The criterion then would have to be selected. Would the egg white be beaten to a specific stiffness? If so, how would that stage be determined? Or would the egg white be beaten to a certain specific gravity? Such questions should be anticipated so that important decisions will not have to be made under stress after the study begins.

If any special forms are to be used for collecting data, as stated previously, they are included in the plan. A score card for sensory evaluation is an example. On occasion, a diagram showing the sampling plan for sensory and/or objective testing is pertinent. Codes for samples to be subjected to sensory evaluation can be worked out in advance. Treatment of the data, such as the statistical analysis, if any, should be predetermined and stated in the plan. At the end of the plan the references cited in the justification and/or procedure are listed.

In a class situation, a market order usually must be included with the plan. The order should clearly differentiate between those staple supplies that are to be obtained only in the beginning and the perishables that are needed each time. Equipment that is needed also should be ordered. It usually is not safe to make any assumptions as to availability, even of equipment that normally is in the laboratory.

Considerably more detail concerning experimental plans and other aspects of planning and conducting experiments may be found in Harrison's presentation in *Food Theory and Applications* (1972).

SUGGESTED EXERCISES

1. List some questions about foods that could be answered through experimental work. Work out a general plan for answering each experimentally.
2. Compile a bibliography on a selected subject, using a serial index or

abstract journal. Locate at least one of the articles and prepare an original abstract.

REFERENCES

Compton, N. H. and Hall, O. A. 1972. "Foundations of Home Economics Research." Burgess Publ. Co., Minneapolis.

Harrison, D. L. 1972. Planning and conducting experiments. In "Food Theory and Applications," Eds. Paul, P. C. and Palmer, H. H. John Wiley and Sons, Inc., New York.

15. Evaluating Food by Sensory Methods

Sensory evaluation of food or evaluation of food quality by a panel of judges is essential to most food experiments because it answers the important questions of how a food looks, smells, feels, and tastes. The Sensory Division of the Institute of Food Technologists (Prell, 1976) defines sensory evaluation as: "a scientific discipline used to evoke, measure, analyze, and interpret reactions to those characteristics of foods and materials as they are perceived by the senses of sight, taste, touch, and hearing."[1]

The answers to the questions posed in sensory evaluation may seem easy, but they are actually difficult because the answers depend on human judgment, which is individual and not always consistent. Many types of sensory tests are used in food research. Panelists may be asked to discriminate among samples, describe or score the quality of a product, rate the acceptability of a product, or describe their preference for a product. The test selected will depend on the objectives of the investigation. In this chapter some of the principles of sensory testing

[1]Reprinted from *Food Technology* Vol. 30, No. 11, p. 40, 1976, Copyright © by Institute of Food Technologists.

are discussed. A fuller treatment of the subject, including the physiological and psychological aspects, is available in *Principles of Sensory Evaluation of Food* (Amerine et al., 1965). Basic procedures for sensory evaluation and interpretation are outlined in *The Manual of Sensory Testing Methods* (ASTM, 1968). Sensory methods frequently are supplemented with chemical and physical methods of evaluation of food quality such as those described in Chapter 16.

PREPARATION AND SAMPLING

Only food that is prepared by a method that can be duplicated is worth the effort of scoring. When, and only when, the conditions of preparation are controlled carefully and defined can differences in quality be attributed to known variables. Careful sampling of the food also is necessary. A well-mixed, homogeneous sample is ideal, but is possible with only a few foods such as applesauce, fruit juices, and mashed potatoes. If canned or frozen foods are being sampled, it is desirable to mix the contents of several cans or packages before sampling for preparation.

Special care is necessary to obtain reliable results with nonhomogeneous materials. For example, one muscle is selected from a slice of meat for scoring. If possible, a judge is given a sample from the same area of the same numbered slice from each roast in a series. Paired roasts are used if possible to minimize animal variation. A judge should be given a sample from the same position in each cake or loaf of bread since these products vary from end to end. Outside slices of meat or baked products are not included as samples. All samples should be equally fresh when judged. It is sometimes possible to keep products baked at different times in satisfactory condition by freezing.

Preparing Samples for Testing

It is important that all samples in a series be at the same temperature when judged. In order to obtain meaningful results, it is logical that the food should be tasted at the customary serving temperature. However, if the usual temperature is very hot or very cold a more moderate temperature is appropriate, since the sense of taste is rendered less acute by extreme temperatures. Temperatures of not lower than 7°C or higher than 77°C should be used (ASTM, 1968). Serving containers for all samples in a series should be identical in size, color, and shape. White or clear containers are preferred so that the color of the food will show clearly. Glass is used most frequently. Plastic or disposable containers may be used if they do not impart flavor to the sample. Sufficient food is required to provide each judge with at least two bites or sips of each sample. Samples are best identified by a code rather than a descriptive name. Codes such as A, B, and C or 1, 2, and 3 are undesirable because the first of a series suggests first choice to the judges. Randomly selected letters, three-digit numbers, geometric shapes, colors, or symbols can be used. When one series of samples is to be evaluated several times, judging will be more accurate if the code and order of presentation are altered each time. In some experiments, judges tend to score the first samples presented higher than others. This effect of position can be minimized by having the judges evaluate an additional sample before the experimental series. In other experiments, judges may base their evaluations on previously presented samples. A good sample preceding a poor sample may result in a lower than normal score for the poor sample, whereas the opposite order of presentation results in a higher than normal score for the good sample (Amerine et al., 1965). Effects such as this, known as the *contrast effect,* can be minimized in experiments with a small number of samples by using every possible order of sample presentation an equal number of times. In larger experiments, randomizing the order of presentation will help to minimize such psychological effects (Larmond, 1973).

Number of Samples

The number of samples that can be judged efficiently in one session is limited. Recommendations for the exact number differ as the type of food and type of test vary (ASTM, 1968; Prell, 1976). More samples can be scored in color or texture evaluation than in flavor evaluation. More bland than strong-flavored samples can be evaluated in one session. Fewer samples can be evaluated when complex score cards are used than when simple scoring systems are used. Judges should not be asked to score so many samples that they will become fatigued and therefore inefficient. As a general guideline, six single samples or pairs of samples or four triangles constitute the maximum number that should be presented to a trained panel for evaluation. Inexperienced consumer panels should be given fewer samples (ASTM, 1968).

ENVIRONMENT FOR TESTING

The environment is an important factor in judging food and should be without distractions so that the judges can concentrate. Judgments should be made independently; therefore, some form of separation of judges is desirable. The best way to satisfy this factor is to seat each judge in an individual compartment, which is properly lighted, painted a neutral color such as white or gray, and temperature controlled. Such an arrangement, which may include sliding doors or a hatch for sample presentation, usually is available only in large food research facilities. A similar effect is accomplished by placing tables against a wall and

Figure 15.1. Portable sensory evaluation booth. (Photo—University of Tennessee, Knoxville.)

separating them with panels about 1.2 m high, or by installing portable carrels on laboratory tables or counters. The carrels should be 45 to 50 cm high. Such an arrangement is shown in Figure 15.1. A room other than the preparation room should be used if available. The room should be well ventilated and free from foreign odors. Convenient, comfortable seating for the judges is essential. Talking during judging is avoided since judges may be influenced by the opinions of others. However, discussion after scoring is desirable because it may lead to better performances by the judges and help to maintain their enthusiasm and interest. The best time for judging is midmorning or midafternoon, when the judges are neither too well fed nor too hungry. Each panelist should be provided with a glass of room-temperature water for rinsing between samples. Unsalted crackers or apple wedges may be used to remove flavors from the mouth. Cold water is avoided because it may dull the sense of taste.

SELECTION AND TRAINING OF JUDGES

Selection of a Panel

Procedures used to select a panel will depend on the type of sensory testing to be done as well as on the circumstances. If few persons are available, it may be necessary to accept anyone who will serve. It is, of course, better to select the most able judges from a larger group. In any case, it is important that the judges be available for the entire period of the experiment, interested in the problem, and willing to serve, and that they not dislike the food to be tasted. They can be either men or women but must be in good health. People with colds are unable to score foods accurately. Judges should be asked to refrain from eating a meal for at least 60 min and from smoking, snacking, or chewing gum for 20 min before the test session (ASTM, 1968).

For discrimination and descriptive testing, some workers have selected judges on the basis of their taste thresholds. However, the results of the tests have not been especially useful in predicting sensitivity to the taste of a food (Boggs and Hanson, 1949). It has been suggested that the ability to recognize the four basic tastes—sweet, sour, salt, and bitter—should be a minimum requirement in flavor work

(Martin, 1973). The ability to recognize duplicate samples in a triangle test is considered important (Martin, 1973) and has been used as a basis for selection of panelists for studies on ground beef patties (Cross et al., 1976).

Another practical method of selecting judges is to present actual test samples to the potential judges, replicating the samples presented to each judge several times. The consistency of each judge is then studied, and those who perform satisfactorily are retained for the panel. Since judges may vary in their ability to score different foods, it is best to select a panel for each food investigated.

Panelists for acceptance-preference testing are selected to represent some segment of the population. Thus, socioeconomic, cultural, and geographic factors rather than ability are used as the basis for selection of the judges.

Size of Panel

The size of the panel is a quesion important to most experimental-foods students and to most research workers. Availability of qualified panelists and differences in those panelists as well as the inherent variability in the product should influence the size of the panel. It should, of course, be as large as possible in order to reduce the experimental error, thus improving the reliability of the results. Several extra judges usually are carried on a panel because of the inevitability of some absences. A minimum of five panelists has been suggested for descriptive and discrimination testing, whereas at least 50 to 100 should be used for acceptance-preference testing with consumer panels (ASTM, 1968). The lack of experience of the consumer panel increases the experimental error and therefore necessitates a large number of judges.

Training the Judges

Judges for discrimination and descriptive testing may be trained before the experiment is started. Important functions of the training period are to show the judges that effort and concentration are essential in evaluation of foods and to develop a common understanding of terminology and procedures among the panelists. The training period may be started by explaining enough of the problem to arouse the interest of the judges. With some problems, however, such as those involving off flavors, it may be necessary to withhold some information to avoid causing prejudice. Interest also can be stimulated by asking the judges to help design the score card, but, in any case, clear instructions on the scoring method are essential. Discussion among the judges and the investigator concerning standards for the food to be tested are desirable. Another helpful device in training judges is to give them a sample that can be labeled to serve as a standard, or coded and presented with unknown samples. If coded, its identity can be revealed later and the findings discussed. The availability of such a sample depends, of course, on the nature of the problem. If it is possible to provide a valid standard throughout the experiment as well as the training period, judging will be improved. However, if its quality is not constant, a standard will be misleading. Judges can be helped by scoring duplicate samples and later learning their identities. Wide variations in the quality of the food should be provided during the training period to give experience in using the extremes of the score card. It can be seen from this discussion that some steps in selecting and training a panel can be combined. Certainly some members of the panel can be omitted on the basis of their performance during the training period if necessary.

TYPES OF TESTS

Sensory tests may be divided into three groups on the basis of the type of information they provide. The three types are discrimination, descriptive, and acceptance-preference tests.

Discrimination Tests

Discrimination tests are used to determine if there is a difference between or among samples. The methods of discrimination testing commonly used are the paired comparison, the triangle, and the duo-trio tests. Tests of this type are especially useful in selecting and training panelists and in quality control. They are relatively easy for the judges as they require only a short memory of food quality. One disadvantage of these methods is that only two samples are compared at a time. Thus, it may be necessary to pair every sample in a series with every other sample. The number of samples that must be evaluated by each judge may be as high as in other methods. A greater quantity of each sample must be prepared. Another disadvantage is that, although these methods indicate whether there is a difference between two samples and may give an indication of the direction of that difference, they do not reveal the degree of difference unless additional questions are asked.

For a *paired comparison test,* two samples are presented together and the judge is asked whether there is a difference in the samples with regard to a specific characteristic. In a directional paired comparison test, the panelist may be asked to identify the sweeter piece of cake or the more tender piece of meat. The judges may be asked to indicate a choice even though they feel there is no difference. In a paired comparison test, the judge has a 50% chance of selecting either sample. The significance of the test results can be determined by statistical analysis or by the use of special tables based on the use of appropriate statistical tests (Roessler et al., 1978).

In the *triangle test,* three samples are presented, two of which are duplicates. The panelist is asked to identify the "odd" sample. In a modified triangle test, the judge may be asked to indicate whether the odd sample or the duplicate samples have a distinguishing feature to a more pronounced degree. The order of sample presentation should be varied because judges tend to choose the central sample of the three as the odd sample (Harries, 1956). The triangle test is especially useful when differences between samples are small. Since the judge is asked to select one sample from a set of three, the chance of guessing correctly is 33.3%. Tables are available for determining the statistical significance of the results (Amerine et al., 1965; ASTM, 1968; Larmond, 1977).

The *duo-trio test* also involves three samples, but the judge is informed that the first sample presented is a control. The panelist is asked to indicate which of the other two samples differs from the control. Therefore, the chance of guessing correctly is 50%. Tables used for deterimining statistical significance are available (Amerine et al., 1965; ASTM, 1968; Larmond, 1977). There is some evidence that the paired comparison and triangle tests are superior to the duo-trio test (Gridgeman, 1955).

Taste threshold tests are also discrimination tests. These tests are conducted to determine the lowest concentration of a substance that can be detected (absolute threshold) or the lowest concentration of a substance required for identification of the substance or taste (recognition threshold). Threshold tests may involve the evaluation of taste acuity for the four basic tastes or variations in concentrations of some constituent of a food. Tests of taste threshold have been made by adding small amounts of evaporated milk to whole milk; salt, sugar, or lemon juice to canned foods; or a sample that has deteriorated to a fresh sample.

Descriptive Tests

The objective of descriptive testing is to characterize or compare samples with respect to one or more specific characteristics. Tests in this category include ranking, scoring, or grading, texture or flavor profiling, quantitative descriptive analysis, and magnitude estimation.

Judges participating in a *ranking test* are asked to rank a series of samples in increasing or decreasing order of a specified characteristic such as a flavor, an odor, a color, or a textural parameter. Numerical values are assigned to the results, with the sample representing the highest degree of the attribute receiving the highest value. Results can be evaluated statistically by consulting tables presented by Kramer and coworkers (1974). This test is especially useful when an entire series of samples is available at one time and is to be ranked for one characteristic. It is not generally satisfactory when products available at different times are to be compared, when there is little difference among samples, or when information is desired on several characteristics at the same time. Ranking gives information regarding the direction of differences but does not give a measure of the degree of difference.

A wide variety of *scoring tests* is used in the evaluation of food quality. Tests vary with the food and with the investigator. Judges are asked to rate products on graphic scales, on scales consisting of a series of numbers representing a range of low to high intensity of a characteristic, or on scales consisting of a series of descriptive words or statements representing successive levels of a characteristic. Scales consisting of numbers may require more intensive training of judges to develop an understanding of the scale. The use of understandable adjectives may make judging relatively easy and precise. For example, juiciness, a textural attribute of meat, might be evaluated on a scale consisting of the terms extremely juicy, moderately juicy, slightly juicy, slightly dry, moderately dry, and extremely dry. To facilitate tabulation and averaging of the scores, a series of descriptive terms is assigned numerical values. The number 1 is assigned to the least desirable and/or least intense term and, in the case of the juiciness scale, 6 would be assigned to the term extremely juicy. These numbers are not included on the score card so that the judges will think in terms of the descriptive adjectives instead of the numerical scores. Similar scales can be devised for other quality attributes. Sample scales are given in the Appendix.

Verbal scales also can be used to evaluate a product in terms of standards for that product. Such a scale might consist of the terms superior, above average, average, below average, and poor. Verbal scales should have five to ten points. Less than five results in limited differentiation among samples. More than ten points may make it difficult for judges to make decisions.

Several problems may be encountered in the development of a score card. Extensive preliminary work on the part of the research worker and the judging panel may be necessary for the development of a score card such as the one in Table 15.1. Various characteristics are listed and points assigned depending on the investigator's estimation of the relative importance of various attributes and defects to the quality of the product. Training of the judges using such a score card would be necessary so that the panel would have a common understanding of the terminology.

In order to obtain useful information about the quality of test products, scales should not reflect just the opinions of the judges. Terms such as desirable, acceptable, and undesirable do only that. These terms are appropriate for acceptance-preference testing but not for descriptive sensory evaluation.

Another way to test the quality of food is the *flavor profile* method, which attempts to analyze and define flavor (Caul, 1957). Flavor factors that are perceived are called *character notes.* A list of these notes is made by each panel member during preliminary work on the food being investigated. The lists are then compared, and agreement is reached on which notes are to be used in future work. The intensity of each character note and the amplitude of the overall aroma and flavor are rated. For example, the character notes listed for one sample of catsup were as follows: sweet, salt, molasses, sour, cooked tomato, and spice complex. A similar technique has been used in the evaluation of the textural characteristics of food (Brandt et al., 1963; Civille and Liska, 1975). A *texture profile* panel evaluates the mechanical, geometrical, fat, and moisture properties of food. Mechanical properties evaluated include hardness, fracturability, springiness, cohesiveness, chewiness,

Table 15.1 Score card for cakes

Qualities	Values used for statistical analysis[a]	Qualities	Values used for statistical analysis[a]
Size of cells		Tenderness	
Large	3	Crumbly	3
Medium	10	Very tender	8
Small	10	Tender	10
Very fine	3	Slightly tender	8
Compact	1	Tough	3
Distribution of cells		Moisture	
Uniform	6	Moist	10
Irregular	4	Dry	5
Tunneled	2	Wet	2
Crumb characteristics		Flavor	
Velvety	10	Well-balanced	10
Slightly harsh	8	Sweet	5
Very harsh	2	Salt	2
		Bitter	2

[a]These values did not appear on the judges' score cards.
SOURCE: Hunter, Briant, and Personius, 1950. Reprinted with permission of the New York State College of Human Ecology, a statutory college of the State University, Cornell University, Ithaca, NY.

gumminess, adhesiveness, and viscosity (Szczesniak, 1975). Foods representing varying degrees of the mechanical properties other than springiness and cohesiveness are described by Szczesniak and coworkers (1963). The definitions of these terms are useful in designing scoring systems for texture evaluations. The profile techniques have been used commercially for comparing competing products, for quality control, and for product development. They are not as generally useful in food research as some of the other methods because they require a highly trained panel and because the results are difficult to interpret or analyze statistically. Results usually are presented graphically (Prell, 1976).

Quantitative descriptive analysis involves the identification of the sensory characteristics of a product. Each identified characteristic is rated on a numerical scale. Data can be analyzed statistically and presented graphically for easy understanding. The technique is most useful in product development (Stone et al., 1974).

A recently described scaling technique is *magnitude estimation.* In this technique, which is a type of ratio scaling, panelists are asked to develop the numerical scale they will use in their own evaluations. For example, if asked to evaluate the sweetness of orange juice, a judge may assign a score of 50 to the first sample. If the judge feels that the second sample is twice as sweet, a score of 100 is assigned; if it is half as sweet, a score of 25 is assigned. The ratio of the scores assigned, not the scores, is the important result. This method obviously gives information that cannot be obtained with more traditional scoring techniques. It is useful in evaluating the feasibility of substituting one ingredient such as a new sweetener for the presently used ingredient (Moskowitz, 1974).

A sensory test that does not seem to fit into the classifications given is the method of *chew counts* suggested by Lowe (1949), in which the judge counts the chews required before a piece of meat of standard size disappears from the mouth without conscious swallowing. This method sometimes is used in addition to a grading chart for meat.

Acceptance-Preference Tests

Acceptance testing is used to determine whether or not a product will be used by consumers, whereas preference testing is used to determine whether or not the panelists like the sample(s). The two types of evaluation may appear to be identical. However, it is possible for a judge to show a strong preference for a sample, but not use it or accept it for reasons other than its likeability. Large panels are used in this type of sensory evaluation and are often called consumer panels because untrained, inexperienced judges are used.

The most commonly used evaluation technique for measuring food preferences is the *hedonic scale.* The word *hedonic* is defined as "pertaining to, or consisting in, pleasure." The phrases used in a nine-point hedonic scale are as follows: like extremely, like very much, like moderately, like slightly, neither like nor dislike, dislike slightly, dislike moderately, dislike very much, and dislike extremely. In addition to determining the likeability or preference for a single sample, difference in responses to foods can be detected with the hedonic scale. It was

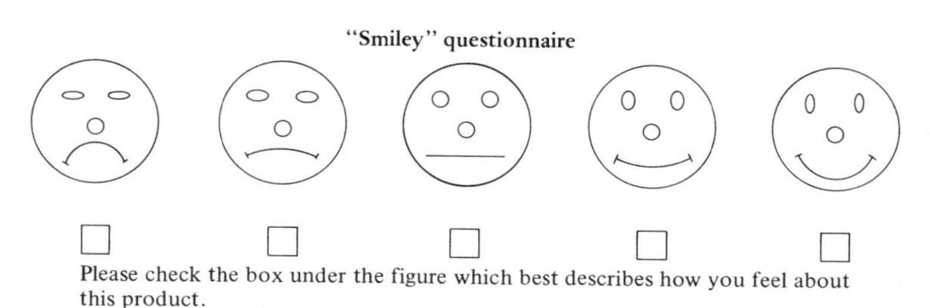

Figure 15.2. Facial hedonic scale developed by the Continental Can Company. (Ellis, 1964)

originated and has been used extensively by the Quartermaster Food and Container Institute of the Armed Forces.

Although the verbal hedonic scale just described has been used frequently, Jones and Thurstone (1955) found that misinterpretations of the term "dislike moderately" often occurred. A modification of the hedonic scale was introduced to minimize misinterpretations of terms. The modification, the *facial hedonic scale,* is useful with young children and others with limited reading ability. It consists of five, seven, or nine faces depicting varying degrees of pleasure and displeasure. One example of a facial hedonic scale is shown in Figure 15.2. Many variations of this basic idea have also been used (Ellis, 1968).

The hedonic scale also may be used for having panelists indicate their like or dislike for a specific characteristic of a sample such as flavor, color, or a textural attribute. It is extremely important to remember that such ratings should be interpreted in terms of likes and dislikes rather than changes in the intensity of the characteristic evaluated.

Schutz (1965) developed a nine-point *food action or attitude rating scale* to determine what actions an individual expects to take with regard to a food product. The scale, which is a measure of overall food acceptance, is shown in Table 15.2. The FACT test, as the food action rating scale is often called, is more sensitive than the hedonic scale. This probably is because panelists tend to be more realistic when evaluating or predicting actions (Schutz, 1965). However, this method should not be used to replace the hedonic scale but should be used to supplement it.

Table 15.2 Food attitude rating form for FACT method

Name ______	Dept.	____	Booth # ____	Date ____
Code	____	____	____	____
I would eat this every opportunity I had.	____	____	____	____
I would eat this very often.	____	____	____	____
I would frequently eat this.	____	____	____	____
I like this and would eat it now and then.	____	____	____	____
I would eat this if available but would not go out of my way.	____	____	____	____
I don't like it but would eat it on an occasion.	____	____	____	____
I would hardly ever eat this.	____	____	____	____
I would eat this only if there were no other food choices.	____	____	____	____
I would eat this if I were forced to.	____	____	____	____

Comments:

Code____ ________________________________

Code____ ________________________________

Code____ ________________________________

Code____ ________________________________

SOURCE: Schutz, 1965. Reprinted from *Journal of Food Science* Vol. 30, No. 2, pp. 365–370, 1965.

Several discrimination and descriptive tests can be adapted for preference testing. Paired comparisons, ranking, and magnitude estimation tests can be used for this type of testing by modification of the score card and instructions to the panelists.

INTERPRETATION OF RESULTS

The results of any of the sensory methods described can be studied by tabulating the data, including the score of each judge for each sample, the averages, and the ranges. Some of the variability in results is due to the samples themselves, and may be a combination of differences in the raw materials and in the method of preparation. Sources of error in the judging include variability in the performance of one judge on duplicate samples as well as variability among several judges on the same sample.

After the data have been tabulated and averaged, the answer to the question posed by the experimenter may be obvious, and further analysis unnecessary. Certainly common sense is a great aid in the interpretation of experimental results. Its application suggests the examination of averages in light of the number of samples in each treatment and the variation among the samples. When there is much variation among replicate samples, differences among averages must be large to be meaningful unless an unusually large number of samples was used.

A study of the data such as described above is not adequate when the investigator wishes to state with confidence that the results obtained are statistically significant. In this case, a statistical analysis of the results is necessary. If an analysis is to be made, the original experiment should have been planned with statistical analysis in mind, because it is difficult, and sometimes impossible, to apply statistics to a completed experiment not appropriately planned. References have been given to tables based on statistical analysis for several discrimination and descriptive tests. The analysis of the data from other tests may be more complex. Various methods of testing the significance of differences among means may be used, depending on the experimental plan. Analysis of variance is very useful in many cases. Correlations or an indication of the relationship between two variables can be found between various sensory measures and between sensory and objective data. Because statistical methods are beyond the scope of this book, the reader is referred to the ASTM *Manual on Sensory Testing Methods* (ASTM, 1968) and to the many textbooks on statistics that are available (Sokal and Rohlf, 1973; Alder and Roessler, 1972; Snedecor and Cochran, 1967).

SUGGESTED EXERCISES

1. Conduct a paired comparison and a triangle test. Analyze the results using tables for determining the statistical significance of the results of paired comparison tests (Roessler et al., 1978) or of triangle tests (Amerine et al., 1965; ASTM, 1968; Larmond, 1977).
2. Add varying amounts of lemon juice to tomato juice (0, 10, 20, and 30% lemon juice on a volume basis). Ask panelists to rank the samples according to the increasing degree of acidity. Analyze the results as suggested by Kramer and coworkers (1974).
3. Develop a scoring system for a food product with the help of several class members. Use the score card several times on samples of varying quality. Discuss the adequacy of the scoring system and make appropriate revisions in the system.

REFERENCES

Alder, H. L. and Roessler, E. B. 1972. "Introduction to Probability and Statistics," 4th ed. W. H. Freeman and Co., San Francisco.

Amerine, M. A., Pangborn, R. M. and Roessler, E. B. 1965. "Principles of Sensory Evaluation of Food." Academic Press, New York.

ASTM. 1968. Manual on sensory testing methods. Spec. Tech. Publ. 434, Amer. Soc. Testing Materials, Philadelphia.

Brandt, M. A., Skinner, E. Z. and Coleman, J. A. 1963. Texture profile method. J. Food Sci. 28: 404.

Boggs, M. M. and Hanson, H. L. 1949. Analysis of foods by sensory difference tests. Adv. Food Res. 2: 219.

Caul, J. F. 1957. The profile method of flavor analysis. Adv. Food Res. 7: 1.

Civille, G. V. and Liska, I. H. 1975. Modifications and applications to foods of the General Foods sensory texture profile technique. J. Texture Studies 6: 19.

Cross, H. R., Green, E. C., Stanfield, M. S. and Franks, W. J. Jr. 1976. Effect of quality grade and cut formulation on the palatability of ground beef patties. J. Food Sci. 41: 9.

Ellis, B. H. 1964. Flavor evaluation as a means of product evaluation. Proceedings 11th Ann. Meeting, p. 181, Society Soft Drink Technologists, Washington, DC.

Ellis, B. H. 1968. Preference testing methodology. Food Technol. 22: 583.

Gridgeman, N. T. 1955. Taste comparisions: two samples or three? Food Technol. 9: 148.

Harries, J. M. 1956. Positional bias in sensory assessments. Food Technol. 10: 86.

Hunter, M. B., Briant, A. M. and Personius, C. J. 1950. Cake quality and batter structure. Cornell Univ. Agric. Exp. Stn. Bull. 860, Ithaca, NY.

Jones, L. V. and Thurstone, L. L. 1955. The psychophysics of semantics: an experimental investigation. J. Appl. Psychol. 39: 31.

Kramer, A., Kahan, G., Cooper, D. and Papavasiliou, A. 1974. A non-parametric ranking method for the statistical evaluation of sensory data. Chem. Senses and Flavor 1: 121.

Larmond, E. 1977. Laboratory methods for sensory evaluation of food. Publ. 1637, Canada Dept. Agric., Ottawa, Ontario.

Larmond, E. 1973. Physical requirements for sensory testing. Food Technol. 27(11): 28.

Lowe, B. 1949. Organoleptic tests developed for measuring the palatability of meat. Proc. 2nd Ann. Recip. Meat Conf., p. 111, Natl. Livestock and Meat Board, Chicago.

Martin, S. L. 1973. Selection and training of sensory judges. Food Technol. 2(11): 22.

Moskowitz, H. R. 1974. Sensory evaluation by magnitude estimation. Food Technol. 28(11): 16.

Prell, P. A. 1976. Preparation of reports and manuscripts which include sensory evaluation data. Food Technol. 30(11): 40.

Roessler, E. B., Pangborn, R. M., Sidel, J. L. and Stone, H. 1978. Expanded statistical tables for estimating significance in paired-preference, paired-difference, duo-trio, and triangle tests. J. Food Sci. 43: 940.

Schutz, H. G. 1965. A food action rating scale for measuring food acceptance. J. Food Sci. 30: 365.

Snedecor, G. W. and Cochran, W. G. 1967. "Statistical Methods," 6th ed. Iowa State Univ. Press, Ames.

Sokal, R. R. and Rohlf, F. J. 1973. "Introduction to Biostatistics." W. H. Freeman and Co., San Francisco.

Stone, H., Sidel, J., Oliver, S., Woolsey, A. and Singleton, R. C. 1974. Sensory evaluation by quantitative descriptive analysis. Food Technol. 28(11): 24.

Szczesniak, A. S., Brandt, M. A. and Friedman, H. H. 1963. Development of standard rating scales for mechanical parameters of texture and correlation between the objective and the sensory methods of texture evaluation. J. Food Sci. 28: 397.

Szczesniak, A. 1975. General Foods texture profile revisited—ten year perspective. J. Texture Studies 6: 5.

16. Evaluating Food by Objective Methods

Food scientists also rely on methods of evaluation that do not depend on human senses. Methods that do not depend on the observations of an individual and can be repeated by using an instrument or a standard procedure are described as *objective methods* (Institute of Food Technologists, 1964). Objective methods include a wide variety of physical and chemical tests. The advantages of these tests are many. They may offer a permanent record of results and invite confidence because they are reproducible and less subject to error than sensory methods of evaluation. Selection of objective tests should involve care. The selected method should give results that are in agreement with the results of sensory evaluation. If the two methods do not correlate, they may not be measuring the same component of quality and hence the chemical or physical method may not be useful for the study. For example, determinations of the alkali insoluble solids content of vegetables give an index of plant tissue maturity but may not be related to overall textural quality (Szczesniak, 1973).

CLASSIFICATION OF OBJECTIVE METHODS

Chemical Methods

Chemical methods include the determination of the nutritive value of foods before and after cooking, as well as of constituents that affect the palatability of the food, such as peroxides in fats. Chemical methods will not be included here except to refer the reader to a publicaton of the Association of Official Analytical Chemists (AOAC, 1975), which gives methods for determining various constituents in specific foods and is kept up to date by continuous study of analytical methods and by periodic revisions.

Physicochemical Methods

Certain *physicochemical determinations* are important in food analysis. Of these, probably the one most frequently used is hydrogen ion concentration, or pH, which was discussed in Chapter 2. Measurement with a refractometer of refractive index, the angle to which light is bent by certain substances, is useful in finding the sugar concentration of sirups and the degree of hydrogenation of fats (Carlson, 1972). The determination with a polariscope of the rotation of plane-polarized light by sugar solutions offers a method for their quantitative analysis.

Microscopic Examination

Some properties of foods depend on the structure or physical arrangement of their components. Microscopic examination of foods such as mayonnaise, whipped cream, fondant, and cake batter may yield valuable information. Fiber diameters and sarcomere lengths of meat samples can be measured easily with the ocular measuring device of a microscope (Tuma et al., 1962; Hegarty and Allen, 1975). Examination of meat, vegetable, and fruit tissue structure is more difficult because it involves making histological sections (Humason, 1972). Although such studies require training and experience, they are essential in some types of food research. *Electron microscopy,* a technique that makes it possible to view a three-dimensional image of the structure of a material, is also being used in food research. Electron microscopy has been used as a research tool in studies of meat (Cheng and Parrish, 1976; Jones et al., 1976), fruits (Mohr, 1973), vegetables (Davis et al., 1976; Haard et al., 1976), starch systems (Miller et al., 1973), and milk and other food gels (Kalab and Harwalkar, 1973). Pomeranz (1976) reviewed the general principles and use of the electron microscope in food science.

Physical Properties

The objective measurements to be discussed in this chapter concern the physical properties of foods. Some physical properties, such as temperature, length, or the amount of liquid drained from the food on standing are simple and used frequently. Others, on which attention will be focused, are of special interest because they are related to the results of sensory tests. These tests are made with special instruments or with improvised devices. Reviews of measurements that are especially applicable to fruits and vegetables (Kramer and Twigg, 1959), meat (Szczesniak and Torgenson, 1965), and baked products (Funk et al., 1969) can be found in the literature. Methods for all foods have been reviewed (deMan et al., 1976).

The advisability of substituting, when possible, an improvised device for a manufactured instrument depends, of course, on the availability of the instrument in question, which in turn may depend on the amount of use that it will receive and the cost. Improvised devices such as those described in this chapter may be used to meet the needs for class projects. The ingenious worker may be able to improvise others.

Many of the objective tests mentioned in this chapter are listed in tables, which indicate whether the equipment is improvised or purchased. Sources for the purchase of equipment are listed for the purpose of information. A price range is included to give a general idea of costs. The ranges are wide to cover price fluctuations but may not be wide enough to include all possible variations in models of instruments. References describing the apparatus or its use also are included in the discussion. Tests in addition to those listed in the tables may be found in the literature. Tests for appearance, color, and volume are listed in Table 16.1.

Table 16.1 Objective tests of appearance, color, and volume of foods

Property measured and name of test or apparatus	Foods	Source of apparatus	Reference
Appearance			
Photography	All	Readily available	Cooley and Davis, 1936
Photocopying	Baked	Readily available	
Color			
Charts			
Color atlases	All	Libraries	Maerz and Paul, 1950; Villalobos-Dominguez and Villalobos, 1947; Kornerup and Wanscher, 1962
Munsell Book of Color[a]	All	Munsell Color 2441 Calvert St Baltimore, MD 21218	Francis and Clydesdale, 1975
Spectrophotometers[b]	Most	Scientific supply companies	Francis and Clydesdale, 1975

Property measured and name of test or apparatus	Foods	Source of apparatus	Reference
Munsell Disc Method			
Improvised[c]	All		Nickerson, 1946
Macbeth-Munsell Colorimeter[d]	All	Munsell Color 2441 Calvert St Baltimore, MD 21218	Francis and Clydesdale, 1975
Gardner Color Difference Meter[e]	All	Gardner Laboratory Bethesda, MD 20014	Francis and Clydesdale, 1975
Volume			
Seed displacement			
Volumeter[c]	Baked	National Manufacturing Lincoln, NE 68502	Cathcart and Cole, 1938
Improvised	Baked		Brown and Zabik, 1967
Index to volume			
Planimeter[c]	Baked	Office supply companies	Funk et al., 1969
Ruler	Baked		Tinklin and Vail, 1946

[a]Cost $300–$400
[b]Cost $600–$1,200
[c]Cost $100–$200
[d]Cost over $2,000
[e]Cost over $5,000

APPEARANCE

The appearance of foods can be recorded by means of photography or, in some cases, photocopying. A photograph furnishes a record of size if scales are included in the picture. The grain of baked products is visible in a photograph when lighting is cntrolled carefully so that an extreme amount of light does not fall on the object (Cooley and Davies, 1936). An example of a photograph is shown in Figure 16.1. A record of the appearance of baked products that is somewhat less clear than a photograph, but is satisfactory for many purposes, is obtained with a photocopy machine. Photocopies furnish a record of the actual size and shape of a sliced or halved baked product and give some record of the grain. They can be made by placing a sheet of clear plastic film on the glass

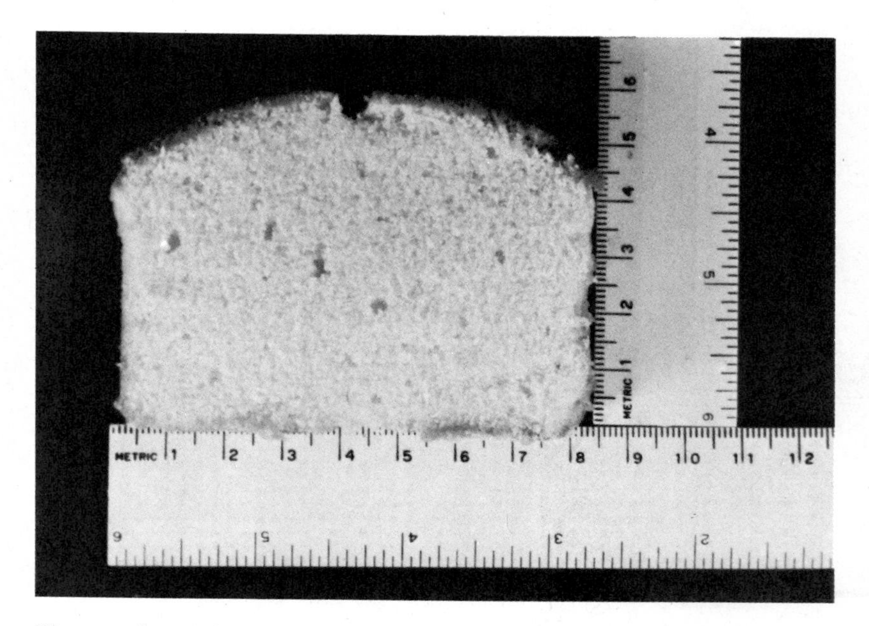

Figure 16.1. Photograph of a cake as a record of size, shape, and grain. (Courtesy of Rogers C. Penfield, Jr.)

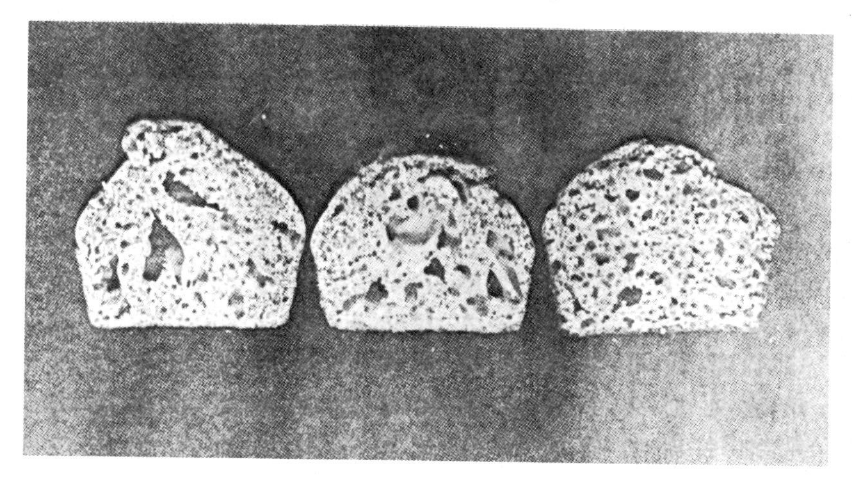

Figure 16.2. Photocopies of muffins as a record of shape and grain.

plate of the copy machine, placing the samples in some logical arrangement on the film, and proceeding as usual to make a copy. Notations of significant characteristics of each sample can be made on the copy. A photocopy of this type is shown in Figure 16.2.

COLOR

Preliminary acceptance or rejection of a food usually is based on the appearance, including the color. Therefore, color, an important quality attribute of food, is studied in many ways. Only a brief introduction to the study of food color can be given here. Further information is available in a book by Francis and Clydesdale (1975) and in several review articles (Clydesdale, 1972, 1976; Noble, 1975).

A rapid method of measuring color that is satisfactory for some purposes is to match the color of the food to a sample of color or a color chip found in a color atlas. One of these books of color samples should be available in most college and university libraries (Maerz and Paul, 1950; Villalobos-Dominguez and Villalobos, 1947; Kornerup and Wanscher, 1962). Although this procedure furnishes a permanent record of the color of a food, it is not entirely satisfactory because it is difficult to match the food with one small block of color on a chart containing many such blocks. These types of data are difficult to tabulate and analyze.

Disc colorimetry was developed to facilitate the use of the Munsell color system in evaluating the color of agricultural products in USDA laboratories (Nickerson, 1946). Two to four discs that have been cut along the radius are interlocked. Selection of the colored discs depends on the color to be matched. The color of the discs is blended by spinning them on a shaft of a small motor so that one color is seen with no flicker. The proportions of the four discs are adjusted until the color of the sample is matched. A method of controlling lighting is necessary and is provided in commercially available disc colorimeters. The color is described in terms of the weighted average of the tristimulus values, X(amber), Y(green), and Z(blue), for each of the discs used (Francis and Clydesdale, 1975).

Two types of instrumental methods of color analysis, spectrophotometry and tristimulus colorimetry, are used in food research. Transmission spectrophotometry is used to measure the intensity of light that passes through a clear or transparent solution. Measurements of this type are used frequently in quantitative analysis of specific compounds. Because most foods are opaque in nature, reflectance spectrophotometry, or, more commonly, tristimulus colorimetry, is used in food research. In a tristimulus system, color is specified by three attributes: hue, value, and chroma in the Munsell system, or dominant wavelength, brightness, and purity in the C.I.E. (Commission Internationale de L'Eclairage) system (Francis and Clydesdale, 1975). The three attributes respectively refer to the actual color, the luminosity, and the strength of the color. Several colorimeters are available for evaluation of these attributes or for the complete specification of color. In some studies, only small differences between like-colored samples are sought. In these cases, either an instrument intended for the complete specification of color or one that is designed to detect small differences in the color of samples or in the color of a sample and a standard-colored tile can be used. The Gardner color difference meter (Francis and Clydesdale, 1975) is one such instrument. Instruments to define color or evaluate color differences are relatively expensive and therefore may not be available for experimental foods classes.

VOLUME

Displacement Method

In the displacement method, the volume of a baked product is found by subtracting the volume of low density seeds such as rapeseeds held in a container with and without the product. The container can be a straight-sided box, a can, or a volumeter, an apparatus built like an hour-glass (Cathcart and Cole, 1938). If a cake is baked in a pan with sides that are higher than the product, its seed displacement can be found by pouring seeds on top of the cake while it is still in the pan. The volume of seeds required to fill the pan then is subtracted from the volume of the empty pan (Brown and Zabik, 1967). After the volume of a baked product has been determined, it may be divided by the weight of the product and reported as specific volume, in units of cubic centimeters per gram. This calculation facilitates comparison of products having different weights. The comparison also is facilitated if the same weight of batter is baked for each product. When the displacement method is used, scoring must be delayed or done on a duplicate product because seed displacement must be done on the entire product at a time when it is not soft, often after 24 hr.

Index to Volume

An index to volume can be found by measuring the area of a slice of food with a *planimeter.* The edges of the slice are traced or a photocopy is used, and the area is found with the planimeter. It is important to use a slice that is representative of the product, such as a center slice, or to use several slices from the same positions in each product and average the areas. Index to volume sometimes is more convenient than seed displacement because a slice can be traced quickly and the area found at any convenient time, making it possible to score the palatability of the product immediately. An alternative system for measuring an index to volume is to measure the height of several slices of the product. Height is measured at the outer edges, the center, and the points halfway between the center and the outer edges. The five measurements are averaged and reported as an index to volume (Tinklin and Vail, 1946).

TEXTURE

The objective measurement of texture is complex because it must reflect the action of the mouth in removing food from an eating utensil, the action of the tongue and jaws in moving the food, and the action of the teeth in cutting, tearing, shearing, grinding, and squeezing food. Textural characteristics include the mechanical properties of hardness, cohesiveness, adhesiveness, fracturability, viscosity , gumminess, springiness, and chewiness, as well as geometrical characteristics, which were previously discussed, and fat and moisture content (Szczesniak, 1963, 1975). Therefore, it is not surprising that there are many ways to evaluate the texture of various foods.

Texture of Meat

Some of the early measurements of the texture of meat were made with a *penetrometer* (Noble et al., 1934). More recently, McCrae and Paul (1974) used a cone penetrometer, similar to the one shown in Figure 16.3, in meat studies, suggesting that the action of the cone was similar to the action of the teeth biting into a piece of meat. The most commonly used device for the evaluation of meat tenderness is the *Warner-Bratzler Shear,* shown in Figure 16.4. The force necessary to shear a cylindrical sample of meat 1.27 or 2.54 cm in diameter against the dull edge of a triangular opening is measured (Bratzler, 1932). Sensory scores usually are highly correlated with Warner-Bratzler shear, possibly because the dull edge against which the meat is torn simulates the grinding surfaces of the teeth, which also are dull. However, some studies have shown poor correlation between Warner-Bratzler shear values and sensory scores for tenderness. Care must be exercised when cylindrical samples or cores are removed from the meat. Meat should be at room or refrigerator temperture to avoid irregularly shaped cores. Kastner and Henrickson (1969) suggested that

Figure 16.3. Universal Penetrometer. (Courtesy of Lab-Line Instruments, Inc., Melrose Park, IL.)

Figure 16.4. Warner-Bratzler Shear. (Courtesy of G-R Electric Mfg. Co., Manhattan, KS.)

a mechanical device should be used to ensure cores of uniform diameter. A recently developed instrument, the Armour Tenderometer, is a ten-probed penetrometer-like instrument that is used to predict tenderness. Values for maximum force required to penetrate the longissimus muscle 1 da after slaughter correlated with panel scores for tenderness of the meat after it was aged for 1 wk and cooked (Hansen, 1972). Other investigators (Campion et al., 1975) have not found this relationship.

Tensile properties of meat samples also have been investigated. The amount of force required to pull apart a small sample of meat with the muscle fibers oriented parallel to the force is a measure of the strength of the muscle fibers, whereas the amount of force required to pull apart a sample with fibers perpendicular to the force is thought to be a measure of the strength of the connective tissue holding the muscle fibers together (Bouton and Harris, 1972). Tests for meat are summarized in Table 16.2.

Texture of Fruits and Vegetables

Puncture testing most frequently is used to evaluate the firmness of fruit and vegetable tissue. The *Magness Taylor pressure tester* is a commonly used apparatus that measures the amount of force required to penetrate the sample to a specific depth (Bourne, 1965). The *Kramer shear press* (one variation of the Texture Test System), which consists of a rectangular box with evenly spaced slits in the bottom, frequently is used to study the tenderness of fruits and vegetables. A series of ten blades is moved through a sample of food in the box. As the blades move the food is compressed, sheared, and extruded through the openings in the box. Many foods have been tested with this system (Szczesniak et al., 1970). Tests for fruits and vegetables are summarized in Table 16.2.

Texture of Liquids and Visco-elastic Foods

Rheology is the science of flow and deformation of materials, both liquid and solid; therefore, an understanding of its principles is important to the study of food texture. Application of rheology to food research was reviewed by Scott Blair (1958) and Finney (1972).

Table 16.2 Objective methods for evaluation of the texture of meat, fruits, and vegetables

Name of test or apparatus	Foods	Source of apparatus	Reference
Penetrometer, universal[a]	Meat	Scientific supply companies	McCrae and Paul, 1974
Shear, Warner-Bratzler[b]	Meat	G-R Electric Manufacturing Co. 1317 Collins Lane Manhattan, KS 66502	Bratzler, 1932; Kastner and Henrickson, 1969
Tensile testing with Instron Universal Testing Machine[c]	Meat	Instron Corp. 2500 Washington St Canton, MA 02021	Bouton and Harris, 1972
Press fluids or expressible moisture index with Harco Hydraulic Press[d]	Meat	Harco Industries 10802 N 21st Phoenix, AZ 85029	Shaffer et al., 1973
Magness Taylor Pressure Tester[e]	Fruits and vegetables	D. Ballauf Co 619 H Street NW Washington, DC 20001	Bourne, 1965
Texture Test System[f]	Most	Food Techology Corp. 12300 Parklawn Drive Rockville, MD 20852	Szczesniak et al., 1970; deMan et al., 1976
Texture profiling with Instron Universal Testing Machine[c]	Fruits and vegetables, meat	Instron Corp. 2500 Washington St Canton, MA 02021	Bourne, 1968; Szczesniak, 1975

[a] Cost \$600–\$700
[b] Cost \$500–\$600
[c] Cost \$8,000–\$12,000
[d] Cost \$400–500
[e] Cost under \$100
[f] Cost \$6,000–\$8,000

Viscosity and consistency All liquids flow; some do not flow readily. Their resistance to flow, or viscosity, is caused by attraction between molecules of the liquid and/or larger particles. In a pure liquid, this attraction or internal friction is greater between large and well-hydrated molecules than between smaller molecules. Temperature must be controlled closely in measurements of viscosity because as the temperature of a pure liquid increases, viscosity decreases. Absolute viscosity is measured in terms of the amount of work required to maintain a certain rate of flow. Its unit is the *poise,* which is defined as a force of 1 dyne per cm^2 to produce a 1-cm per sec difference in velocity of two planes separated by 1 cm of liquid. *Absolute viscosity* is a characteristic of Newtonian fluids or homogeneous liquids such as sugar sirups, oils, very dilute fruit juices, and skimmilk. The term *apparent viscosity* should be used to refer to the flow characteristics of non-Newtonian fluids. *Relative viscosity* is found by comparing the rate of flow of a liquid with that of a reference liquid, usually water. Some of the tests used to study viscosity are listed in Table 16.3. The rate of flow through a tube in comparison to the rate of flow of water through the same tube is a simple measurement of viscosity. An ordinary pipet with a part of the tip cut off, if necessary, for more rapid flow can be used. Pipets especially designed for viscosity measurements, like the Ostwald pipet, are available.

Apparent viscosity or consistency can be evaluated with several tests. The time required for cake batter to flow between two marks on the stem of a funnel is a simple measurement of this type (Tinklin and Vail, 1946). Grawemeyer and Pfund (1943) described the *line-spread test,* which is suitable for foods like white sauce, starch puddings, applesauce, and cake batters. For this test, food is placed in a hollow cylinder. (See Experiment 16-II.) The cylinder is lifted when the food is at the desired temperature and the product is allowed to spread for a specified period of time (30 sec to 2 min). Consistency is reported in distance (cm) spread in the designated time period.

The *Adams consistometer* (Adams and Birdsall, 1946) is a commercially available apparatus, which was first designed for use with creamed corn. It consists of a funnel-like reservoir and a metal plate with concentric rings and works on the same principle as the line-spread. The *Bostwick consistometer* measures the distance that a food flows under its own weight down a trough in a given period of time. It is used in an official National Canners Association consistency test.

Table 16.3 Objective methods for evaluation of the texture of liquid and visco-elastic foods

Property measured and name of test or apparatus	Foods	Source of apparatus	Reference
Viscosity			
Funnel	Cake batter	Improvised	Tinklin and Vail, 1946
Line-spread	White sauce, cake batter, etc.	Improvised	Grawemeyer and Pfund, 1943
Consistometer, Bostwick[a]	Ketchup, preserves, etc.	Central Scientific Cenco Center 2600 Kostner Ave Chicago, IL 60623	Daoud and Luh, 1971
Steel ball, falling	Batter, puddings	Improvised	Morse et al., 1950
Viscometer, Brookfield[b]	Liquids	Brookfield Engineering Laboratories, Inc. 240 Cushing St Stoughton, MA 02072	Balmeceda et al., 1973
Amylograph, Brabender[c]	Starch-water suspensions	C. W. Brabender Instruments 50 E Wesley St South Hackensack, NJ 07606	Mazurs et al., 1957
Elasticity and gel strength			
Ridgelimeter, Exchange[a]	Pectin gels	Sunkist Growers Ontario, CA 91764	Ehrlich, 1968
Penetrometer, universal[d]	Baked custards	See Table 16.2	

[a]Cost $100–$200
[b]Cost $500–$600
[c]Cost over $5,000
[d]Cost $600–$700

Figure 16.5. Brookfield Viscometer. (Courtesy of Brookfield Engineering Laboratories, Stoughton, MA.)

Several other methods can be used for viscosity and consistency measurements. For example, the length of time required for a steel or glass ball to fall through a column of test material is found by using improvised equipment (Morse et al., 1950) or a special device, and is an indicator of viscosity. Several rotational viscometers are available. A spindle attached to the Brookfield viscometer (Figure 16.5) rotates in the test material and the amount of drag on the spindle is measured. The Brookfield viscometer has been used with many foods, including sirups (Collins and Dincer, 1973), protein slurries (Fleming et al., 1975), and gums (Balmaceda et al., 1973). For very viscous materials, the Brookfield can be mounted on a helipath stand that gradually lowers the spindle through the test material for testing of undisturbed material. In the MacMichael viscometer, a horizontal disc is suspended in the liquid by means of a steel wire, and the force required to hold the disc stationary is measured while the outside cup is rotated at a constant speed. In the *Stormer viscometer,* measurements are based on the rate of rotation of a cylinder immersed in a sample and impelled by a uniform force. For some problems, change in consistency must be followed over a period of time, as in heating and cooling of a starch paste. Measurements that are important in the flour and starch industries can be made with a *Brabender Amylograph.* A starch-and-water suspension is rotated and stirred in a cup that can be heated or cooled. Changes in consistency are recorded in graphic form in an arbitrary unit of consistency called *Brabender units* (BU). Interpretation of curves from this instrument was explained by Mazurs and coworkers (1957).

Elasticity The science of rheology includes the study of elastic solids as well as viscous liquids. These solids cannot be said to flow, but they can be deformed by force and recover when the force is removed. Many of the objective tests listed in Table 16.3 are for use with gels such as pectin gels, baked custards, rennet curds, gelatin gels, and starch gels. The firmness of gels is indicated by the extent to which they retain their height when turned from a container or by their resistance to penetration by a variety of instruments.

The percentage sag of a gel can be measured simply by determining the height of the gel before and after its removal from its container. This can be done by inserting the probe of a vernier caliper into the gel at the appropriate times. Sag then is expressed as a percentage of the height before unmolding. The *Exchange Ridgelimeter* is a simple instrument used for direct measurement of percentage of sag. It was designed specifically for evaluation of pectin gels that will sag in the range of 10 to 40% (Ehrlich, 1968).

Various methods are used to measure the resistance of a gel to penetration. The universal penetrometer (Figure 16.3) can be adapted for measurements on baked custards, starch gels, and pectin gels by substituting a light weight cone for the heavier cones available from the maufacturer. A device improvised from common laboratory equipment and operating on the same principle as the universal penetrometer was described by Hanning and coworkers (1955).

Texture of Doughs and Baked Products

Texture measurements frequently are made on batters and doughs as a way of predicting the quality of the final baked product. The consistency of batters has been measured with line-spread apparatus. Hunter and coworkers (1950) suggested that very viscous batters of a given specific gravity are indicative of fine dispersion of incorporated air, whereas a high line-spread reading in association with the same specific gravity is indicative of the dispersion of air in larger units. Other viscosity and consistency measuring devices also can be used to evaluate batters. The rheological properties of dough are related to the quality of the finished product. The consistency and stability of doughs can be measured with a *farinograph,* which is designed to measure the force required to turn mixer blades at a constant speed during mixing of the dough (Locken et al., 1972). The force increases as gluten is being developed and later decreases as gluten is broken down with continued mixing. Similar information can be obtained with a mixograph (Shuey, 1975).

The evaluation of the texture of baked products is approached in several ways. The breaking strength of pastries, cookies, and crackers can be determined with a *shortometer,* developed by Bailey (1934) and shown in Figure 16.6. The pastry or other wafer (approximately 4.5 × 9.0 cm) is placed across two horizontal bars; a single upper bar is brought down by means of a motor until it breaks the wafer, and the force is recorded. Shortometer values were highly correlated with sensory values for tenderness, and values obtained with a Kramer shear press also were highly correlated with sensory scores for tenderness (Stinson and Huck, 1969).

Compression testing is used frequently to evaluate crumb firmness or softness of baked products. The Baker compressimeter can be used to measure these textural characteristics (Platt and Powers, 1940). Softness was defined as the degree of compression under a constant load, whereas firmness was defined as the amount of force required to attain a specified compression (Babb, 1965). Firmness values correlated well with sensory scores for firmness (Crossland and Favor, 1950). A penetrometer fitted with a flat disc (Funk et al., 1969; Paul et al., 1954) and a Bloom gelometer fitted with a modified plunger (Edelmann and Cathcart, 1949) also were used for measurements of softness. Uniform slices for these and other tests are obtained with a cutting box like the one shown in Figure 16.7. Compressibility measurements seem to indicate both the aeration of a cake and its structural rigidity (Hunter et al., 1950).

Shearability, a measure of the force required to cut through a sample of a baked product, was measured with the Warner-Bratzler Shear by

Figure 16.6. Bailey Shortometer. (Courtesy of Donald B. McIntyre, St. Paul, MN.)

Matthews and Dawson (1963, 1966). They used the instrument to evaluate the tenderness of pastry, biscuits, and white cakes. Statistically significant correlations between shear values and panel scores for tenderness were reported, but the relationship is not clear because of the type of scale used for the panel evaluations of tenderness. Tests for baked products are listed in Table 16.4.

Texture as Measured by Multipurpose Instruments

Texture as perceived by humans is not a simple characteristic but a composite of characteristics. Thus, it is understandable that methods for evaluating more than one characteristic have been developed. Texture

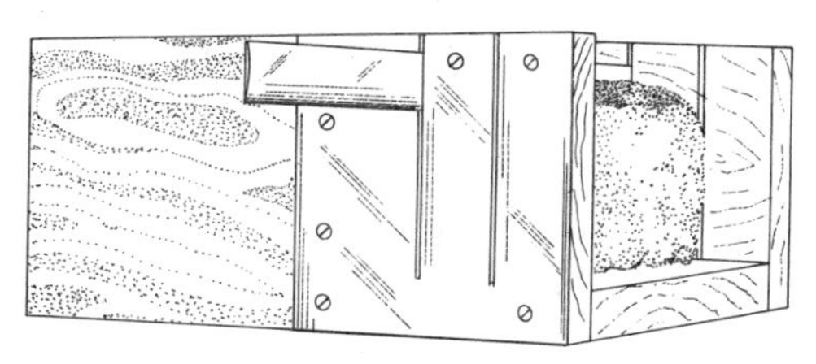

Figure 16.7. Cutting box used to obtain slices of uniform thickness.

profile analysis by sensory panels was mentioned in Chapter 15. It also is possible to use instrumental methods to arrive at a texture profile for most foods. The *General Foods Texturometer,* a texture-measuring device that imitates chewing motions of the mouth, was developed to give values to the parameters of the General Foods Texture Profile (Friedman et al., 1963). Parameters evaluated included hardness, springiness, gumminess, fracturability, chewiness, adhesiveness, and cohesiveness. Viscosity, the eighth profile parameter, usually is evaluated with a viscometer such as the Brookfield viscometer. The *Instron Universal Testing Machine* pictured in Figure 16.8 was first used by Bourne (1968) for the objective texture profile analysis of a food. The crosshead of the universal testing machine moves up or down at a constant rate of speed. As it moves, the force that is required to compress a sample of food is continuously recorded. In addition to the compression type of attachments used in texture profile analysis, other test cells may be attached to the universal testing machine including probes for puncture testing, extrusion cells, Warner-Bratzler Shear attachments, and Kramer shear cells. Many others have used this instrument to study textural properties of food materials. Breene (1975) reviewed the use of instrumental methods for texture profile analysis.

Texture-Moisture Components

Press fluid measurements *Press fluid measurements* involve the mechanical removal of fluid from a sample. Measurements of this type most frequently are used in meat studies, although that amount of fluid

Table 16.4 Objective methods for evaluation of the texture of batters and doughs and baked products

Property measured and name of test or apparatus	Foods	Source of apparatus	Reference
Batter consistency		See Table 16.3	
Dough consistency			
Farinograph[a]	Doughs	C. W. Brabender Instruments 50 E Wesley St South Hackensack, NJ 07606	Locken et al., 1972
Compressibility			
Compressimeter, Baker[b]	Baked products	Watkins Corp. P.O. Box 445 Caldwell, NJ 07006	Crossland and Favor, 1950
Penetrometer, Universal with improvised disc[b]	Baked products	See Table 16.2	Funk et al., 1969
Shearability			
Shortometer, Bailey[c]	Pastry	D. McIntyre Dept of Physics University of Minn Minneapolis, MN 55455	Stinson and Huck, 1969
Kramer shear cell of Texture Test System[d]	Cake	See Table 16.2	Stinson and Huck, 1969; Szczesniak et al., 1970
Shear, Warner-Bratzler[e]	Cake, biscuits	See Table 16.2	Matthews and Dawson, 1963, 1966

[a]Cost over $3,000
[b]Cost $600–$700
[c]Cost $900–$950
[d]Cost $6,000–$8,000
[e]Cost $500–$600

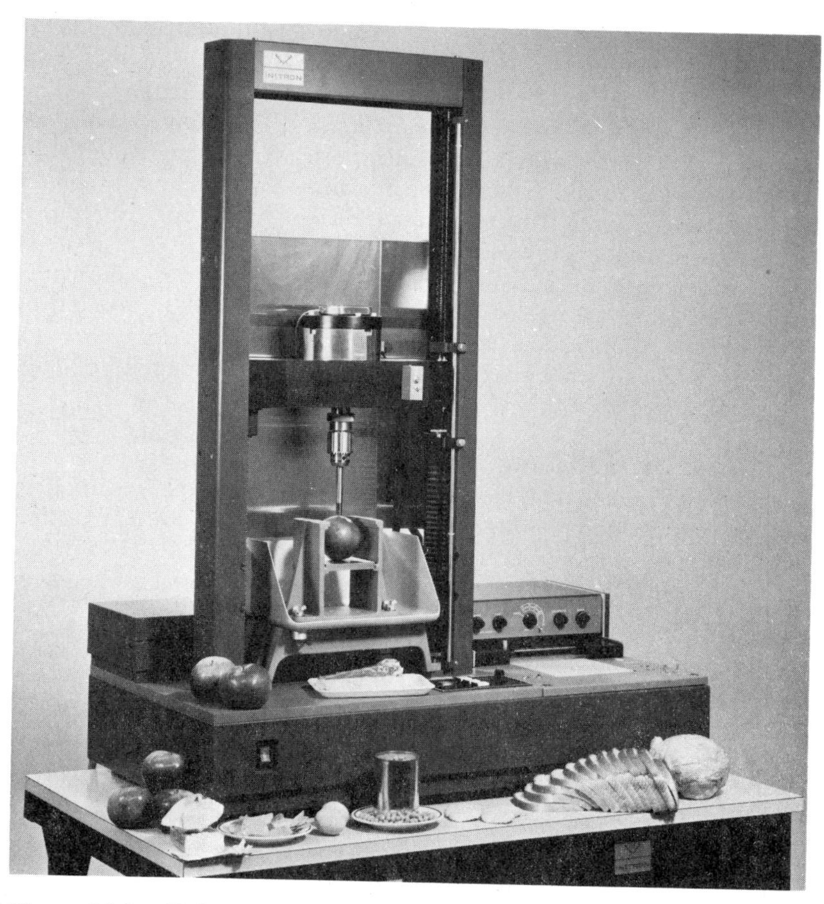

Figure 16.8. Universal Testing Machine. (Instron Model 1132 Food Testing Instrument, Courtesy of Instron Corp., Canton, MA.)

that can be pressed from a vegetable such as corn is used as an index of maturity and succulence. Press fluid values for meat are indicators of water-binding capacity. Fluids may be pressed from a sample by packing it into a Carver cell, a device resembling a cylinder and piston, and exerting high force with a hydraulic press. An alternative method involves placing small samples of meat on filter paper and pressing them between acrylic sheets. The meat area is divided by the fluid area to give an expressible moisture index. Subtracting this ratio from 1 gives an index of water-holding capacity (Shaffer et al., 1973). An alternative method for objective evaluation of juiciness is outlined by Bouton and coworkers (1975). A significant relationship between sensory scores for juiciness and the amount of fluid expressed from a sample of meat centrifuged at 100,000 × G was reported.

Moisture determinations The moisture content of food materials may be determined as an indicator of the quality of food. Traditionally used methods involve a long, slow drying process in an oven or vacuum oven. Methods for specific foods are outlined by AOAC (1975). Lee and Latham (1976) reported a procedure for using the microwave oven for more rapid moisture determinations.

MISCELLANEOUS TESTS

Specific gravity and the similar measurement, density, indicate the amount of air incorporated into products such as whipped cream, egg white foams, creamed shortening, and cake batter. The determination of specific gravity, described at the end of this chapter, involves dividing the weight of the food packed into a small, even-rimmed cylindrical container by the weight of the water held by the same container. Density would be determined by dividing the weight of the food packed into the container by the volume of the container. Low specific gravity or density values for batter, which indicate that a large amount of air is present, are associated with good volume of cakes in some reports (Hunter et al., 1950) but not in others (Jooste and Mackey, 1952).

EXPERIMENTS

I. Specific Gravity

Specific gravity is a useful measure of the lightness of products such as egg white foams, creamed shortening and sugar, or cake batter. The greater the amount of air incorporated into such a mixture, the lower will be its specific gravity. Specific gravity is found by dividing the weight of the material by the weight of an equal volume of water. The result, being a ratio, carries no units.

A. *Equipment for test*

Small container, such as a crystallizing dish, having a capacity of about 50 ml or a diameter of about 5 cm, a smooth even rim, and a cylindrical or nearly cylindrical shape.

B. *Procedures*

1. Weigh the dry container to the nearest 0.1 g.
2. Partially fill the container with boiled distilled water that is at room temperature. Place on a balance and add more water until the container is completely full. Fullness should be judged at eye level. Weigh to the nearest 0.1 g.
3. Fill the dry cup with the material to be tested in such a way that no air pockets are left. If egg white foam or creamed shortening and sugar are being tested, pack carefully to avoid large air cells. If cake batter is tested, fill the cup half full and tap on a folded towel 12 times, rotating frequently. Then add excess batter and again tap 12 times. Remove the excess batter with a spatula, starting in the center. Wipe the outside of the container and weigh to the nearest 0.1 g.
4. Calculate the specific gravity as follows:

$$\text{specific gravity} = \frac{\text{weight of filled container} - \text{weight of container}}{\text{weight of water-filled container} - \text{weight of container}}$$

II. Line-spread Test

The line-spread test is used to measure the consistency of foods in terms of the distance that they spread on a flat surface in a given period of time. It is suitable for foods such as white sauce, soft custard, applesauce, starch puddings, cake batters, and cream filling.

A. *Equipment for test*

1. A hollow cylinder having a diameter of 5 to 8 cm. This can be made by removing the handle from a cookie cutter, by cutting both ends out of a small flat tin can, or by cutting heavy copper tubing into cylinders approximately 7 cm high. It also is possible to purchase plumbing fittings that are suitable for line-spread rings.
2. Diagram of concentric circles drawn 1 cm apart, the smallest having a diameter equal to the inside edge of the line-spread ring. The smallest ring has no number, but the second one is numbered 1, and others are numbered consecutively. Alternatively, the circles other than the inner one may be omitted and four lines drawn at right angles to one another outward from the circle. The numbers representing centimeters are placed along these lines. Readings of spread are facilitated by marking the lines into millimeters as well as centimeters.
3. A flat glass plate or a large glass pie plate.
4. A spirit level.

B. *Procedures*

1. Place the glass plate or pie plate over the diagram and check for evenness with the spirit level.
2. Place the cylinder directly over the smallest circle and fill with the food to be tested. Level off with a spatula. Remove any food that falls onto the plate. Suspend a thermometer into the test material.
3. When the material reaches the desired temperature, lift the cylinder and allow the food to spread for exactly 2 min.

4. Quickly take readings (at four equally spaced axes) on the limit of spread of the substance.
5. The line-spread of the sample is found by averaging the four readings. The value represents the distance in centimeters that the material spreads in 2 min.

III. Volume by Seed Displacement

The volume of many foods, including cakes, breads, biscuits, and muffins, can be measured by seed displacement. Unless equal portions of batter or dough have been weighed for the samples, it is best to use an entire baking. These directions describe the use of improvised apparatus that can be assembled readily at little expense. The volume of the seed is measured in a large graduated cylinder. If desired, the seed displaced by the product can be weighed instead of measured and the volume of the seed calculated. To do this, the volume of a certain weight of seed is found by calculating the volume of the box containing the seed from carefully determined internal measurements and weighing the seed contained within it. Dividing the seed volume (cm^3) by its weight (g) provides a conversion factor.

A. *Equipment for test*

1. Sturdy box, made either of wood or metal, or a straight-sided can, larger than products to be measured. The selected container should have a flat, open top.
2. Shallow box or pan.
3. Light seed, such as rapeseed.
4. Ruler or vernier caliper for measuring dimensions of container.
5. Graduated cylinder, 1- or 2-liter.

B. *Procedure*

1. Method 1
   a. To find the volume of the empty box or can, put it into the shallow pan or box and pour seed into it until it

overflows. Always pour in the same way and from the same height to avoid variations in packing. Avoid shaking the box. Level the seed by passing a ruler across the top of the box once. Using a graduated cylinder, measure the volume of seed held by the box. Avoid shaking.

b. The product should not be too soft when the determination is made. Cake is more easily handled if it is about 24 hr old. If desired, it can be dusted lightly with corn starch so that seeds will be less likely to stick during the determination. Place the product in the box or can, pour seed on it until the box overflows, and level the seed as before. Measure the volume of the seed.
c. Find the volume of the product by subtracting the volume of the seed that was around the product from the volume of the empty container.
d. Find the specific volume of the product by dividing its volume by its weight. If each product in a series is made from the same weight of dough or batter, the volumes can be compared directly.

2. Method 2
This method can be used if the sides of the pan extend well above the top of the product baked in it, and is especially useful for delicate cakes such as sponge or angel food. The cake is cooked in the pan and its top is dusted with cornstarch. Seeds are added and leveled off as in Method 1. The volume of the cake is found by subtracting the volume of seeds held by the pan containing the product from the volume of the empty pan.

IV. Index to Volume

In this method, the area of a slice of baked product as found with a polar planimeter is used as an index to the volume of the product. This index furnishes a valuable comparison of products that are similar as to shape. It is easily recognized that the comparison would be less useful if the products were dissimilar, such as two loaves of bread, one of which had risen evenly, the other of which was high in the center and low on the sides. Products should be made from equal amounts of batter or dough if comparison of values obtained is to be meaningful.

A. *Equipment for test*

1. Compensating polar planimeter measuring in square centimeters or in square inches. The simplest instrument available is satisfactory.
2. Flat, level surface that will not be damaged by needle point of the pole arm or by tape. A large board such as a bread board is suitable.

B. *Procedure*

1. Baked products of uniform age should be used for testing. Cut a slice 1.5–2 cm thick directly from the center of the product using a cutting box to assure uniformity. Put the surface that came directly from the center of the product on a large piece of paper and draw around it with a sharp-pointed, soft lead pencil. Be careful not to press the product out of shape and to hold the pencil at the same angle for the entire sample and for all samples to be compared. If a large piece of paper is used, several samples can be traced on it. Mark across each tracing at any one point. This becomes the starting and finishing point for the measurement.
2. Place the paper on the flat surface and tape or tack in place.
3. The planimeter is in two pieces. Assemble by placing the small ball of the pole arm into the socket on the tracer arm. Place the assembled planimeter on the paper with the tracer arm a little to your right and perpendicular to the front of the table.
4. Test the placement of the planimeter by quickly moving the tracer point around the area to be traced. The planimeter should be placed so that the tracer arm moves freely and stays on the paper. The angle between the two arms must stay between 15 and 165°. Reposition the planimeter if the conditions are not met.
5. Prior to using the planimeter for the first time, practice reading it. The first digit of the reading is the number on

the dial at which the pointer is standing or which it has just passed. The pointer usually will fall between two numbers. The smaller of the two numbers is recorded. The second and third digits are read from the measuring wheel. Observe the line that is opposite or has just been passed by the zero line on the vernier scale. The second digit is the number, or major division on the measuring wheel, and the third is found from the single divisions on the measuring wheel. The last digit is the graduation on the vernier that makes a straight line with a wheel graduation. The value of the units depends on the calibration of the instrument.

It may be advantageous to practice finding the area of a known square or rectangle. Place the tracer point on the starting line. Record the reading. Follow the outline of the figure carefully by moving the tracer point in a clockwise direction. Avoid any counterclockwise movement. If the tracer point leaves the line slightly, the error can be compensated for by moving the tracer arm an equivalent amount away from the line in the other direction. Do not use a guide in tracing straight lines. When the tracer has returned to the starting point, make a second reading. The difference between the two readings, multiplied by the appropriate factor (0.1 for instrument calibrated in centimeters or 0.01 for instrument calibrated in inches), is the area of the figure. The area found for the known square or rectangle should agree with the calculated area within 2%.

6. Measure the area of the outlined baked product as described in step 5. Repeat the measurement at least once. The areas found should agree within 2%. Average the figures.

REFERENCES

Adams, M. C. and Birdsall, E. L. 1946. New consistometer measures corn consistency. Food Ind. 18: 844.

AOAC. 1975. "Official Methods of Analysis," 12th ed. Assoc. Official Analytical Chemists, Washington, DC.

Babb, A. T. S. 1965. A recording instrument for the rapid evaluation of the compressibility of bakery goods. J. Sci. Food Agric. 16: 670.

Bailey, C. H. 1934. An automatic shortometer. Cereal Chem. 11: 160.

Balmaceda, E., Rha, C-.K. and Huang, F. 1973. Rheological properties of hydrocolloids. J. Food Sci. 38: 1169.

Bourne, M. C. 1965. Studies on punch testing of apples. Food Technol. 19: 413.

Bourne, M. C. 1968. Texture profile of ripening pears. J. Food Sci. 33: 223.

Bouton, P. E. and Harris, P. V. 1972. The effects of cooking temperature and time on some mechanical properties of meat. J. Food Sci. 37: 140.

Bouton, P. E., Ford, A. L., Harris, P. V. and Ratcliff, D. 1975. A research note—objective assessment of meat juiciness. J. Food Sci. 40: 884.

Bratzler, L. J. 1932. Measuring the tenderness of meat by means of a mechanical shear. Master's thesis. Kansas State College, Manhattan.

Breene, W. M. 1975. Application of texture profile analysis to instrumental food texture evaluation. J. Texture Studies 6: 53.

Brown, S. L. and Zabik, M. E. 1967. Effect of heat treatments on the physical and functional properties of liquid and spray-dried egg albumen. Food Technol. 21: 87.

Campion, D. R., Crouse, J. D. and Dikeman, M. E. 1975. A research note—the Armour tenderometer as a predictor of cooked meat tenderness. J. Food Sci. 40: 886.

Carlson, D. 1972. Critical angle refractometer. Food Technol. 26(5): 84.

Cathcart, W. H. and Cole, L. C. 1938. Wide-range volume-measuring apparatus for bread. Cereal Chem. 15: 69.

Cheng, C. S. and Parrish, F. C. Jr. 1976. Scanning electron microscopy of bovine muscle: effect of heating on ultrastructure. J. Food Sci. 41: 1449.

Clydesdale, F. M. 1972. Measuring the color of foods. Food Technol. 26(7): 45.

Clydesdale, F. M. 1976. Instrumental techniques for color measurement of foods. Food Technol. 30(10): 52.

Collins, J. L. and Dincer, B. 1973. Rheological properties of syrups containing gums. J. Food Sci. 38: 489.

Cooley, L. and Davies, J. R. 1936. The technique of photography for permanent records of baking studies. Cereal Chem. 13: 613.

Crossland, L. B. and Favor, H. H. 1950. A study of the effects of various techniques on the measurement of the firmness of bread by the Baker compressimeter. Cereal Chem. 27: 15.

Davis, E. S., Gordon, J. and Hutchinson, T. E. 1976. Scanning electron microscope studies on carrots: effects of cooking on the xylem and phloem. Home Econ. Res . J. 4: 214.

Daoud, H. N. and Luh, B. S. 1971. Effect of partial replacement of sucrose by corn syrup on quality and stability of canned applesauce. J. Food Sci. 36: 419.

deMan, J. M., Voisey, P. W., Rasper, V. F. and Stanley, D. W. 1976. "Rheology and Texture in Food Quality." Avi Publ. Co., Westport, CT.

Edelmann, E. C. and Cathcart, W. H. 1949. Effect of surface-active agents on the softness and rate of staling of bread. Cereal Chem. 26: 345.

Ehrlich, R. M. 1968. Contolling gel quality by choice and proper use of pectin. Food Prod. Dev. 2(1): 74.

Finney, E. E. Jr. 1972. Elementary concepts of rheology relevant to food texture studies. Food Technol. 26(2): 68.

Fleming, S. E., Sosulski, F. W. and Hamon, N. W. 1975. Gelation and thickening phenomena of vegetable proteins. J. Food Sci. 40: 805.

Francis, F. J. and Clydesdale, F. M. 1975. "Food Colorimetry: Theory and Applications." Avi Publ. Co., Westport, CT.

Friedman, H. H., Whitney, J. E. and Szczesniak, A. S. 1963. The texturometer—a new instrument for objective texture measurement. J. Food Sci. 28: 390.

Funk, K., Zabik, M. E. and ElGidaily, D. A. 1969. Objective measurements for baked products. J. Home Econ. 61: 119.

Grawemeyer, E. A. and Pfund, M. C. 1943. Line-spread as an objective test for consistency. Food Res. 8: 105.

Haard, N. F., Medina, M. B. and Greenhut, V. A. 1976. Scanning electron microscopy of sweet potato root tissue exhibiting "hardcore." J. Food Sci. 41: 1378.

Hanning, F., Bloch, J. deG. and Siemers, L. L. 1955. The quality of starch puddings containing whey and/or non-fat milk solids. J. Home Econ. 47: 107.

Hansen, L. J. 1972. Development of the Armour tenderometer for tenderness evaluation of beef carcasses. J. Texture Studies 3: 146.

Hegarty, P. V. J. and Allen, C. E. 1975. Thermal effects on the length of sarcomeres in muscles held at different tensions. J. Food Sci. 40: 24.

Humason, G. L. 1972. "Animal Tissue Techniques," 3rd ed. W. H. Freeman and Co., San Francisco.

Hunter, M. B., Briant, A. M. and Personius, C. J. 1950. Cake quality and batter structure. Cornell Univ. Agric. Exp. Stn. Bull. 860, Ithaca, NY.

Institute of Food Technologists. 1964. Sensory testing guide for panel evaluation of foods and beverages. Food Technol. 18: 1135.

Jones, S. B., Carroll, R. J. and Cavanaugh, J. R. 1976. Muscle samples for scanning electron microscopy: preparative techniques and general morphology. J. Food Sci. 41: 867.

Jooste, M. E. and Mackey, A. D. 1952. Cake structure and palatability as affected by emulsifying agents and baking temperatures. Food Res. 17: 185.

Kalab, M. and Harwalkar, V. R. 1973. Milk gel structure. I. Application of scanning electron microscopy to milk and other food gels. J. Dairy Sci. 56: 835.

Kastner, C. L. and Henrickson, R.L. 1969. Providing uniform meat cores for mechanical stress force measurement. J. Food Sci. 34: 603.

Kornerup, A. and Wanscher, J. H. 1962. "Reinhold Color Atlas." Reinhold Publ. Co., New York.

Kramer, A. and Twigg, B. A. 1959. Principles and instrumentation for the physical measurement of food quality with special reference to fruit and vegetable products. Adv. Food Res. 9: 153.

Lee, J. W. S. and Latham, S. D. 1976. A research note—rapid moisture determination by a commercial-type microwave oven technique. J. Food Sci. 41: 1487.

Locken, L., Loska, S. and Shuey, W. 1972. "The Farinograph Handbook." American Association of Cereal Chemists, St. Paul, MN.

Maerz, A. J. and Paul, M. R. 1950. "A Dictionary of Color," 2nd ed. McGraw-Hill Book Co., New York.

Matthews, R. H. and Dawson, E. H. 1963. Performance of fats and oils in pastry and biscuits. Cereal Chem. 40: 291.

Matthews, R. H. and Dawson, E. H. 1966. Performance of fats in white cake. Cereal Chem. 43: 538.

Mazurs, E. G., Schoch, T. J. and Kite, F. E. 1957. Graphic analysis of the Brabender viscosity curves of various starches. Cereal Chem. 34: 141.

McCrae, S. E. and Paul, P. C. 1974. Rate of heating as it affects the solubilization of beef muscle collagen. J. Food Sci. 39: 18.

Miller, B. S., Derby, R. I. and Trimbo, H. B. 1973. A pictorial explanation for the increase in viscosity of a heated wheat starch-water suspension. Cereal Chem. 50: 271.

Mohr, W. P. 1973. Applesauce grain. J. Texture Studies 4: 263.

Morse, L. M., Davis, D. S. and Jack, E. L. 1950. Use and properties of non-fat dry milk solids in food preparation. I. Effect on viscosity and gel strength. II. Use in typical foods. Food Res. 15: 200.

Nickerson, D. 1946. Color measurement and its application to the grading of agricultural products. A handbook on the method of disk colorimetry. Misc. Publ. 580, USDA, Washington, DC.

Noble, A. C. 1975. Instrumental analysis of the sensory properties of food. Food Technol. 29(12): 56.

Noble, I. T., Halliday, E. G. and Klaas, H. K. 1934. Studies on tenderness and juiciness of cooked meat. J. Home Econ. 26: 238.

Paul, P., Batcher, O. M. and Fulde, L. 1954. Dry mix and frozen baked products. I. Dry mix and frozen cakes. J. Home Econ. 46: 249.

Platt, W. and Powers, R. 1940. Compressibility of bread crumb. Cereal Chem. 17: 601.

Pomeranz, Y. 1976. Scanning electron microscopy in food science and technology. Adv. Food Res. 22: 205.

Scott Blair, G. W. 1958. Rheology in food research. Adv. Food Res. 8: 1.
Shaffer, T. A., Harrison, D. L. and Anderson, L. L. 1973. Effects of end point and oven temperatures on beef roasts cooked in oven film bags and open pans. J. Food Sci. 38: 1205.
Shuey, W. C. 1975. Practical instruments for rheological measurements of wheat products. Cereal Chem. 52(3II): 42r.
Stinson, C. G. and Huck, M. B. 1969. A comparison of four methods for pastry tenderness evaluation. J. Food Sci. 34: 537.
Szczesniak, A. S. 1963. Classification of textural characteristics. J. Food Sci. 28: 385.
Szczesniak, A. S. 1973. Indirect methods of objective texture measurements. In "Texture Measurements of Food," Eds. Kramer, A. and Szczesniak, A. S. Reidel Publ. Co., Boston.
Szczesniak, A. S. 1975. General Foods texture profile revisited—ten year perspective. J. Texture Studies 6: 5.
Szczesniak, A. S., Humbaugh, P. R. and Block, H. W. 1970. Behavior of different foods in the standard shear compression cell of the shear press and the effect of sample weight on peak area and maximum force. J. Texture Studies 1: 356.
Szczesniak, A. S. and Torgenson, K. W. 1965. Methods of meat texture measurement viewed from the background of factors affecting tenderness. Adv. Food Res. 14: 33.
Tinklin, G. L. and Vail, G. E. 1946. Effect of method of combining the ingredients upon the quality of the finished cake. Cereal Chem. 23: 155.
Tuma, H. J., Venable, J. H., Wuthier, P. R. and Henrickson, R. L. 1962. Relationships of fiber diameter to tenderness and meatiness as influenced by bovine age. J. Animal Sci. 21: 33.
Villalobos-Domínguez, C. and Villalobos, J. 1947. "Villalobos Color Atlas." Stechert-Hafner, New York.

17. Preparing the Report

The main sections of the report are the following: introduction, review of the literature, materials and methods, results and discussion, summary and conclusions, and references. The headings are subject to some variation and the sections are subject to some combining.

THE ORGANIZATION OF THE REPORT

Introduction, Review of Literature

The introduction includes a statement of purpose and provides justification for the work. The review of the literature frequently is combined with the introduction, providing the background for the statement of the purpose of the study and justification for the work. In this case, the review will have been written during the planning phase. If the collection of data has covered a period of even a month, some new related information may have appeared in the research literature; therefore, current issues of research journals should be consulted and new material incorporated.

The review of the literature may be a separate section if it is long and if the introduction does not depend on it. A long review of the literature should not consist merely of a series of separate summaries but should be organized into a cohesive exposition of the subject. Preparation of an outline is helpful in achieving organization and in excluding irrelevant and repetitious material.

Materials and Methods

The purpose of this section is to state clearly what was done. Decisions concerning the materials and methods will of necessity have been made during the planning stage; therefore, much of the writing of this section could have been completed before the data were collected. Possibly some revision and/or elaboration will be necessary because of the course of events during the study. For example, the actual number of replications might have differed from the number planned. Unforeseen difficulties might have necessitated changes in materials or methods. Early results might even have brought about a change in direction of the study.

The materials and methods section should include a clear presentation of the overall experimental plan, an adequate description of the materials used, and the details of the procedure in chronological sequence. Methods for which references can be cited need not be detailed, but any modifications that are made in published methods should be described.

Judgment is needed as to the detail in which methods are described. Woodford (1968) agreed with the frequently stated rule of thumb that the detail should be sufficient to permit the reader to repeat the experiment. He went on to caution, however, that "who the reader is" should be defined. Unless some background on the part of the reader is assumed, this section might be intolerably long. Woodford suggested a constant self-questioning: "Are these details essential to the success of the experiment?" Peterson (1961) also cautioned the writer not to "suffocate your reader with minutiae."

Results and Discussion

The results and discussion may be either combined or presented as separate sections. In a short report they are likely to be combined.

Tables Tables should be prepared before any statements of results are written. Eventually the tables will be integrated into the text, but the text should be fitted to the tables rather than the tables to the text. Tables, in addition to showing the results, should describe an experiment and its purpose in an abbreviated way (Woodford, 1968).

The content of the tables requires some thought. A decision that must be made is the extent to which the raw data are to be included along with the averages and other calculated values. Consider a study that involves sensory evaluation. The complete data would show the extent of variation among the judges but would add considerable bulk. Another decision might involve inconclusive data: Should they be presented in tabular form or should they be covered by a statement in the text? Judgment must be applied when answering such questions in individual cases.

The amount of information to be included in a single table is another important consideration. There is an advantage in presenting the results of more than one type of evaluation in the same table because doing so makes certain relationships apparent. For example, presenting the results of the objective measurements and the sensory evaluation side by side could reveal a relationship between a physical characteristic and acceptance of a product.

Tabular form varies somewhat with publications, but there are some principles to be observed. Tables are placed after their first mention in the text. They are numbered consecutively through the report. Each title, placed above the table, should be as concise as possible while giving the reader a comprehensive idea of the content. Side headings and column headings are used as needed to identify the treatments and the nature of the values. Units of measurement must be given for all values but not individually. They can be at the top of the columns or at the left side or stub, depending on what is convenient in a given situation. Sometimes all of the values in a single table have the same unit, such as percentage, which can be conveniently specified in the title.

Accuracy in entering figures in a table is of obvious importance. Decimal points should be aligned. A zero should precede a decimal fraction. Consideration should be given to significant digits. In general, the calculations should be carried to as many digits as practicality permits, and then the values rounded. For a given measurement, the number of significant digits should reflect the accuracy of the measurement.

Figures Figures such as instrument tracings and photographs represent primary results. Other figures, for example, graphs such as plotted curves and histograms, may be derived from the numerical data. Figures contribute, along with tables, to the description of the experiment. Their uniqueness lies in their ability to record natural appearances, as in the case of light and electron micrographs, or to reveal trends and relationships, as with graphs.

Figures, like tables, are numbered consecutively through the report. Like tables, they should be self-explanatory. Titles, normally placed below the figures, should indicate the exact nature of the information presented. Axes should be clearly and accurately labeled and legends should be complete. Additional information concerning figures is found in the style guide of the Conference of Biological Editors (CBE, 1964).

Conclusions The results and discussion section might end with conclusions, or the section might be followed by a separate summary and statement of conclusions. When conclusions are drawn, they should be stated in relation to the specific experimental conditions. Sometimes suggestions are made for future work. If this is done, the suggestions should be realistic.

References

The list of references at the end of the report should follow a definite form consistently, as discussed in Chapters 1 and 14. Again, frequent use of research reports as a guide is encouraged. Because of compilation of a reference list during the preliminary review of literature (Chapter 14), only insertion of later references should be necessary when the final report is written. The list should include all of the references that have been cited and none that have not.

THE LANGUAGE OF THE REPORT

The language of the report begins with the title, which should be concise and yet indicative of the content. Beginning a title with a key word, as opposed to "A Comparison of . . . ," "A Study of . . . ," or "The Effect of . . . ," is recommended. For example, "The Effect of Triticale

Flour on the Volume of Bread" might better be written "Bread Volume as Affected by Triticale Flour" or "Triticale Flour in Bread: Effect on Loaf Volume."

Plagiarism should be avoided throughout the report. The review of literature probably is the section in which it is most likely to occur unless care is taken.

The passive voice, which is used commonly in scientific writing, predominates in the methods and materials section. A hazard of the passive voice, particularly in this section, is the dangling participle, which occurs all too commonly in scientific writing. A participle functions as an adjective, but when the passive voice is used, the noun or pronoun to be modified is likely to be absent. This is not really an argument against the passive voice but against dangling participles. An example of incorrect usage is the following: "Viscosity was measured using a Brookfield viscometer." *Viscosity* did not use a viscometer; the *worker* did. In this case the problem is easily avoided by writing, "Viscosity was measured with a Brookfield viscometer."

Elimination of writing frills throughout the report contributes to control of length and to readability. Examples given in the CBE style manual (CBE, 1964) include the following: "conducted inoculation experiments on" versus the more concise "inoculated"; "due to the fact that" versus "because"; "the treatment having been performed" versus "after treatment"; and "a method, which was found to be expedient and not very difficult to accomplish and which possessed a high degree of accuracy in its results" versus "an easy, accurate way."

The data presented in tables and figures should be discussed but not belabored. The writer should compare the results with the work of others if references are available and should attempt to explain apparent discrepancies. Limitations of the study should be recognized.

Caution is needed in comparing treatment means when statistical tests of significance are not conducted. Small differences may or may not be real differences, as was discussed in Chapter 1. Statistical analysis sometimes is feasible for an individual problem. Harrison's (1972) discussion of analysis of data in relation to various experimental plans is applicable to some problems undertaken in a class.

A common misuse of terms involves words that frequently appear in the results and discussion section: *affect* and *effect*. *Affect* is used only as a verb; for example, "Oven temperature affects optimal baking time." *Effect* most frequently is used as a noun; for example, "The effect of oven temperature on optimal baking time is inverse and approximately linear." *Effect* is used as a verb only in the sense of *to bring about;* for example, "An increase in oven temperature effects a decrease in optimal baking time." Frequent use of the classic treatise by Strunk and White (1959) on English usage, principles of composition, and writing style can benefit anyone, student or not. Most freshman English textbooks also are useful references.

Portions of a sample report are presented below. Locations of omitted portions are represented by (. . .). See Exercise 2b before reading further. While reading the following excerpts, look for illustrations of the points that have been made above.

Storage Stability of Cake Batters as Affected by the Leavening System

Today's time-conscious consumer seeks means of increasing the ease of food preparation while maintaining some of the advantages of home cooking. The practice of preparing foods in quantity for use over an extended period of time is feasible for many foods; however, batters containing conventional leavening agents have not represented a particularly successful application of the procedure.

Glucono-delta-lactone (GDL) has been available for some time for commercial use in products in which leavening stability is particularly important. Its use has not been extended to the consumer level, however. This study was an investigation of the feasibility of utilizing GDL in the development of storage-stable plain cake batter.

REVIEW OF LITERATURE

GDL is sold in commercial quantities as a white crystalline solid. Chemically it is an inner ester of gluconic acid with the formula:

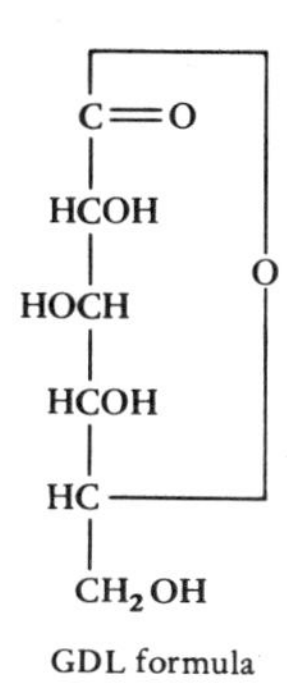

GDL formula

GDL is not an acid, but in water solution it is hydrolyzed to gluconic acid. Just as gas production by a baking powder involves a reaction between an acid salt and sodium bicarbonate, the gluconic acid produced by hydrolysis of GDL reacts with sodium bicarbonate to produce carbon dioxide when heat is applied. One part GDL neutralizes 0.472 part by weight of sodium bicarbonate (Feldberg, 1959).

The rate of hydrolysis of GDL is slow at room temperature and below but increases with heating (Anon., 1957). This combination of stability at low temperatures and reactivity at elevated temperatures suggests potential usefulness of GDL in a batter to be subjected to low temperature storage.

. . .

PROCEDURE

The study of plain cakes containing two leavening systems was conducted according to a factorial plan that included two variables, each at two levels:

1. Leavening system: a. sulfate-phosphate baking powder
   b. GDL–sodium bicarbonate system
2. Oven temperature: a. 171°C (340°F)
   b. 190°C (375°F)

Cakes for all four treatment combinations in each of four replications were tested on the zero day and on day 21 of freezer storage at −15°C.

Cake batters were prepared according to the following formula and mixing procedure:

. . .

Table 1. Schedule for preparing cake batters and testing cakes after 0 and 21 days of storage (4 replications)

		Replication			
Week	Day	1	2	3	4
I	A	batter prepn. 0-da cake test			
	B		batter prepn. 0-da cake test		
II	A			batter prepn. 0-da cake test	
	B				batter prepn. 0-da cake test
III	A				
	B				
IV	A	21-da cake test			
	B		21- da cake test		
V	A			21-da cake test	
	B				21-da cake test

Table 2. Volumes of cakes leavened with sulfate-phosphate baking powder or a GDL system, stored as batter for 0 or 21 days and baked at 171 or 190°C[a]

Storage days	Oven temperature °C	Leavening	
		Baking powder	GDL system
		Volume, cm^3	
0	171	290	401
0	190	324	406
21	171	188	405
21	190	199	412

[a]Each value is an average for four replications.

Ten class periods, covering a 5-wk time span, were available after completion of the preliminary work. The scheduling of these periods to permit freezer storage for 21 da is shown in Table 1. On each day during the first two weeks, two batters were prepared for a replication. Each batter was weighed into four pans, two of which were frozen and stored. The other two were baked, one at each oven temperature, and evaluated. On each day during the last two weeks, the four pans of batter that had been frozen 3 wk previously were thawed in the refrigerator (4 hr), baked as on the zero storage day, and evaluated.

Cakes from freshly prepared and stored batters were evaluated by volume measurement by rapeseed displacement (Cathcart and Cole, 1938) and by sensory evaluation of volume, shape, grain, odor, taste, and mouth feel.

. . .

RESULTS

The average cake volumes are presented in Table 2. The cakes containing GDL consistently had greater average volumes than those containing baking powder. The effect of the leavening system was much greater after 21 da of freezer storage than when batters were freshly mixed. Stored samples containing the baking powder had much lower average volumes than did fresh samples; volumes of the samples containing GDL were unaffected by storage. The higher oven temperature resulted in the greater average volume for cakes from unstored batters containing the baking powder; otherwise, oven temperature had essentially no effect on cake volume.

. . .

Under the conditions of this study, GDL showed promise as a leavening system component in cake batter subjected to freezer storage. Future work should involve longer storage periods. Quick bread batters might be studied also, as well as the effect of thawing batters versus initiating baking from the frozen state.

REFERENCES

Anon. 1957. Glucono-delta-lactone in food products. Pfizer Tech. Bull. 93. Chas. Pfizer and Co., Inc., Brooklyn, NY.

Cathcart, W. H. and Cole, L. C. 1938. Wide-range volume-measuring apparatus for bread. Cereal Chem. 15: 69.

Feldberg, C. 1959. Glucono-delta-lactone. Cereal Sci. Today 4: 96.

ORAL REPORTS

Oral reports of experimental work frequently are made in classes. Such reports are deadly if read or if presented with poor visual aids. Talking from an outline eliminates the disadvantages of either reading or memorizing the report.

An oral report is organized similarly to the written report of a problem. A major difference between oral and written reports involves the tables and figures. They must be relatively simple for oral reports to enable the members of the audience to digest the information. An overhead projector usually is available for showing transparencies. If one is not available, and if the group is small, a large pad of paper on an easel can serve the purpose. It is important that the entire group be able to read any visual material shown and to hear the speaker.

The oral report should be well prepared and carefully timed. A speaker who exceeds the time limit risks boring the audience.

SUGGESTED EXERCISES

1. Select one or more research articles that are of interest to you and do the following:
   a. Without reading any text, select some tables and/or figures and for each write a summary of the results. Now read the pertinent portions of text and compare your statements with those of the author(s). Are your interpretations of the information in the tables and/or figures in agreement? If not, are you at fault or are the authors? To what extent did the design of the tables and figures make this exercise unnecessarily difficult?
   b. Make a list of obvious errors. These may include anything from dangling participles and lack of agreement between subjects and verbs to obviously incorrect values in tables, interchanged legends, or mislabeled information in graphs. The purpose of this exercise is not to find fault with others' writing but to increase the student's awareness of some pitfalls that one should attempt to avoid.
2. Study the sample report excerpts and do the following:
   a. Identify specific applications of points made earlier in this chap-

ter. This exercise might include some suggestions for improvement.

b. Look at the gaps in the report and list specific kinds of information that would need to be included.

REFERENCES

CBE. 1964. "Style Manual for Biological Journals," 2nd ed. Conference of Biological Editors, Committee on Form and Style. American Institute of Biological Sciences, Washington, DC.

Harrison, D. L. 1972. Planning and conducting experiments. In "Food Theory and Applications," Eds. Paul, P. C. and Palmer, H. H. John Wiley and Sons, Inc., New York.

Peterson, M. S. 1961. "Scientific Thinking and Scientific Writing." Reinhold Publ. Co., New York.

Strunk, W. and White, E. B. 1959. "The Elements of Style." The Macmillan Co., New York.

Woodford, F. P. 1968. "Scientific Writing for Graduate Students." The Rockefeller University Press, New York.

APPENDIX

Scales for sensory evaluation

Scale	Possible uses
6 Excellent 5 Very good 4 Good 3 Fair 2 Poor 1 Very poor	To evaluate quality attributes such as: color (crumb, crust, meat, etc.) flavor overall acceptability volume cell structure odor consistency
8 Extremely ____ 7 Very ____ 6 Moderately ____ 5 Slightly ____ 4 Slightly ____ 3 Moderately ____ 2 Very ____ 1 Extremely ____	To evaluate such quality attributes as those represented by the polar terms below, replace blanks 1–4 with the first term and blanks 5–8 with the second term. thin/thick tough/tender sour/sweet weak/strong undesirable/desirable dry/moist dry/juicy fine/coarse
4 High 3 Moderately high 2 Moderately low 1 Low	To evaluate volume
9 Like extremely 8 Like very much 7 Like moderately 6 Like slightly 5 Neither like nor dislike 4 Dislike slightly 3 Dislike moderately 2 Dislike very much 1 Dislike extremely	To evaluate preference. This scale should not be used for evaluation of quality.
5 Superior 4 Above average 3 Average 2 Below average 1 Poor	To evaluate products in relation to perviously determined standards

INDEX

*The letter *t* after a page number refers to a table.